ADVANCES IN MICROFLUIDICS

Edited by **Ryan T. Kelly**

Advances in Microfluidics
Edited by Ryan T. Kelly

Published by InTech
Janeza Trdine 9, 51000 Rijeka, Croatia

Publishing Process Manager Martina Durovic

Technical Editor Teodora Smiljanic

Cover Designer InTech Design Team

Additional hard copies can be obtained from orders@intechopen.com

Advances in Microfluidics, Edited by Ryan T. Kelly
p. cm.
978-953-51-0106-2

For EU product safety concerns:
IN TECH d.o.o., Prolaz Marije Krucifikse Kozulić 3, 51000 Rijeka, Croatia,
info@intechopen.com or visit our website at intechopen.com.

Meet the editor

Dr Kelly currently serves as lead of the microfabrication facility in the Environmental Molecular Sciences Laboratory at Pacific Northwest National Laboratory (PNNL). He has a broad background in the design and fabrication of microfluidic devices and other microsystems for a diverse array of applications. His prior research focused on the development of novel separation and analysis methods for proteins and peptides within microfluidic devices. He also developed new fabrication methods for creating microdevices in plastic substrates and interfacing microchannels with semipermeable membranes. While at PNNL, Dr. Kelly developed approaches for increasing the sensitivity of electrospray ionization mass spectrometry using both capillary based and microfabricated electrospray sources. His current interests involve coupling microfluidic systems with mass spectrometry for single cell and other sample-limited bioanalyses.

Contents

Preface

When the field of microfluidics emerged in the early 1990s, it was primarily focused on the development of analytical microdevices. Since then, microfluidics has expanded its influence into virtually every branch of science and engineering. There are many driving forces behind this explosive growth. To name a few:

- Scaling properties afforded by miniaturization are desirable for many applications. For example, enhanced mass transfer and heat dissipation enable faster chemical separations without sacrificing separation performance.
- Sample and reagent requirements can be greatly reduced.
- The unique properties of fluids when confined to small channels (e.g., laminar flow) make novel applications possible.
- Photolithographic patterning provides tremendous design flexibility. Rather than manually coupling different components and capillaries to create a microsystem, microfluidic design relies on the creation of photomasks that are drawn using computer aided design software.

These favorable conditions have led to a positive feedback loop in which new applications drive additional technology development and vice versa. Of course, such developments are ongoing, and we will undoubtedly continue to see brisk growth in both the research environment and in commercial settings for many years to come.

This book provides a current snapshot of the field of microfluidics as it relates to a variety of sub-disciplines. The chapters have been divided into three sections: Fluid Dynamics, Technology, and Applications, although a number of the chapters contain aspects that make them applicable to more than one section. It is hoped that this book will serve as a useful resource for recent entrants to the field as well as for established practitioners.

Ryan T. Kelly, Ph.D.
Senior Research Scientist
Pacific Northwest National Laboratory,
Richland, Washington,
USA

Part 1

Fluid Dynamics

1

Microfluidic Transport Driven by Opto-Thermal Effects

Matthieu Robert de Saint Vincent and Jean-Pierre Delville
Univ. Bordeaux, LOMA, UMR 5798, F-33400 Talence,
CNRS, LOMA, UMR 5798, F-33400 Talence
France

1. Introduction

Microfluidic applications to biology and chemistry rely on precise control over the transport of (bio-)molecules dissolved in tiny volumes of fluid. However, while the rigid environment of a microfluidic chip represents a convenient way to impose flows at the micrometer scale, an active control of transport properties usually requires the action of an external field (Squires & Quake, 2005).

Can light provide such control? Light indeed has several specific assets. First, as optical methods are contact-free, they are intrinsically sterile. Second, light fields can be tightly focused, providing by the way a very local and selective action. Third, light excitation can be totally disconnected from the chip (even though integration is possible (Monat et al., 2007)), therefore no microfabrication or specific treatment of the chip are required. This also provides a high degree of reconfigurability and versatility. The interest of applying optical fields to lab-on-a-chip devices is therefore evident.

Optical forces, which rely on the exchange of momentum between a light beam and a material object at a refractive index discontinuity (Ashkin, 1970), have led to the development of optical tweezers (Ashkin et al., 1986), themselves having opened a huge field of applications (Jonáš & Zemánek, 2008). However, the use of optical forces in the scope of microfluidic transport is limited by their very weak amplitude—typically, in the picoNewton range.

To circumvent this limitation, several alternatives have been proposed. The basic idea is to use light to induce hydrodynamic forces. A convenient means of doing this is to use a light source as a localized heater. The light beam thus provides a direct transfer of energy, rather than a transfert of momentum. Indeed, as the photon momentum equals its energy divided by the velocity of light, the total impulsion which can be communicated to an object is weak at given energy per photon. A direct transfer of energy therefore appears more favorable than a transfer of momentum to provide mechanical effects.

Besides the assets mentioned above, the use of focused light as a heating source has two extra advantages. On the one hand, it allows for producing very strong temperature gradients with a moderate heating. On the other hand, the possible disconnection from the chip, and the ability to duplicate or displace at will a laser beam (through galvanometric mirrors or holographic methods) provide two complementary ways of using the heating source: (i) a 'remote controlled' mode, in which the source is static, and (ii) a 'writing' mode, involving a continuously moving source. This complementarity opens the way to various opportunities.

How can the heating affect the transport properties of a fluid, or of a solute carried by this fluid? A first method consists in directly tuning the concentration of the solute, providing that the thermally-induced transport is strong enough to overcome the natural Brownian diffusion. Alternatively, the manipulation of the carrier fluid provides another way to control the transport of reagents. Such a manipulation can be achieved by tuning the fluid properties, density and viscosity, which are both temperature-dependent. On the other hand, diphasic flows are particularly relevant in lab-on-a-chip applications since they allow for the manipulation of calibrated volumes of reagents, while preventing from potential cross-contamination according to the immiscible character of the fluids involved (Song et al., 2006; Theberge et al., 2010). From the viewpoint of fluid manipulation, diphasic flows add another degree of freedom, namely, the interfacial tension, toward the control of fluid transport. Finally, a last possibility consists in performing phase changes, involving liquid-gas or gas-liquid transitions.

These two families of flows—mono- or diphasic flows—build the structure of the present chapter, basically constituting its two main parts, inside which we overview the main approaches developed in the literature. The scope of this review includes the transport of fluids and macromolecules of biological interest in the view of—proven or potential—lab-on-a-chip applications. Our purpose is not to give an exhaustive overview of the literature (especially, the manipulation techniques of small molecules, colloids, and nanoparticles, are not included in the present chapter), but rather to give a comprehensive survey, centered on the main physical mechanisms, and then to bridge the gap between the highly diverse opto-thermal approaches.

2. One-fluid flows

This section reviews the main transport phenomena involved in monophasic flows. We will first remind the main principles involved, then we will show two major research directions combining these methods, namely, the generation of channel-free microfluidic flows, and the manipulation of biological molecules.

2.1 Basic principles and methods

Three basic mechanisms, as summarized in Fig. 1, are involved in monophasic solutions: thermophoresis, thermal convection, and thermoviscous expansion. A more anecdotic alternative, involving a thermally-induced sol-gel transition, will also be briefly presented.

2.1.1 Thermophoresis

Thermophoresis, also called thermodiffusion, or Ludwig-Soret effect, takes place in solutions submitted to a temperature gradient (Piazza & Parola, 2008; Würger, 2010). The macroscopic effect is the creation, at steady state, of a concentration gradient overtaking the natural smoothing due to the Brownian diffusion (Fig. 1(a)). While this effect has been discovered in the mid-nineteenth century (independently by Ludwig and Soret), its theoretical

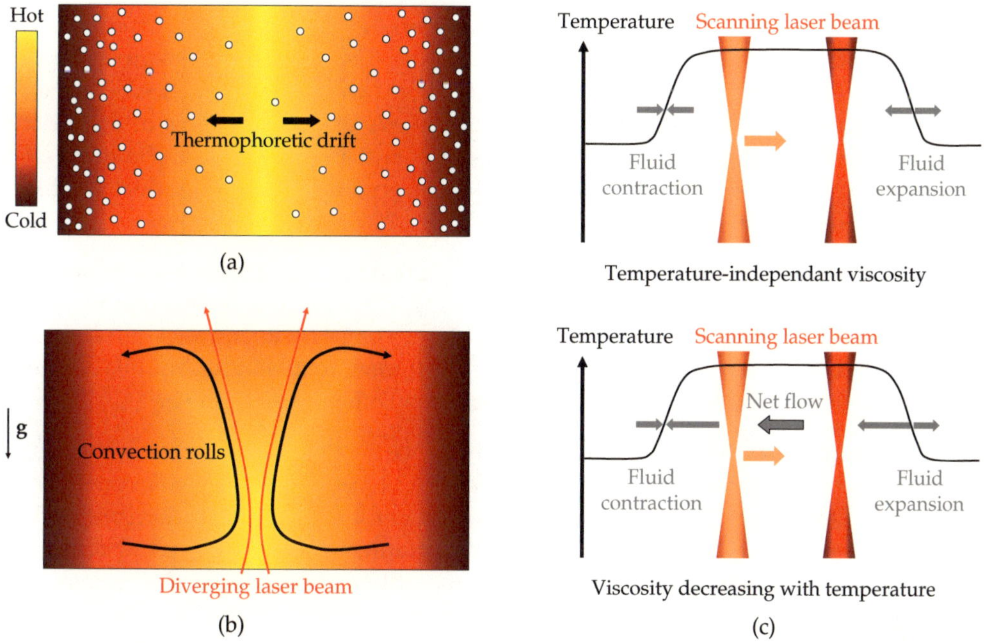

Fig. 1. Schematic illustration of the three main mechanisms involved in monophasic opto-thermal transport. (a) Thermophoresis of (here thermophobic) molecules, (b) laser-induced convection, and (c) thermoviscous expansion (adapted after Weinert & Braun (2008b)).

understanding is still controversial. The recent review by Würger (2010) provides significant insight on the different mechanisms which can be involved.

From a phenomenological point of view, the motion of particles submitted to a temperature gradient can be described as a thermophoretic drift of velocity

$$\mathbf{u}_{\text{Soret}} = -D_T \boldsymbol{\nabla} T, \tag{1}$$

with D_T the thermophoretic mobility. Note that the word 'particle' should be understood here in a generic meaning, including both molecules, nanoparticles, microbeads, etc. Indeed, while biomolecules will mainly be considered in the following, thermodiffusion applies to a broad range of systems. Even though fundamental differences exist in the involved physical mechanisms (more details can be found in (Würger, 2010)), the phenomenological description we provide here keeps its generality.

Comparing thermophoresis to the Brownian diffusion leads to the definition of the Soret coefficient,

$$S_T = \frac{D_T}{D}, \tag{2}$$

with D the Brownian diffusivity. This coefficient has the dimension of the inverse of a temperature. It can be either positive or negative and then determines both the direction and amplitude of the overall particle drift. To date, no unified theory is able to predict either the sign or the order of magnitude of the Soret coefficient, which have been observed to usually depend on both solute and solvent parameters, as well as external conditions such as temperature (Piazza & Parola, 2008; Würger, 2010). The theoretical background aiming at describing the fluidic thermophoresis is built upon two main approaches. On the one hand, hydrodynamic descriptions rely on the hypothesis of quasi-slip flow at the particles boundary (Weinert & Braun, 2008a; Würger, 2007). On the other hand, at the microscopic scale, thermodynamic approaches assume the local thermodynamic equilibrium to account for solvent diffusivity and fluctuations (Duhr & Braun, 2006b; Würger, 2009).

A positive value of the Soret coefficient thus corresponds to a migration toward the colder regions ('thermophobic' behavior, as shown on Fig. 1(a)). Conversely, a solute with a negative Soret coefficient will be said 'thermophilic'. For DNA in aqueous buffer solution Braun & Libchaber (2002) measured $S_T = 0.14\ \mathrm{K}^{-1}$ at room temperature, but this coefficient has been observed to change of sign with temperature (Duhr & Braun, 2006b).

From an experimental point of view, the study of thermophoresis requires (i) to apply a temperature gradient to the test cell, and (ii) to detect and measure the resulting concentration distribution. Optical methods are indeed well suited to fulfill these two requirements. First, as already pointed out, a much higher temperature gradient can be produced by direct laser heating of the fluid than by externally heating the cell boundary. Second, the same laser beam can also be used to characterize the concentration gradient. One possible method relies on the thermal lensing effect: as the concentration gradient created by thermal diffusion modifies locally the refractive index of the solution, the transmitted beam is either focused or spread (effect called 'Soret lens'), depending on the direction of the solute migration (Giglio & Vendramini, 1974). An alternative method makes use of a fluorescent marker grafted to the particles of interest, or of the particles fluorescence themselves if applicable, to reconstruct the concentration profile in real time by microscope imaging (Duhr et al., 2004). Moreover, the temperature profile can also be monitored by using a temperature-dependant fluorescent marker.

2.1.2 Thermoconvection

Thermoconvection relies on the difference of density of an homogeneous fluid heated inhomogeneously. As density usually decreases with temperature, the local heating of a fluid leads to its dilatation. Considering the heating induced by a collimated laser beam with radial symmetry, the thermal expansion would also be axisymmetric, and no net flow would appear even in the case where the laser beam moves. Inducing a net flow in this case would require to break the heating symmetry. This can be done if the laser beam is divergent, as shown in Fig. 1(b): the fluid more heated at the bottom side raises up by buoyancy, then loses its heat and falls down, creating convective rolls (Boyd & Vest, 1975). This mechanism, generally known as Rayleigh-Bénard convection, is involved in many processes at the macroscopic scale, ranging from the cooking of pasta to atmospheric currents. However, in the micrometer-scale, gravity is not the predominant force, and the heating symmetry breakup induced by gravity is rather limited because the Rayleigh number, which controls the convection onset, behaves as the cube of the heated layer width. In that sense, thermal convection is usually not relevant at this scale. Microfluidic applications of thermoconvection can nevertheless be developed provided that the sample is thick enough, or that the other forces (essentially, viscous or capillary) can be efficiently reduced.

2.1.3 Thermoviscous expansion

Another elegant means of breaking the heating symmetry to induce a net flow has recently been proposed by Weinert & Braun (2008b). This method relies on the temperature dependance of viscosity of the fluid submitted to a scanning heating beam (Fig. 1(c)).

Let us consider a confined fluid, in which the influence of gravity is negligible. We first assume that the fluid viscosity does not depend on temperature, as shown on the top part of Fig. 1(c). As the laser beam moves, the fluid at the front of the spot scanning expands due to its decrease in density while, on the other hand, the fluid at the rear of the spot scanning contracts as well. As this thermal expansion is a linear process, expansion and contraction balance, and no net fluid flow is produced.

Let us now add the temperature dependence of fluid viscosity. As viscosity usually decreases with temperature, the expansion and contraction processes will be favored in the heated regions, as shown on the bottom part of Fig. 1(c). This dissymmetry results in a net flow, directed in the direction opposite to the scanning.

As thermal diffusion is faster, by several orders of magnitude, than the fluid flow, the fluid warms and cools down in milliseconds, so scanning can be operated at rates in the kiloHertz range. The resulting pump velocity can be expressed in a simple manner, dropping a numerical prefactor of order unity, as (Weinert & Braun, 2008b)

$$\begin{aligned} u_{\text{thermoviscous}} &= \text{scanning velocity} \times \text{thermal expansion} \times \text{thermal decrease of viscosity} \\ &= -f \ell_{\text{th}} \alpha \beta T^2. \end{aligned} \tag{3}$$

In this expression f is the scanning rate, ℓ_{th} the heating spot length scale, T the temperature rise, $\alpha = (1/\rho)(\partial\rho/\partial T)$ and $\beta = (1/\eta)(\partial\eta/\partial T)$ the thermal expansion coefficient and temperature dependance with temperature, respectively. For water, $\alpha = -3.3 \times 10^{-4}\ \text{K}^{-1}$ and $\beta = -2.1 \times 10^{-2}\ \text{K}^{-1}$, then considering a heating spot size of 30 μm, a scanning rate of 5 kHz and a temperature rise of 10 K lead to a pump velocity of 104 $\mu\text{m}\,\text{s}^{-1}$.

2.1.4 Thermally-induced sol-gel transition

An alternative way, inducing a local phase transition in the fluid, should also be mentioned. As the thermoviscous expansion presented above, this approach relies on a thermally-induced change in the fluid viscosity, but in the framework of a phase change. Krishnan et al. (2009) used a thermorheological fluid (water containing 15 % w/w of Pluronic F127, a tribloc copolymer) flowing in a channel including an absorbing substrate. The laser heating induced a reversible gelation of the fluid, resulting in the interruption of the flow. A flow switch without any moving part was then achieved. A similar approach was also used to perform fluorescence-activated cell sorting (Shirasaki et al., 2006).

2.2 Channel-free microfluidic flows

The direct manipulation of volumes of fluid allows for the controlled creation of arbitrary flows without the need of a rigid microfluidic channel. In particular, Weinert & Braun (2008b) have shown that flows can be driven along complex patterns by thermoviscous pumping (Fig. 2). As illustrated in Fig. 2(a), an infrared laser beam writing the words 'LASER PUMP' can

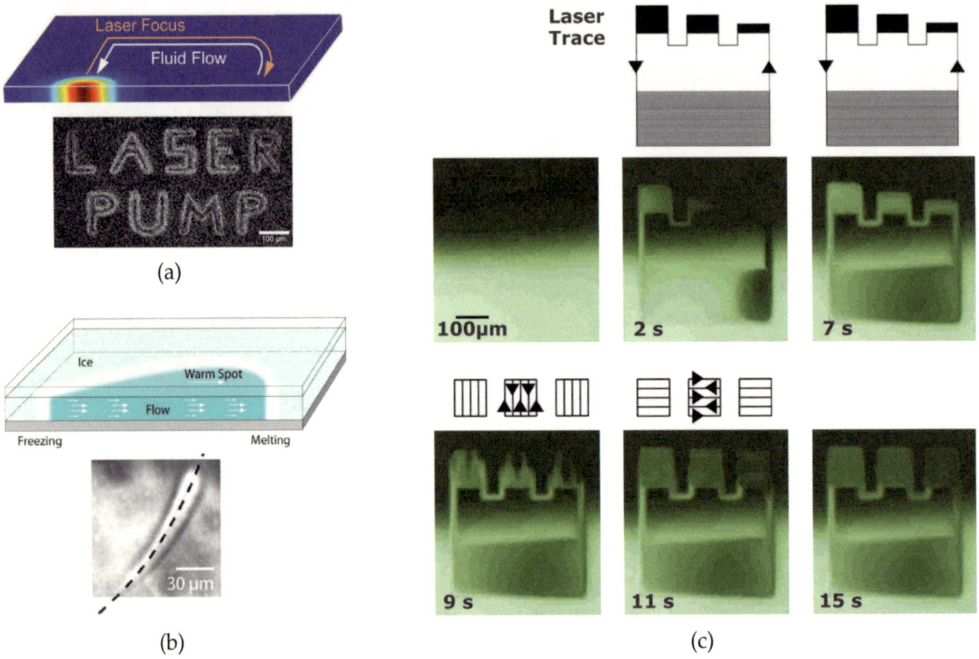

Fig. 2. Examples of channel-free flows performed by thermoviscous expansion. (a) A water flow is induced by a laser beam scanning in the opposite direction, and visualized by fluorescent tracer particles. From Weinert & Braun (2008b). (b) Thermoviscous motion in an ice sheet: the ice melts in front of the heating spot, and moves in the same direction due to the thermal expansion. The image is a superposition of frames of the molten spot along a curved path, shown in dashed line. From Weinert et al. (2009). (c) Optical creation of a dilution series of biomolecules. The system is composed of two neighboring gels, one of them (at the bottom) containing fluorescein-marked biomolecules. The laser beam first creates a liquid channel including three chambers of distinct volumes (upper panel), then mixes the content of these chambers with the ambiant fluid by pumping along a pattern alternating horizontal and vertical stripes. From Weinert & Braun (2008b).

produce a flow, in a 10-µm-thick water layer, in the direction opposite to the laser scanning. Due to the very small thickness of fluid involved, the thermoconvection cannot be invoked as a driving mechanism in this case.

The thermoviscous paradigm has also been extended to the case of melting ice (Weinert et al., 2009). In that particular case, the scanning laser first melts the ice, the liquid motion is then driven by thermal expansion, and finally the liquid refreezes (Fig. 2(b)). The motion can be described as a thermoviscous pumping in the case where the water does not freeze in the channel, when the chamber is cooled above 0°C. However, as the water density increases with temperature below 4°C the fluid flow takes same direction as the scanning. Pumping velocities of several cm s^{-1} can be reached.

The creation of fluid flows along arbitrarily complex patterns can, in principle, provide an alternative to the design of rigid dedicated channels. To highlight the potentialities of the method for (bio)chemical applications, Weinert & Braun (2008b) created a dilution series by thermoviscous expansion. To this aim, they used a drop of agarose gel, gelated at room temperature, and molten by moderate heating (Fig. 2(c)). Biomolecules (30 kDa dextran marked with fluorescein) were added at the bottom part of the drop only, with a large amount of saccharose in order to avoid diffusion across the interface between the two halves of the gelated drop. The laser first draws a liquid channel along the two parts of the gel, creating in particular three liquid chambers of 65, 40, and 20 pL, respectively, in the upper part (initially without biomolecules). This step is represented in the upper row of Fig. 2(c). In a second step (lower row of Fig. 2(c)), the laser scans the gelated zones surrounding these chambers, along successive crossing lines. This scan enlarges the actual chambers, and dilute the biomolecules by mixing them with the molten agarose gel. As a result, a dilution series is created, with volume ratios of 4:1, 1:1, and 1:4 in equal volumes.

2.3 Manipulation of biological molecules: Diluting, trapping, replicating, and analyzing

Besides setting in motion a fluid, manipulating directly molecules of biological interest which might be dissolved in it is also of particular relevance. Such direct manipulation should indeed allow for precise tuning of the molecule concentration, and, further, for inducing particular reactions (especially, DNA replication).

2.3.1 DNA dilution or accumulation

As stated above, DNA exhibit a thermophobic behavior at room temperature (Braun & Libchaber, 2002). Therefore, the laser heating of a buffer solution of DNA deplete the zone at the vicinity of the spot due to the DNA thermophoretic drift, as illustrated in the right image of Fig. 3(a). Thermophoresis is therefore a convenient way of locally diluting a DNA solution. However, the most relevant issue is rather to concentrate molecules at a given point. Duhr & Braun (2006b) observed that the thermophoretic behavior of DNA could be reversed by simply cooling the sample: at 3°C, the DNA molecules become thermophilic and can therefore be trapped at the hot spot (left image of Fig. 3(a)). Besides this very simple method, several alternatives exist to perform effective DNA trapping.

One elegant way consists in opposing a liquid flow to the thermophoretic drift (Duhr & Braun, 2006a). This method seems particularly relevant in the lab-on-a-chip context due to its easy integrability into microfluidic channels. A 16-fold increase in DNA concentration was reached, at about 10 μm upstream from the beam axis, with a peak flow velocity of 0.55 $\mu m\,s^{-1}$. However, the time required to reach the equilibrium concentration profile is about 15 min, which limits the potentiality of the method for high-throughput applications.

By increasing the vertical temperature gradient effects, thermal convection can become significant. Figure 3(b) represents the effective DNA trapping by the interplay between these two mechanisms, as observed by Braun & Libchaber (2002). They considered a 50-μm-thick chamber, in the center of which a heating beam was focused. The top and bottom walls of the chamber were cooled to enhance the axial thermal gradient. The trapping mechanism is made of four main steps. First, the lateral thermophoresis drives the DNA molecules away from the heating spot (step 1). Then, the convection rolls carry the molecules downward, as the upward part of the rolls occur in the depleted zone close to the beam axis (step 2). The axial thermophoresis holds the DNA molecules at the chamber floor (step 3), where they finally accumulate at a radial position which result from the balance between lateral thermophoresis

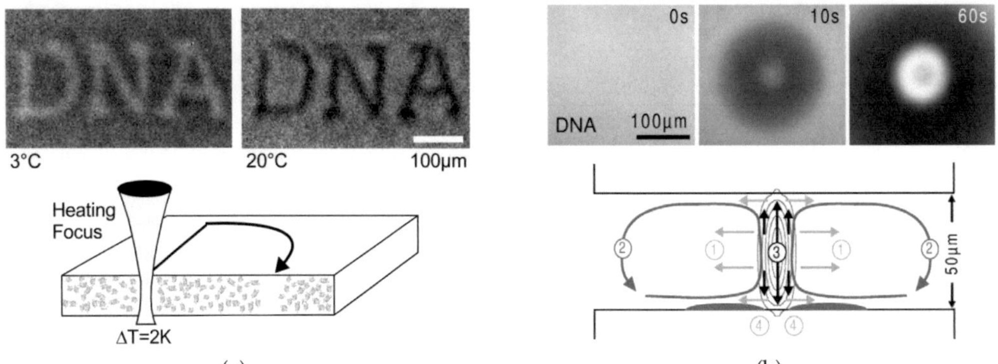

Fig. 3. Optothermal dilution and trapping of DNA. (a) Use of the temperature dependance of the Soret coefficient to write complex DNA-enhanced or DNA-depleted patterns. When the microfluidic chamber is cooled down to 3°C the DNA molecules are thermophilic (left picture), while they are thermophobic at room temperature (right picture). From Duhr & Braun (2006b). (b) DNA trapping by a combination of thermophoresis and thermoconvection. From Braun & Libchaber (2002).

and convection (step 4). The DNA molecules are therefore trapped in a ring-shaped pattern around the laser beam axis. Braun & Libchaber (2002) observed a 60-fold local increase in concentration at steady state, which is reached within 60 s, for a mean temperature of about 80°C in the chamber. They even increased significantly the trapping efficiency, by using a thicker chamber (500 μm) and a divergent laser beam, in a scheme comparable to that represented in Fig. 1(b). The enhancement of the convection effect compared to the lateral thermophoresis leads to a point-like trapping pattern along the beam axis. After 180 s, a concentration increase by a factor of 2,450 was measured (Braun & Libchaber, 2002).

Another interesting trapping mechanism involves the combination of thermophoresis and a bidirectional flow induced by thermoviscous expansion in a very thin (2 μm) liquid layer (Weinert & Braun, 2009). Let us consider a vertical slice of this liquid layer, along the scanning path of the laser beam. As the laser scans, say, from the right to the left, then the lower part of the fluid will flow from the left to the right. However, the mass conservation applied together with lateral boundary conditions impose a symmetric counterflow in the upper part of the slice. Let us now reproduce at high rate the same scanning pattern, in such a way that each slice draws a radius of a circular fluid pancake, as illustrated in Fig. 4(a). The resulting flow pattern is therefore a toroidal roll, with a centrifugal flow at the bottom of the cell, and its centripetal counterpart at the top. In the meantime, the vertical temperature gradient drives the DNA molecules upward by thermophoresis. It means that the centrifugal flow concerns rather DNA-depleted fluid, while the centripetal flow advects more DNA molecules toward the center, where they accumulate.

As the scanning pattern is radial, the average trapping position is stationary. However, it can be still moved by displacing the average position of the scanning laser, allowing to collect particles over longer ranges. The so-called optothermal conveyor has been demonstrated efficient with small beads as well as DNA molecules (Fig. 4(b)).

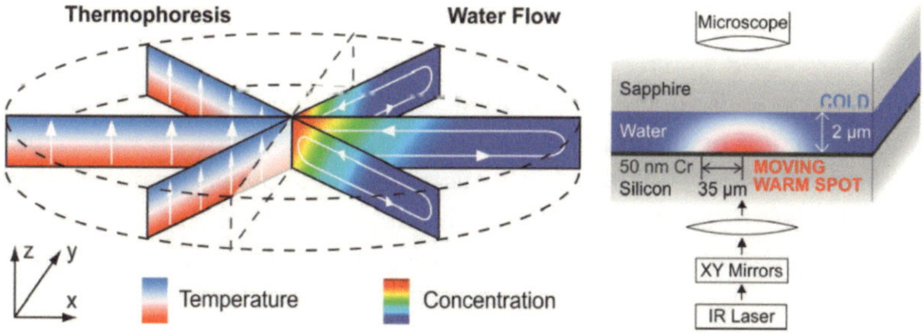

(a)

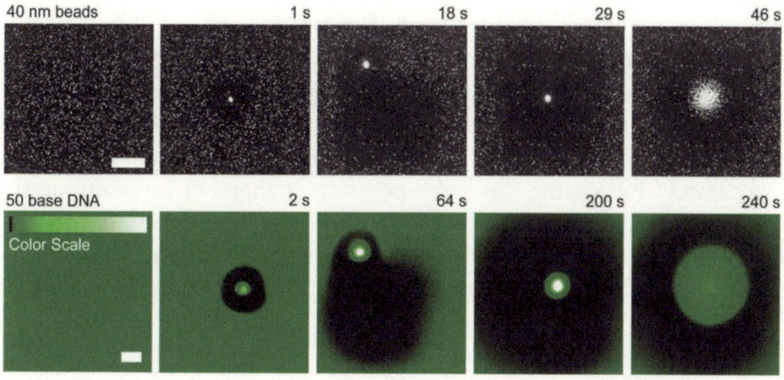

(b)

Fig. 4. Optothermal molecule conveyor (Weinert & Braun, 2009). (a) A radial centripetal laser scanning leads to the efficient trapping of molecules by a combination of thermophoresis and thermoviscous flow. The heating is provided at the bottom surface of the chamber by the laser absorption by a thin chromium coating. Typical warm spot radius is 35 μm. (b) Optical conveyor: the optothermal trap is moved along arbitrary patterns in order to collect particles (40 nm polystyrene beads, top row) or DNA molecules (bottom row, DNA concentration is given in color scale). After the laser has been switched off, the trapped objects are released and diffuse freely (right frames). The scale bars are 100 μm.

2.3.2 DNA replication

The investigations reported above highlight the possibility of DNA accumulation by purely thermal mechanisms. On the other hand, thermal convection has been shown to be able to drive the replication of DNA. The most popular DNA replication mechanism used by molecular biologists is polymerase chain reaction (PCR). Basically, each DNA molecule first dissociates into two single strands when heated at 95°C. In a second step (in the 55-65°C temperature range), pairs of short DNA fragments called primers, exhibiting sequences complementary to that to be amplified, anneal at the end of each single strand. Each of the two so-generated double helices finally elongate, by enzyme-activated replication of the complementary part of the initial strand (72°C). The process is then repeated cyclicly, each reaction cycle doubling the concentration of DNA exhibiting the targeted sequence. As pointed out by Braun et al. (2003), DNA molecules carried along a circular convection streamline experience a cycling change in temperature which mimic the temperature pattern of a PCR. Mast & Braun (2010) then combined the trapping (by thermoviscous expansion and thermophoresis) and convective replication mechanisms in a capillary. Beyond the biotechnological interest of such a combination, it opens new perspectives for fundamental studies on the molecular evolution of life. Indeed, the two pillars on which the Darwinian evolutional theory relies, namely, the duplication of genetic material, and its storage against molecular diffusion, are retrieved. Moreover, inhomogeneously heated microfluidic chambers can be viewed as model systems reproducing the pores of hydrothermal rocks in the deep oceanic floor, in which life could have originated (Braun & Libchaber, 2004).

2.3.3 Analysis of biomolecular binding

Understanding the interactions between biomolecules, or between a particular biomolecule and its environment, is of crucial relevance for medical applications. Very recently, Braun and co-workers have proposed the use of thermophoresis as a probe of these interactions. As the thermophoretic properties of a solute depend on its interactions with the solvent (or, more generally, with its environment), they indeed proposed to quantify the biomolecular binding by accurately measuring the corresponding changes in thermophoretic depletion. This method has thus been used to quantify the aptamer-target interactions in a buffer solution (Baaske et al., 2010), and then generalized to various protein-protein and protein-ion binding in buffer solutions as well as in more complex biological liquids (Wienken et al., 2010).

2.3.4 Single molecule stretching

Besides the transport and analysis of DNA samples, several studies investigated the stretching of individual DNA molecules under the action of a laser-induced thermal gradient. Ichikawa et al. (2007) characterized the elongation of long DNA chains by the hydrodynamic stresses arising from thermal convection. Jiang & Sano (2007) anchored a DNA molecule by one or two ends and observed its deformation when located at a given distance from a heating laser. One-end-anchored molecules appear elongated along the direction opposite to the temperature gradient, while the two-end-anchored exhibit an arc-like shape when the laser is approached between the two bonded points. As the authors ensured the convection to be negligible, they interpreted the deformations to result from the thermophoretic drift of the movable parts of the molecule. From their observations, they calculated a tension force of about 70 fN for a 3 $\mathrm{K\,\mu m^{-1}}$ temperature gradient. These methods for studying the physical properties of biopolymers compete with others, such as AFM, optical or magnetic tweezers, by their contact- and probe-free characters.

3. Two-fluid flows

Let us now turn to the case of diphasic flows. We consider here, more generally, the case of a liquid droplet (a possible microreactor) immersed in, or floating on, another fluid, both of them being not miscible with each other. Controlling the microfluidic transport should here mean, essentially, controlling the motion of these microreactors. Then, hereafter is an overview of techniques to push, pull, divert, sort, or broadly speaking, manipulate droplets.

This section is divided into two parts. The first part deals with the direct manipulation of droplets through interfacial tension effects. The second one gathers different approaches involving one (or several) change(s) of state, in which at least two fluid phases are involved.

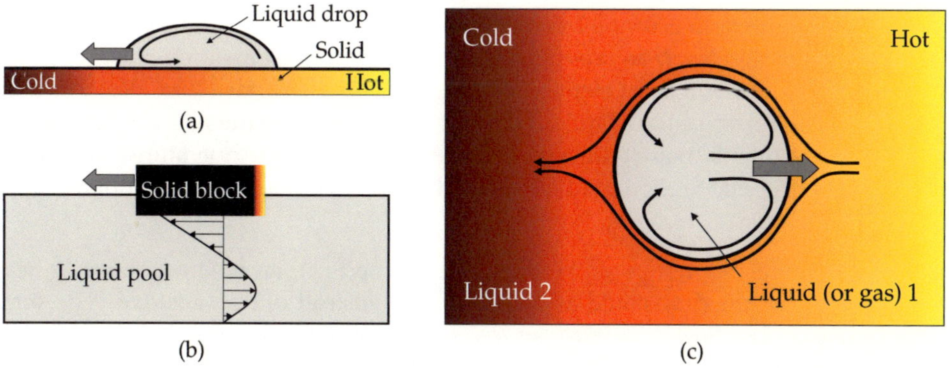

Fig. 5. Principle of thermocapillary migration, considering a surface (or interfacial) tension decreasing with temperature. (a) A liquid partly wetting a heated solid substrate is driven toward the colder regions by 'rolling' on the surface (as for a tracked vehicle). (b) A solid block floating on a liquid pool and heated at its end is propelled in the opposite direction (the schematic velocity field is in the laboratory frame). (c) An immersed droplet (or bubble) migrates to the warm by a 'swimming' motion.

3.1 Optocapillary effect

A spatial unbalance of surface tension along a liquid surface creates surface stresses. This effect, responsible for example for the phenomenon of 'wine tears', which can be observed above the free surface of wine in a clean glass, has been evidenced by Marangoni in 1871 and named after him. Over the past decade, this effect has known a renewed interest as it opens important vistas on the fluid manipulation at small scales. These potentialities are justified by the increased surface-to-volume ratio at small scales, which tends to favor interfacial effects compared to volume effects.

Basically, Marangoni effect appears as soon as a gradient of interfacial tension[1] is created, which can be achieved essentially through gradients of chemical composition or temperature. The resulting effect is then related as 'solutocapillary' or 'thermocapillary', respectively. For example, wine tears result from a gradient in ethanol concentration due to the excess evaporation in the rising wetting film, which increases the surface tension in these zones. Inhomogeneities in surfactant interfacial concentration can play the same role. Even though solutocapillary effect can be particulary relevant in microfluidic systems, this section will focus on thermocapillary effect—more precisely, 'optocapillary', i.e., thermocapillary effect in the special case where the temperature gradient is optically induced. Nevertheless, the role of surfactants can be significant, as we will see hereafter, to understand experimental features.

3.1.1 Thermocapillary migration: Basic principles

Creating an interfacial stress allows for setting in motion a fluid element (bubble, drop, liquid film), which we call thermocapillary migration, providing that a condensed surrounding medium can support this motion. For example, what follows would never happen if

[1] The term 'interfacial tension' is related to a liquid-liquid interface, when 'surface tension' is used for liquid-gas free surfaces. However, the phenomena presented in the following under the denomination 'interfacial tension' can be generalized to the free surface case.

considering a free levitating droplet in vacuum. Thermocapillary migration can therefore occur in the following cases, as reported in Fig. 5: a liquid element on a solid substrate (Fig. 5(a)), or conversely a solid element floating on a liquid surface (Fig. 5(b)), or a fluid element (either liquid or gas) in a liquid (Fig. 5(c)). In all cases considered here, the interfacial tension is supposed to decrease with temperature, so the driving temperature gradient leads to an inverted interfacial tension gradient—and therefore, interfacial stresses opposed to the temperature gradient.

In the first case (Fig. 5(a)), a partly wetting liquid drop is on an inhomogeneously heated substrate—note that the drop could as well be heated instead of the substrate. The surface stresses drive a surface flow at the droplet free surface, an opposite counterflow then results from mass conservation. Due to the no-slip condition at the solid-liquid boundary, this flow sets the drop in motion toward the high surface tension regions (Cazabat et al., 1990; Darhuber et al., 2003). Note that the particular case of liquids on solid surfaces offers several additional degrees of freedom to spatially modulate the surface tension, including surface texturation, chemical patterning, or electrocapillarity. Further improvements can be found in the review by Darhuber & Troian (2005).

The symmetric case, less documented in the literature, involves an inhomogeneously heated solid, floating on a liquid surface (Fig. 5(b)). The surface stresses symmetrically drive the fluid on either sides of the hot point (creating a deeper counterflow as well). However, as the solid is heated from one side, the surface flow carries it toward the less heated direction.

The case of immersed bubbles (Fig. 5(c)) has been first treated half of a century ago by Young et al. (1959), then extended to droplets (Barton & Subramanian, 1989; Hähnel et al., 1989). It seems to be the most relevant to droplet microfluidic systems as no solid part is involved, avoiding any difficulty related to the liquid wetting (Chen et al., 2005). Here, the interfacial stresses drive an interfacial flow, in both sides of the interface, toward the colder part of the droplet. In the droplet, this flow creates internal rolls.[2] In the surrounding medium, this flow drags the bulk fluid, which in turn propels the droplet in the opposite direction according to the action-reaction principle. A simple analogy can be made with a swimmer, who drags fluid backward and then moves forward. Young et al. (1959) quantified the thermocapillary migration velocity of a bubble, and extended their calculation to droplets, in an infinite medium:

$$\mathbf{u}_{\text{thermocapillary}} = -\frac{2}{3\eta_1 + 2\eta_2}\frac{R}{2 + \Lambda_1/\Lambda_2}\frac{\partial\sigma}{\partial T}\boldsymbol{\nabla}T. \tag{4}$$

Here, R is the drop radius, Λ the thermal conductivity, and σ the interfacial tension. The subscripts 1 and 2 denote the droplet and surrounding phases, respectively. Considering a water droplet ($\eta_1 = 1$ mPa s and $\Lambda_1 = 0.6$ W K^{-1} m^{-1}), of radius 100 µm, in silicone oil ($\eta_2 \simeq 10$ mPa s and $\Lambda_2 \simeq 0.1$ W K^{-1} m^{-1}), with $\partial\sigma/\partial T \sim -0.1$ mN m^{-1} K^{-1}, a temperature gradient of 1 K mm^{-1} leads to a migration velocity of 110 µm s^{-1}. This value rises up to 5 mm s^{-1} in the case of a gas bubble in water.

Beyond this ideal background, two main disturbing effects must be taken into account in microfluidic environments. First, the confining effect of channel walls would quantitatively

[2] The bubble case significantly differs: as gas is inviscid, the interfacial flow does not diffuse in the bulk gas phase, and then no flow is produced.

or qualitatively modify the physics of thermocapillary migration. Considering a squeezed bubble in a Hele-Shaw cell, Bratukhin & Zuev (1984) showed that the friction to walls reduces the migration efficiency, but the scaling $u_{\text{thermocapillary}} \sim -R\nabla T$ remains valid. The case of elongated bubbles in capillary tubes has then been treated both theoretically (Mazouchi & Homsy, 2000; 2001) and experimentally (Lajeunesse & Homsy, 2003). In the case of a polygonal cross-sectioned channel, they evidenced the strong influence of liquid flow through both corners and wetting films on the migration velocity. Especially, for a rectangular cross section the migration velocity then varies non-trivially with the channel aspect ratio, and no longer depends on the bubble size, due to the fluid recirculation through the corners. Furthermore, the recirculating flows through the wetting films add nonlinear corrections.

On the other hand, diphasic microfluidics often involves surfactants. Several investigations, both experimental (Barton & Subramanian, 1989; Chen et al., 1997) and theoretical (Kim & Subramanian, 1989a;b), have shown that insoluble surfactants tend to reduce the thermocapillary migration efficiency. A recent theoretical study by Khattari et al. (2002) has proposed to account for the surfactant effect by writing an 'effective' variation of interfacial tension with temperature, which for insoluble surfactants reads as

$$\left(\frac{\partial\sigma}{\partial T}\right)_{\text{eff}} = \left(\frac{\partial\sigma}{\partial T}\right)_{\Gamma} + \frac{\left(\frac{\partial\sigma}{\partial\Gamma}\right)_T}{\left(\frac{\partial^2\sigma}{\partial\Gamma^2}\right)_T}\left[\frac{1}{T}\left(\frac{\partial\sigma}{\partial\Gamma}\right)_T - \frac{\partial^2\sigma}{\partial T\partial\Gamma}\right], \tag{5}$$

with Γ the interfacial concentration. Although difficult to link with measurable quantities, the corrective term accounting for surfactant effect is expected to be positive as it reduces the effective variation of interfacial tension.

What was described above is related to thermocapillary effect in general. Let us now focus more specifically on optocapillary effect.

3.1.2 Optocapillary propelling

The three configurations depicted in Fig. 5 have been experimentally explored in an optocapillary scheme, as illustrated in Figs. 6 and 7.

The optocapillary migration of a liquid object (a film) on a heated solid surface is represented in Fig. 6(a). Garnier et al. (2003) considered an horizontal film, inhomogeneously heated by a light pattern which superimposes a gradient of intensity perpendicular to the contact line, and a sinusoidal intensity fluctuation along it. As a result, the contact line, mostly the less enlighten part, moves toward the darker edge, drawing the wavy-like pattern shown on Fig. 6(a). By varying the spatial periodicity of the illumination, the authors studied quantitatively the contact line instability.

Besides this study, relatively few publications are related to the optocapillary migration in the 'rolling droplet' configuration. However, this configuration is well adapted to the optical heating through surface plasmon decay (Farahi et al., 2005; Passian et al., 2006).

Thermocapillary migration of a solid floating object, within a scheme as represented in Fig. 5(b), is practically difficult to achieve since this requires the heating of a small movable object—heating the pool itself is however possible. Laser-induced heating is therefore a

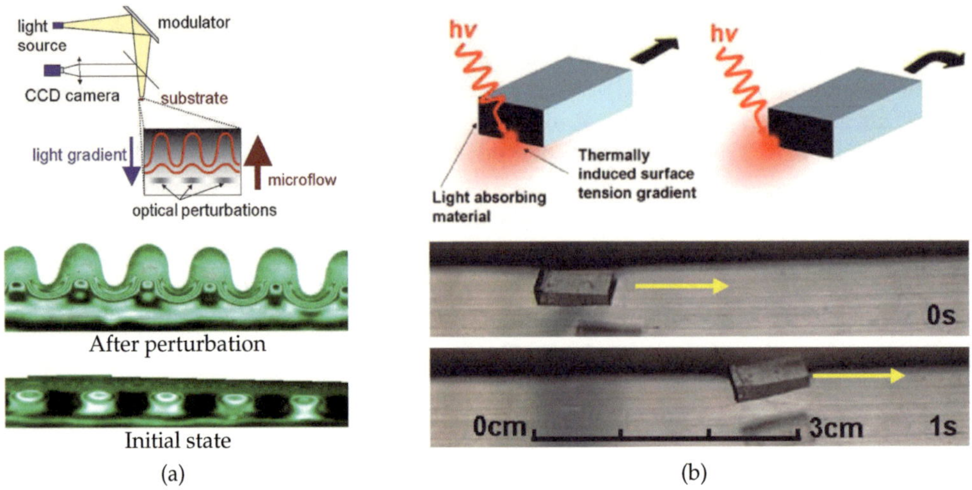

Fig. 6. Optocapillary migration at the liquid-solid boundary. (a) Migration of the contact line of an horizontal thin film induced by optical heating. The light intensity is spatially modulated in order to superimpose an horizontal light gradient perpendicular to the contact line and a sinusoidal fluctuation along it. Adapted from Garnier et al. (2003). (b) A millifluidic boat: a floating PDMS block, coated by an absorbent material at its back, is set in motion by light. Straight as well as turning motions can be achieved, depending on the irradiated point position compared to the object main axis. After Okawa et al. (2009).

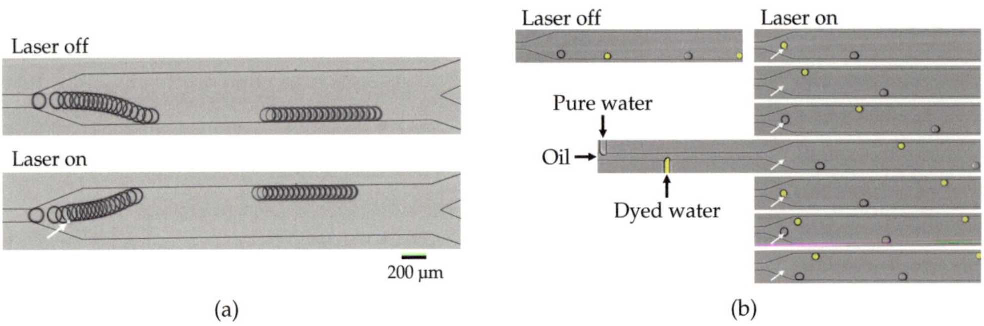

Fig. 7. Reverse optocapillary migration of immersed liquid droplets, in the presence of surfactants. (a) Superposition of frames showing droplet switching. Water droplets flowing in oil are optocapillary pushed away, and forced to flow through the upper part of the channel, of higher hydrodynamic resistance. (b) Application to droplet sorting: as water is transparent at the laser wavelength, droplets containing dye (colored on the frames) are diverted while pure water droplets are not. The white arrows depict the laser position. Adapted after Robert de Saint Vincent et al. (2008).

good solution as the focal point can follow the object motion. Okawa et al. (2009) realized a 'millifluidic boat' (Fig. 6(b)) by enlighting the back face of a PDMS block, coated with an absorbent material. The solid block then flow, at velocities up to the cm s^{-1} range, along remote-controlled trajectories, which can be either linear or curved depending on the relative location of the light spot in respect with the solid main axis.

As pointed out by the representation of Fig. 5(c), an immersed bubble should be attracted by the hot point. In the case of a laser heating, the bubble is trapped (Berry et al., 2000; Marcano & Aranguren, 1993), which implies that setting a bubble in permanent motion requires the laser to write this motion by moving the spot in the absorbing medium (Ohta et al., 2007). Rybalko et al. (2004) considered a slightly different case, in which an hemispherically-shaped absorbing droplet of nitrobenzene, floating on water, is alternatively shined at its front or at its back interface by simply changing the optical path from the above to the bottom part of the droplet. The resulting remote-controlled motion is alternatively forward and backward. Finally, another interesting alternative scheme has been presented recently by Nagy & Neitzel (2008), who set in levitation a laser-heated droplet above a solid substrate by using the thermocapillary flow to continuously feed the lubrication film. They were able to impose a translating motion to the drop, at velocities in the $\mathrm{mm\,s^{-1}}$ range.

The reversal of optocapillary motion has been reported in several recent studies (Baroud et al., 2007a; Dixit et al., 2010). While poorly understood yet, this reversal is expected to be related to the coupling between thermo- and solutocapillary effects. Indeed, eq. 5 suggests that the effective variation of interfacial tension with temperature could become positive provided that the term describing the influence of interfacial concentration is positive and larger than the pure temperature variation. However, answering decently this question would require much more complex calculations, involving coupled equations of heat and surfactant transport, which remain to date a numerical challenge.

As a consequence of this reversal, the thermocapillary migration is now oriented in such a way that a droplet is repelled by the heating light. This is illustrated in Fig. 7. Water droplets are carried by an oil flow in an enlarged channel with two outlets, and naturally flow through the lower outlet which has a lower hydrodynamic resistance. These droplets are passively diverted and then forced to flow through the other outlet by optocapillarity (Fig. 7(a)). This optocapillary effect is provided by heating the water droplets by a visible laser beam, a dye being added to the water to ensure light absorption. Such a scheme can be used to passively sort droplets on the basis of their optical properties (transparent or absorbent), as shown in Fig. 7(b). Considering alternate droplets of pure water and droplets containing dye (emphasized in color on the pictures), only the dyed droplets are diverted, while the others, which do not feel the laser heating, continue through the lower channel (Robert de Saint Vincent et al., 2008). Alternative applications, such as optocapillary pinball and conveyor, have also been demonstrated in the same view (Cordero et al., 2008).

3.1.3 Optocapillary blocking

In the case where $(\partial\sigma/\partial T)_{\text{eff}} > 0$, the laser beam can be viewed as a sort of 'soft wall' of tuneable stiffness. This ability to optically repel a droplet opens the way to the realization of optofluidic components (Baroud et al., 2007b). As represented in Fig. 8, the optocapillary blocking applied to a growing droplet in a flow focusing geometry (Anna et al., 2003) can interrupt the motion of the leading interface during several seconds. During the meantime, the water flow feeds the growing droplet. When eventually released, the blocked droplets are therefore larger than the unblocked ones, all the more so the blocking time has been long. Besides the ability to instantaneously interrupt a droplet flow, the optocapillary blocking of growing droplets provides a means of tuning in real time the droplet volume, by adjusting the blocking time via the beam power, without any action on the imposed flow parameters.

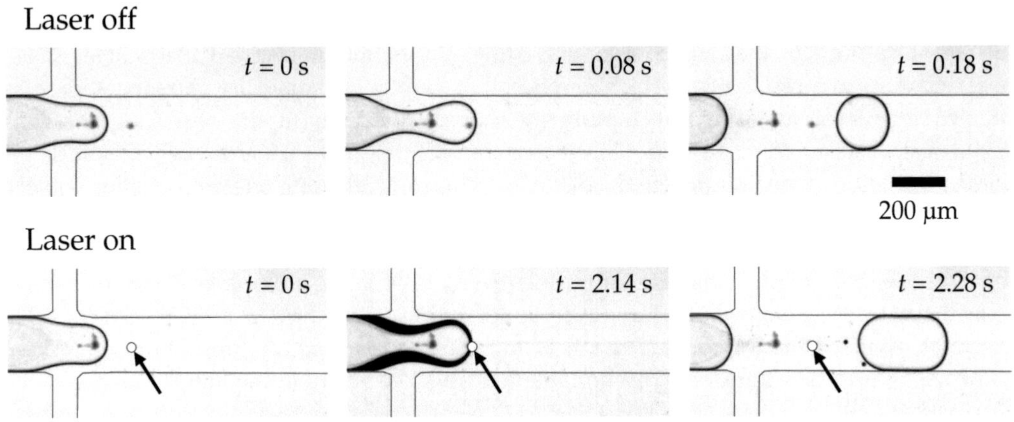

Fig. 8. Optocapillary blocking of the leading interface of an absorbent water droplet in oil during its formation. The picture representing the blocking process (lower row, center) is a superposition of 100 successive frames. The arrows depict the laser position. Adapted from Baroud et al. (2007a).

The optocapillary force exerting on a droplet during its blocking has been estimated theoretically (Baroud et al., 2007a) and experimentally (Verneuil et al., 2009) to be in the range of 0.1 µN with typical laser powers of 100 mW.

3.1.4 Splitting, merging and mixing

An important issue in the scope of application of droplets as microreactors is the ability of further controlling their volume and composition. This includes the calibration of droplets, to accurately adjust the quantity of reagents to be mixed, and the droplet coalescence. Both these operations have been performed optically (Fig. 9). First, calibrating the droplets can be done during their formation, as shown above (see Fig. 8). A subsequent modification of droplet volume would then require to split the droplet. Link et al. (2004) proposed a purely geometrical passive splitting scheme: a droplet arriving at a diverging junction splits into two equal or unequal parts depending on the hydrodynamic resistance (the length, in that work) of the diverging arms. The channel asymmetry can be reproduced, in a tunable way, by optocapillary repelling the droplet as it arrives at the T junction (Fig. 9(a)). A controllable part of the incoming droplet, depending on the beam power, is then forced to flow through the opposite channel (Baroud et al., 2007b). It has also been shown that, above a threshold power, the droplet is totally diverted and no longer splits.

Merging droplets is the symmetrical counterpart of the splitting operation. Due to the presence of surfactants, droplets are prevented from spontaneous coalescence. Interestingly, two very different approaches manage this issue—considering a $(\partial\sigma/\partial T)_{\mathrm{eff}} > 0$ case in both. One consists in pushing a droplet toward a neighboring one, until contact, as represented on the top row of Fig. 9(b). The subsequent 'remote-controlled' merging is interpreted through the formation of a metastable bilayer surfactant film (Dixit et al., 2010; Kotz et al., 2004). The alternative approach consists in placing the laser spot at the interface that separates two contacting droplets (Baroud et al., 2007b). The bottom panel of Fig. 9(b) represents a train of contacting, noncoalescing droplets, flowing from the left to the right. Droplets then merge as

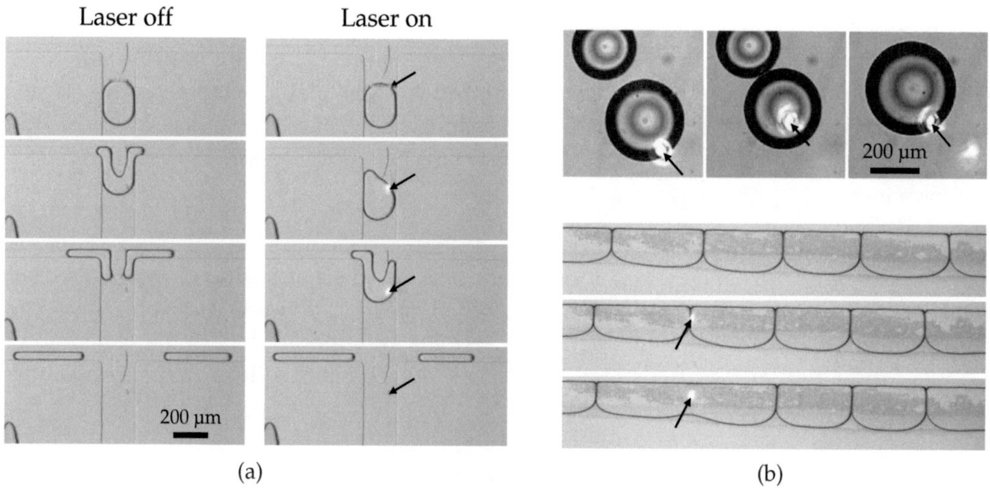

Fig. 9. Droplet splitting and merging. (a) Controlled splitting of droplet arriving at a symmetric T junction. The laser breaks the symmetry by reducing the water flow through the right channel. Adapted from Baroud et al. (2007b). (b) Optocapillary merging of droplets. Top row: the laser beam pushes a droplet away up to another droplet, then the two droplets coalesce (Dixit et al., 2010). Bottom panel: successive contacting droplets, flowing from the left to the right, are repeatedly merged when the interface is shined (Baroud et al., 2007b). The arrows depict the laser position.

the interface crosses the laser beam. Despite its reproducibility, this coalescing process remains misunderstood to date. In fact, Dell'Aversana et al. (1996) have shown, in an experimental scheme which can be viewed as comparable to that presented here, that thermocapillary flows have a preventing influence from coalescence, as the interfacial flow feeds the lubrication film which separates two contacting droplets. Therefore, while the film drainage could be possible in the 'remote-controlled' coalescence scheme where interfacial flows, directed toward the laser beam, drain the lubrication film off, the droplet coalescence in the flowing scheme is rather surprising.

Even though fusing droplets is a prerequisite to perform reactions in droplets, mixing the reagents after the merging is also necessary. However, in the microfluidic world mixing is unfavored due to the laminar character of the flows (low Reynolds number), which limits the advective mass transfers transversally to the main flow. Mixing would thus be driven by diffusion alone, requiring by the way unacceptably long times. Efficient mixing therefore requires a chaotic, 'stretching and folding', flow (Ottino & Wiggins, 2004).

As observed in Fig 5(c), thermocapillary stresses create rolls inside a droplet. Such rolls could be good candidates to perform effective mixing in microfluidic droplets (Grigoriev, 2005). However, the dipolar flow pattern created at steady state by a single heating source is not sufficient to induce mixing, as the streamlines do not intercross. A combination of dipolar and higher-order flows is thus required. By scanning a nanoliter droplet with a heating laser beam along a two-dimensional pattern, Grigoriev et al. (2006) induced chaotic mixing inside the droplet, through a combination of a bulk thermoconvective flow and an interface-driven thermocapillary flow. Cordero et al. (2009) used an optocapillary microdroplet blocking scheme (see Fig. 8), combined with spatial and temporal light modulation techniques. They compared the mixing efficiency resulting from two stationary or one rapidly alternating heating beams and demonstrated that, despite the fact the spatial patterns are equivalent, only the non-stationary flow pattern produces mixing.

3.2 Flows induced by liquid-gas phase transitions

Up to now, we have essentially considered continuous changes on the fluid properties. Besides, we have also mentioned the specific use of phase transitions (namely, sol-gel transitions) to provide or prevent fluid motion when one fluid phase is involved. This section deals with phase transitions involving liquid and gas systems, applied to the generation or control of fluid motion. The two main approaches reported in the literature are presented in Fig. 10.

3.2.1 Successive evaporation-condensation cycles

A first approach consists in displacing tiny volumes of fluid by inducing successive cycles of evaporation-condensation-coalescence. Liu et al. (2006) considered a dilute solution of photothermal nanoparticles, which absorb light from a laser beam. As schematized on Fig. 10(a), the laser beam close to the leading edge of the liquid film first provides liquid evaporation. As the evaporated liquid cools down, it condensates and forms tiny droplets ahead the liquid film contact line. These droplets eventually coalesce and merge into the initial liquid film. This process results in an advance of the contact line: by repeating it after laser translation, a continuous flow can be obtained along the beam path. This flow can be guided laterally when the manipulation takes place in straight channels. Then, the laser can drive the fluid motion along a selected path, as illustrated in the experimental pictures in the right part of Fig. 10(a). Finally, the authors demonstrated the transport of Jurkat T-cells embedded in the solution.

Liu et al. (2006) experimentally obtained flows at velocities up to several hundreds of $\mu m\,s^{-1}$. According to the time required by the different mechanisms involved (heating, evaporation, condensation, coalescence, film advance), they estimated a maximal possible flow velocity of $1\,mm\,s^{-1}$.

Boyd et al. (2008) have recently proposed an alternative scheme involving the transfer of mass across a bubble in a partly filled microfluidic channel. A gas bubble is formed in the liquid phase, and a heating laser beam is focused in the liquid just behind the bubble. By slightly increasing the temperature at the rear of the bubble, the laser induces evaporation and the vapor then condenses at the front interface. This leads to a net fluid transfer across the bubble. The authors then applied this method to the distillation of a dye solution, the transferred solution being dye-free while the fluorescence enhancement can be observed in the untransferred one.

3.2.2 Flow actuation through bubble nucleation

The second phase transition-related approach is based on the nucleation of bubbles by laser. The simplest way of nucleating bubbles is to heat an absorbent fluid enough to reach the boiling point. Alternatively, a nonlinear process, called laser-induced cavitation, can be used

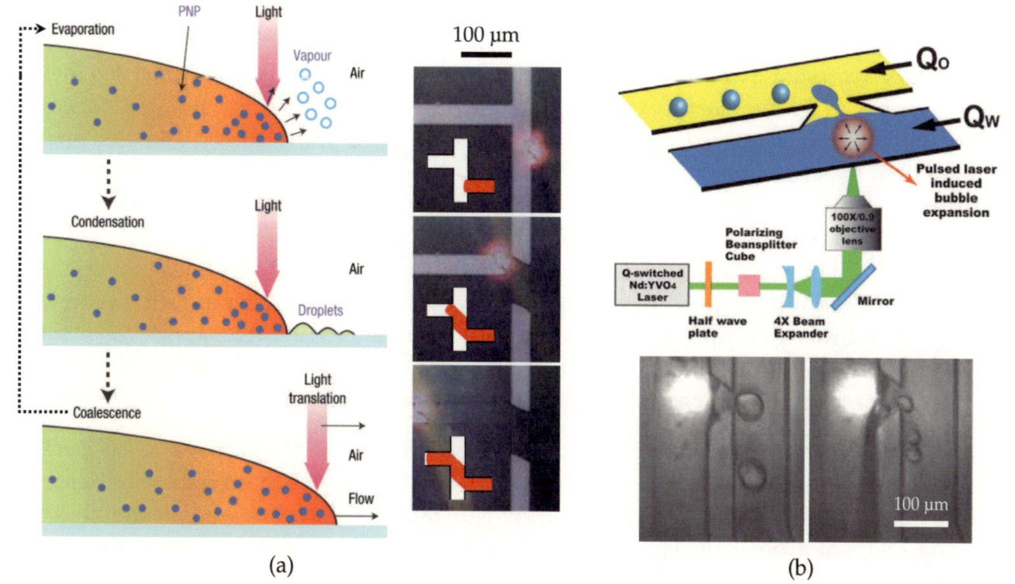

Fig. 10. Phase transition-driven microfluidic flows. (a) A laser beam induces successive evaporation-condensation-coalescence cycles by heating a diluted photothermal nanoparticle (PNP) solution. The resulting fluid motion can follow complex paths inside a microfluidic channel (right pictures). Adapted from Liu et al. (2006). (b) Cavitation-induced droplet generation: a cavitation bubble created by a nanosecond pulsed laser expands and pushes a controlled volume of water into a parallel oil flow. Bottom pictures illustrate the produced bubbles, for pulse repetition rates of 0.5 (left) and 10 kHz (right, the three droplets shown here result from three successive pulses). Adapted from Park et al. (2011).

in transparent liquids (Vogel et al., 1989). The basic mechanism can be summarized as follows. A high-intensity light pulse is absorbed by an impurity contained in the transparent liquid, triggering optical breakdown. This creates a localized high-temperature and high-pressure plasma, which rapidly expands, resulting in shock wave emission and bubble formation. The bubble then grows and collapses in times in the millisecond range.

This bubble nucleation can be used to actuate fluid flows. Very recently, Park et al. (2011) used laser-induced bubble cavitation to trigger droplet formation in an oil flow in microfluidic channels. As represented in Fig. 10(b), water and oil flow in separated parallel channels, connected by a straight junction channel. A nanosecond pulsed laser, tightly focused in water close to this junction, induces cavitation bubble formation, which pushes water as it expands. The volume of water pushed in the oil channel is then dragged by the oil flow. By varying the pulse repetition rate, the authors produced monodisperse droplets at rates ranging from 0.5 up to 10 kHz. A comparable actuation process (while not involving the cavitation process) has also been proposed to eject particles or cells trapped in microfluidic traps, which is an elegant way of resetting trap-and-release-based microarray systems (Tan & Takeuchi, 2007; 2008). Finally, while not directly connected to fluid manipulation, biophotonics applications of laser-induced cavitation such as phototransfection (Stevenson et al., 2010) represent a current active research field.

4. Conclusions and prospects

Light fields represent a convenient and versatile tool to drive fluid transport by thermal effects. As seen in the present chapter, very different effects can be induced, involving either continuous changes in the fluid properties or phase transitions. The efficiency of opto-thermal effects, in addition to the specific assets of optical methods, set them as serious alternatives to non-optical microfluidic manipulation techniques—e.g. electric fields or microfabrication-based techniques. One key point in favor of these opto-thermal approaches is that they keep most of the advantages associated to the conventional optical manipulation techniques, without falling short of the high-throughput requirements in terms of force or velocity. The heating of the medium, sometimes relatively strong, can appear as a severe restriction, especially for biological application. However, as optical heating can be very localized, the inconvenience caused to the sample can be circumvent—or, at least, limited.

In addition, the diversity of opto-thermal approaches opens perspectives to cooperative effects. A good example is given by the optical conveyor, in which thermophoresis depletion of DNA and thermoviscous pumping, brought together, ultimately lead to the ability of trapping DNA at a position which can be changed at will. In the same way, the diverse complementary applications of optocapillary effect—blocking and propelling, splitting and merging—represent a good example of the high degree of versatility which can be reached by the same effect, when used in complementary means. Therefore, one can imagine that ultimately, light would be able to perform all operations relevant to lab-on-a-chip devices, ranging from sample preparation (dilution, concentration enhancement) to the final analysis (imaging and spectroscopy, which is beyond the scope of this review), including all steps associated with transport (sampling, carrying, sorting) and reactions (mixing). The optical lab-on-a-chip paradigm is furthermore compatible with droplet microfluidics.

From a more fundamental point of view, opto-thermal fluid manipulation techniques provide tools opening new perspectives in a broad field spectrum. One of the most exciting basic research directions for the next future is the investigation of prebiotic evolution of life, as suggested by the works of Braun and co-workers on thermally-driven DNA concentration and replication in microfluidic porous media. Still in the life sciences field, investigating biomolecular interactions, what has been proven a promising development for thermophoresis, is highly relevant for public health questions, as poorly-understood antigene-antibody interactions play a major role in medical treatments. Another direction is the investigation of biological and soft matter properties. This has been suggested, for example, by the investigations performed on single DNA molecule stretching by thermal convection, or submolecular thermodiffusion. Likely, optocapillarity provides a tool to investigate the behavior of interfaces submitted to strongly inhomogeneous stresses. Connexions could be found in relation with the study of breakup phenomena, which remains an active field of research since the second half of the nineteenth century.

5. References

Anna, S. L., Bontoux, N. & Stone, H. A. (2003). Formation of dispersions using "flow focusing" in microchannels, *Appl. Phys. Lett.* 82: 364–366.
URL: *http://link.aip.org/link/?APL/82/364/1*

Ashkin, A. (1970). Acceleration and trapping of particles by radiation pressure, *Phys. Rev. Lett.* 24: 156–159.
URL: *http://link.aps.org/doi/10.1103/PhysRevLett.24.156*

Ashkin, A., Dziedzic, J. M., Bjorkholm, J. E. & Chu, S. (1986). Observation of a single-beam gradient force optical trap for dielectric particles, *Opt. Lett.* 11: 288–290.
URL: *http://ol.osa.org/abstract.cfm?URI=ol-11-5-288*

Baaske, P., Wienken, C. J., Reineck, P., Duhr, S. & Braun, D. (2010). Optical thermophoresis for quantifying the buffer dependence of aptamer binding, *Angew. Chem. Int. Ed.* 49: 2238–2241.
URL: *http://dx.doi.org/10.1002/anie.200903998*

Baroud, C. N., Delville, J.-P., Gallaire, F. & Wunenburger, R. (2007a). Thermocapillary valve for droplet production and sorting, *Phys. Rev. E* 75: 046302.
URL: *http://link.aps.org/doi/10.1103/PhysRevE.75.046302*

Baroud, C. N., Robert de Saint Vincent, M. & Delville, J.-P. (2007b). An optical toolbox for total control of droplet microfluidics, *Lab Chip* 7: 1029–1033.
URL: *http://dx.doi.org/10.1039/B702472J*

Barton, K. D. & Subramanian, R. S. (1989). The migration of liquid drops in a vertical temperature gradient, *J. Colloid Interface Sci.* 133: 211–222.
URL: *http://dx.doi.org/10.1016/0021-9797(89)90294-4*

Berry, D. W., Heckenberg, N. R. & Rubinsztein-Dunlop, H. (2000). Effects associated with bubble formation in optical trapping, *J. Mod. Opt.* 47: 1575–1585.
URL: *http://www.tandfonline.com/doi/abs/10.1080/09500340008235124*

Boyd, D. A., Adleman, J. R., Goodwin, D. G. & Psaltis, D. (2008). Chemical separations by bubble-assisted interphase mass-transfer, *Anal. Chem.* 80: 2452–2456.
URL: *http://dx.doi.org/10.1021/ac702174t*

Boyd, R. D. & Vest, C. M. (1975). Onset of convection due to horizontal laser beams, *Appl. Phys. Lett.* 26: 287–288.
URL: *http://link.aip.org/link/?APL/26/287/1*

Bratukhin, Y. K. & Zuev, A. L. (1984). Thermocapillary drift of an air bubble in a horizontal Hele-Shaw cell, *Fluid Dyn.* 19: 393–398.
URL: *http://dx.doi.org/10.1007/BF01093902*

Braun, D., Goddard, N. L. & Libchaber, A. (2003). Exponential DNA replication by laminar convection, *Phys. Rev. Lett.* 91: 158103.
URL: *http://link.aps.org/doi/10.1103/PhysRevLett.91.158103*

Braun, D. & Libchaber, A. (2002). Trapping of DNA by thermophoretic depletion and convection, *Phys. Rev. Lett.* 89: 188103.
URL: *http://link.aps.org/doi/10.1103/PhysRevLett.89.188103*

Braun, D. & Libchaber, A. (2004). Thermal force approach to molecular evolution, *Phys. Biol.* 1: P1–P8.
URL: *http://stacks.iop.org/1478-3975/1/i=1/a=P01*

Cazabat, A.-M., Heslot, F., Troian, S. M. & Carles, P. (1990). Fingering instability of thin spreading films driven by temperature gradients, *Nature* 346: 824–826.
URL: *http://dx.doi.org/10.1038/346824a0*

Chen, J. Z., Troian, S. M., Darhuber, A. A. & Wagner, S. (2005). Effect of contact angle hysteresis on thermocapillary droplet actuation, *J. Appl. Phys.* 97: 014906.
URL: *http://link.aip.org/link/?JAP/97/014906/1*

Chen, Y. S., Lu, Y. L., Yang, Y. M. & Maa, J. R. (1997). Surfactant effects on the motion of a droplet in thermocapillary migration, *Int. J. Multiphase Flow* 23: 325–335.
URL: *http://dx.doi.org/10.1016/S0301-9322(96)00066-3*

Cordero, M. L., Burnham, D. R., Baroud, C. N. & McGloin, D. (2008). Thermocapillary manipulation of droplets using holographic beam shaping: Microfluidic pin ball, *Appl. Phys. Lett.* 93: 034107.
URL: *http://link.aip.org/link/?APL/93/034107/1*

Cordero, M. L., Rolfsnes, H. O., Burnham, D. R., Campbell, P. A., McGloin, D. & Baroud, C. N. (2009). Mixing via thermocapillary generation of flow patterns inside a microfluidic drop, *New J. Phys.* 11: 075033.
URL: *http://stacks.iop.org/1367-2630/11/i=7/a=075033*

Darhuber, A. A. & Troian, S. M. (2005). Principles of microfluidic actuation by modulation of surface stresses, *Annu. Rev. Fluid Mech.* 37: 425–455.
URL: *http://www.annualreviews.org/doi/abs/10.1146/annurev.fluid.36.050802.122052*

Darhuber, A. A., Valentino, J. P., Davis, J. M., Troian, S. M. & Wagner, S. (2003). Microfluidic actuation by modulation of surface stresses, *Appl. Phys. Lett.* 82: 657–659.
URL: *http://link.aip.org/link/?APL/82/657/1*

Dell'Aversana, P., Banavar, J. R. & Koplik, J. (1996). Suppression of coalescence by shear and temperature gradients, *Phys. Fluids* 8: 15–28.
URL: *http://link.aip.org/link/?PHF/8/15/1*

Dixit, S. S., Kim, H., Vasilyev, A., Eid, A. & Faris, G. W. (2010). Light-driven formation and rupture of droplet bilayers, *Langmuir* 26: 6193–6200.
URL: *http://dx.doi.org/10.1021/la1010067*

Duhr, S., Arduini, S. & Braun, D. (2004). Thermophoresis of DNA determined by microfluidic fluorescence, *Eur. Phys. J. E* 15: 277–286.
URL: *http://dx.doi.org/10.1140/epje/i2004-10073-5*

Duhr, S. & Braun, D. (2006a). Optothermal molecule trapping by opposing fluid flow with thermophoretic drift, *Phys. Rev. Lett.* 97: 038103.
URL: *http://link.aps.org/doi/10.1103/PhysRevLett.97.038103*

Duhr, S. & Braun, D. (2006b). Why molecules move along a temperature gradient, *Proc. Natl. Acad. Sci. USA* 103: 19678–19682.
URL: *http://www.pnas.org/content/103/52/19678.abstract*

Farahi, R. H., Passian, A., Ferrell, T. L. & Thundat, T. (2005). Marangoni forces created by surface plasmon decay, *Opt. Lett.* 30: 616–618.
URL: *http://ol.osa.org/abstract.cfm?URI=ol-30-6-616*

Garnier, N., Grigoriev, R. O. & Schatz, M. F. (2003). Optical manipulation of microscale fluid flow, *Phys. Rev. Lett.* 91: 054501.
URL: *http://link.aps.org/doi/10.1103/PhysRevLett.91.054501*

Giglio, M. & Vendramini, A. (1974). Thermal lens effect in a binary liquid mixture: A new effect, *Appl. Phys. Lett.* 25: 555–557.
URL: *http://link.aip.org/link/?APL/25/555/1*

Grigoriev, R. O. (2005). Chaotic mixing in thermocapillary-driven microdroplets, *Phys. Fluids* 17: 033601.
URL: *http://link.aip.org/link/?PHF/17/033601/1*

Grigoriev, R. O., Schatz, M. F. & Sharma, V. (2006). Chaotic mixing in microdroplets, *Lab Chip* 6: 1369–1372.
URL: *http://dx.doi.org/10.1039/B607003E*

Hähnel, M., Delitzsch, V. & Eckelmann, H. (1989). The motion of droplets in a vertical temperature gradient, *Phys. Fluids A* 1: 1460–1466.
URL: *http://link.aip.org/link/?PFA/1/1460/1*

Ichikawa, M., Ichikawa, H., Yoshikawa, K. & Kimura, Y. (2007). Extension of a DNA molecule by local heating with a laser, *Phys. Rev. Lett.* 99: 148104.
URL: *http://link.aps.org/doi/10.1103/PhysRevLett.99.148104*

Jiang, H.-R. & Sano, M. (2007). Stretching single molecular DNA by temperature gradient, *Appl. Phys. Lett.* 91: 154104.
URL: *http://link.aip.org/link/?APL/91/154104/1*

Jonáš, A. & Zemánek, P. (2008). Light at work: the use of optical forces for particle manipulation, sorting, and analysis, *Electrophoresis* 29: 4813–4851.
URL: *http://dx.doi.org/10.1002/elps.200800484*

Khattari, Z., Steffen, P. & Fischer, T. M. (2002). Migration of a droplet in a liquid: effect of insoluble surfactants and thermal gradient, *J. Phys.: Condens. Matter* 14: 4823–4828.
URL: *http://stacks.iop.org/0953-8984/14/i=19/a=309*

Kim, H. S. & Subramanian, R. S. (1989a). Thermocapillary migration of a droplet with insoluble surfactant: I. Surfactant cap, *J. Colloid Interface Sci.* 127: 417–428.
URL: *http://dx.doi.org/10.1016/0021-9797(89)90047-7*

Kim, H. S. & Subramanian, R. S. (1989b). Thermocapillary migration of a droplet with insoluble surfactant: II. General case, *J. Colloid Interface Sci.* 130: 112–129.
URL: *http://dx.doi.org/10.1016/0021-9797(89)90082-9*

Kotz, K. T., Noble, K. A. & Faris, G. W. (2004). Optical microfluidics, *Appl. Phys. Lett.* 85: 2658–2660.
URL: *http://link.aip.org/link/?APL/85/2658/1*

Krishnan, M., Park, J. & Erickson, D. (2009). Optothermorheological flow manipulation, *Opt. Lett.* 34: 1976–1978.
URL: *http://ol.osa.org/abstract.cfm?URI=ol-34-13-1976*

Lajeunesse, E. & Homsy, G. M. (2003). Thermocapillary migration of long bubbles in polygonal tubes. II. Experiments, *Phys. Fluids* 15: 308–314.
URL: *http://link.aip.org/link/?PHF/15/308/1*

Link, D. R., Anna, S. L., Weitz, D. A. & Stone, H. A. (2004). Geometrically mediated breakup of drops in microfluidic devices, *Phys. Rev. Lett.* 92: 054503.
URL: *http://link.aps.org/doi/10.1103/PhysRevLett.92.054503*

Liu, G. L., Kim, J., Lu, Y. U. & Lee, L. P. (2006). Optofluidic control using photothermal nanoparticles, *Nature Mater.* 5: 27–32.
URL: *http://dx.doi.org/10.1038/nmat1528*

Marcano, A. O. & Aranguren, L. (1993). Laser-induced force for bubble-trapping in liquids, *Appl. Phys. B: Lasers and Optics* 56: 343–346.
URL: *http://dx.doi.org/10.1007/BF00324530*

Mast, C. B. & Braun, D. (2010). Thermal trap for DNA replication, *Phys. Rev. Lett.* 104: 188102.
URL: *http://link.aps.org/doi/10.1103/PhysRevLett.104.188102*

Mazouchi, A. & Homsy, G. M. (2000). Thermocapillary migration of long bubbles in cylindrical capillary tubes, *Phys. Fluids* 12: 542–549.
URL: *http://link.aip.org/link/?PHF/12/542/1*

Mazouchi, A. & Homsy, G. M. (2001). Thermocapillary migration of long bubbles in polygonal tubes. I. Theory, *Phys. Fluids* 13: 1594–1600.
URL: *http://link.aip.org/link/?PHF/13/1594/1*

Monat, C., Domachuk, P. & Eggleton, B. J. (2007). Integrated optofluidics: A new river of light, *Nature Photon.* 1: 106–114.
URL: *http://dx.doi.org/10.1038/nphoton.2006.96*

Nagy, P. T. & Neitzel, G. P. (2008). Optical levitation and transport of microdroplets: proof of concept, *Phys. Fluids* 20: 101703.
URL: *http://link.aip.org/link/?PHF/20/101703/1*

Ohta, A. T., Jamshidi, A., Valley, J. K., Hsu, H.-Y. & Wu, M. C. (2007). Optically actuated thermocapillary movement of gas bubbles on an absorbing substrate, *Appl. Phys. Lett.* 91: 074103.
URL: *http://link.aip.org/link/?APL/91/074103/1*

Okawa, D., Pastine, S. J., Zettl, A. & Fréchet, J. M. J. (2009). Surface tension mediated conversion of light to work, *J. Am. Chem. Soc.* 131: 5396–5398.
URL: *http://dx.doi.org/10.1021/ja900130n*

Ottino, J. M. & Wiggins, S. (2004). Introduction: mixing in microfluidics, *Phil. Trans. R. Soc. Lond. A* 362: 923–935.
URL: *http://rsta.royalsocietypublishing.org/content/362/1818/923.abstract*

Park, S.-Y., Wu, T.-H., Chen, Y., Teitell, M. A. & Chiou, P.-Y. (2011). High-speed droplet generation on demand driven by pulse laser-induced cavitation, *Lab Chip* 11: 1010–1012.
URL: *http://dx.doi.org/10.1039/C0LC00555J*

Passian, A., Zahrai, S., Lereu, A. L., Farahi, R. H., Ferrell, T. L. & Thundat, T. (2006). Nonradiative surface plasmon assisted microscale Marangoni forces, *Phys. Rev. E* 73: 066311.
URL: *http://link.aps.org/doi/10.1103/PhysRevE.73.066311*

Piazza, R. & Parola, A. (2008). Thermophoresis in colloidal suspensions, *J. Phys.: Condens. Matter* 20: 153102.
URL: *http://stacks.iop.org/0953-8984/20/i=15/a=153102*

Robert de Saint Vincent, M., Wunenburger, R. & Delville, J.-P. (2008). Laser switching and sorting for high speed digital microfluidics, *Appl. Phys. Lett.* 92: 154105.
URL: *http://link.aip.org/link/?APL/92/154105/1*

Rybalko, S., Magome, N. & Yoshikawa, K. (2004). Forward and backward laser-guided motion of an oil droplet, *Phys. Rev. E* 70: 046301.
URL: *http://link.aps.org/doi/10.1103/PhysRevE.70.046301*

Shirasaki, Y., Tanaka, J., Makazu, H., Tashiro, K., Shoji, S., Tsukita, S. & Funatsu, T. (2006). On-chip cell sorting system using laser-induced heating of a thermoreversible gelation polymer to control flow, *Anal. Chem.* 78: 695–701.
URL: *http://dx.doi.org/10.1021/ac0511041*

Song, H., Chen, D. L. & Ismagilov, R. F. (2006). Reactions in droplets in microfluidic channels, *Angew. Chem. Int. Ed.* 45: 7336–7356.
URL: *http://dx.doi.org/10.1002/anie.200601554*

Squires, T. M. & Quake, S. R. (2005). Microfluidics: fluid physics at the nanoliter scale, *Rev. Mod. Phys.* 77: 977–1026.
URL: *http://link.aps.org/doi/10.1103/RevModPhys.77.977*

Stevenson, D. J., Gunn-Moore, F. J., Campbell, P. & Dholakia, K. (2010). Single cell optical transfection, *J. R. Soc. Interface* 7: 863–871.
URL: *http://rsif.royalsocietypublishing.org/content/7/47/863.abstract*

Tan, W.-H. & Takeuchi, S. (2007). A trap-and-release integrated microfluidic system for dynamic microarray applications, *Proc. Natl. Acad. Sci. USA* 104: 1146–1151.
URL: *http://www.pnas.org/content/104/4/1146.abstract*

Tan, W.-H. & Takeuchi, S. (2008). Dynamic microarray system with gentle retrieval mechanism for cell-encapsulating hydrogel beads, *Lab Chip* 8: 259–266.
URL: *http://dx.doi.org/10.1039/B714573J*

Theberge, A. B., Courtois, F., Schaerli, Y., Fischlechner, M., Abell, C., Hollfelder, F. & Huck, W. T. S. (2010). Microdroplets in microfluidics: An evolving platform for discoveries in chemistry and biology, *Angew. Chem. Int. Ed.* 49: 5846–5868.
URL: *http://dx.doi.org/10.1002/anie.200906653*

Verneuil, E., Cordero, M. L., Gallaire, F. & Baroud, C. N. (2009). Laser-induced force on a microfluidic drop: Origin and magnitude, *Langmuir* 25: 5127–5134.
URL: *http://dx.doi.org/10.1021/la8041605*

Vogel, A., Lauterborn, W. & Timm, R. (1989). Optical and acoustic investigations of the dynamics of laser-produced cavitation bubbles near a solid boundary, *J. Fluid Mech.* 206: 299–338.
URL: *http://dx.doi.org/10.1017/S0022112089002314*

Weinert, F. M. & Braun, D. (2008a). Observation of slip flow in thermophoresis, *Phys. Rev. Lett.* 101: 168301.
URL: *http://link.aps.org/doi/10.1103/PhysRevLett.101.168301*

Weinert, F. M. & Braun, D. (2008b). Optically driven fluid flow along arbitrary microscale patterns using thermoviscous expansion, *J. Appl. Phys.* 104: 104701.
URL: *http://link.aip.org/link/?JAP/104/104701/1*

Weinert, F. M. & Braun, D. (2009). An optical conveyor for molecules, *Nano Lett.* 9: 4264–4267.
URL: *http://dx.doi.org/10.1021/nl902503c*

Weinert, F. M., Wühr, M. & Braun, D. (2009). Light driven microflow in ice, *Appl. Phys. Lett.* 94: 113901.
URL: *http://link.aip.org/link/?APL/94/113901/1*

Wienken, C. J., Baaske, P., Rothbauer, U., Braun, D. & Duhr, S. (2010). Protein-binding assays in biological liquids using microscale thermophoresis, *Nat. Commun.* 1: 100.
URL: *http://dx.doi.org/10.1038/ncomms1093*

Würger, A. (2007). Thermophoresis in colloidal suspensions driven by Marangoni forces, *Phys. Rev. Lett.* 98: 138301.
URL: *http://link.aps.org/doi/10.1103/PhysRevLett.98.138301*

Würger, A. (2009). Molecular-weight dependent thermal diffusion in dilute polymer solutions, *Phys. Rev. Lett.* 102: 078302.
URL: *http://link.aps.org/doi/10.1103/PhysRevLett.102.078302*

Würger, A. (2010). Thermal non-equilibrium transport in colloids, *Rep. Prog. Phys.* 73: 126601.
URL: *http://stacks.iop.org/0034-4885/73/i=12/a=126601*

Young, N. O., Goldstein, J. S. & Block, M. J. (1959). The motion of bubbles in a vertical temperature gradient, *J. Fluid Mech.* 6: 350–356.
URL: *http://dx.doi.org/10.1017/S0022112059000684*

2

Hydrodynamic Focusing in Microfluidic Devices

Marek Dziubinski
Department of Chemical Engineering, Lodz Technical University, Poland

1. Introduction

In the last decade one of the quickest developing trends in fluid mechanics and chemical engineering has been microfluidics, covering the issues of heat, mass and momentum transfer in microscale. This corresponds directly to intensive research of nano- and microscale technology, as in such scales the system behavior shows significant deviations, compared to macroscale. That is mainly due to a drastically different surface-to-volume ratio and a minor role of buoyancy and inertia forces compared to surface forces like surface tension and adhesion. Due to the characteristic dimensions of microchannels, the flow of liquid is characterized by parallel streamlines, Reynolds number is small and only molecular diffusion is responsible for the inter-diffusion of a reagent.

One of the phenomena involved in a growing number of applications within the microfluidics area is hydrodynamic focusing. Hydrodynamic focusing is a technique relying on squeezing one of the streams in a four-microchannel intersection by two side streams and reshaping it downstream into a thin sheathed film (Domagalski, 2011; Dziubinski and Domagalski, 2007; Mielnik and Saetran, 2006). As can be seen in Fig. 1, the stream of interest, Q_C, is focused and sheathed downstream by streams Q_B and Q_A.

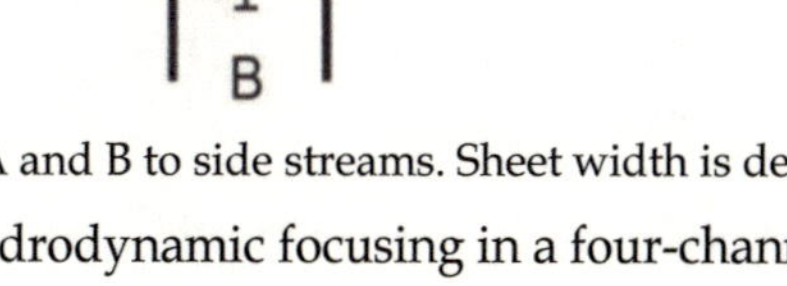

Index C refers to central inlet, A and B to side streams. Sheet width is denoted by δ_S

Fig. 1. Schematic view of hydrodynamic focusing in a four-channel intersection

By manipulating flow rates of the focusing flows, location of the focused sheet can be deformed and moved out of the symmetry plane. Achieving a precise control of the focused stream width is crucial in various applications of the flow focusing systems.

2. Applications of hydrodynamic focusing

Due to specific features it has been successfully involved in several microfluidic applications ranging from ultra-fast mixers and microreactors via flow addressed in Lab-on-a-Chip applications and cytometry, two-phase system generators, rheometry and flow visualization to microfabrication. Chemical synthesis in microscale is faster, small volumes and high area-to-volume ratios reduce risks and can improve economics, short diffusion lengths enable fast mixing, generally showing a way for process intensification.

Hydrodynamic focusing is a well known phenomenon in the area of fluid mechanics thanks to Osborne Reynolds, who first used it for flow visualization in his break-through experiment and it is widely utilized as a pipe mixer in chemical technology. However, the first 'non-academic' microfluidic application of hydrodynamic focusing was in the area of flow cytometry, a technique for counting, examining and sorting microscopic particles suspended in a stream of fluid. Hydrodynamic focusing, where the core flow of investigated sample is sheathed by an inert fluid, is used in flow cytometry as a way to deliver the sample of suspended cells to the analyzed region in an appropriate form. Such technique is used to precisely align optical detection system giving the possibility of high speed, high through-output analysis easily integrated with sorting, which makes the hydrodynamic focusing the main principle of flow cytometric hardware up to day (Donguen et al., 2005; Givan, 2011; Shapiro, 2003).

Focused stream residing in a channel centre gives a new possibility - to control the focused sheet position by changing the ratio of side streams, which was quickly utilized in the area of μ-TAS (micro-total-analysis systems). In such systems of reactors, mixers and detectors, a precise control of fluid flow is essential. This can be achieved by means of hydrodynamic focusing presenting several advantages as the characteristic switching time being in the order of magnitude of millisecond and near zero dead volume (see the example in Fig. 2).

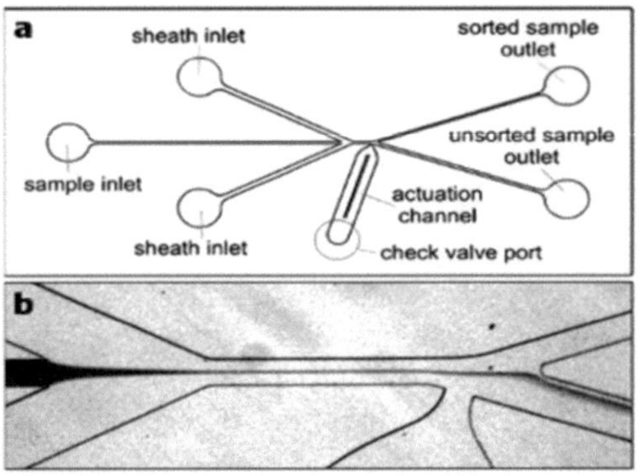

Fig. 2. Flow addressing: overall channel design (a), CCD image of focused sample stream (Lee et al.,2005b), visible focused sample stream

This idea was developed experimentally, theoretically and by CFD means by several authors (Bang et al., 2006; Brody et al., 1996; Chein and Tsai, 2004; Dittrich and Schwille, 2003; Hyunwoo et al., 2006; Kruger et al., 2002; Lee et al., 2001a; Lee et al., 2005b; Stiles et al., 2005; Vestad et al., 2004;) and it can be used in conjunction with electrokinetic effects (Dittrich and Schwille, 2003; Schrum et al., 1999; Yamada and Seki, 2005). All proposed applications used initial pre-focusing prior to precise spatial manipulation of the stream making it possible to integrate a whole system of mixers, reactors and separators at one chip (Chung et al., 2003; Goranovic et al., 2001; Klank et al., 2002; Sundararajan et al., 2004). Such approach is gaining a lot of enthusiasm in analytical science and engineering society as it possesses unquestionable advantages - low sample consumption, possibility of in situ analysis, low cost of single analysis due to mass production lying among the most important ones.

The next area where hydrodynamic focusing is used lies directly within chemical technology field of interest, namely in mixing. The geometrical setup used in cytometry was adopted in a continuous flow mixer leading to laminar, diffusion based mixer which used the hydrofocusing geometry (Knight et al., 1998). The proposed mixer consists of a system of four 10 μm wide channels of rectangular cross section, intersecting in the middle and micromachined in silicon wafer by photolithographic technique. The chip is covered with a glass slip providing the possibility of direct observation of the fluorescent quenching reaction by means of fluorescent and confocal scanning microscopy. Mixing between the inlet and side streams in such a system occurs at the interphase and is fully controlled by diffusion. As the time scale for diffusion changes with the square of a characteristic length, the micron dimensions of focused sheet (down to 50 nm) provides the efficient mixing. In a mixer of such construction, the obtained mixing times are less than 10 μs and reagents consumption is 3 orders of magnitude lower (5 nl/s compared to typical 10 μl/s) than in turbulent continuous flow mixers, which was a big achievement bearing in mind mixing is a challenging issue by itself in micro world.

Such a novel concept of mixer was very flexible and became a subject of many development researches. The speed and efficiency was enhanced by flow segmentation (Nguyen and Huang, 2005), side streams oscillations leading to focused film folding (Tabeling et al., 2004), preventing the slow speed reaction stage which takes place in the intersection, before the focusing process finishes (Park et al., 2006), working on slow reactions requiring steady pumping system (Stiles et al., 2005) or by forcing the turbulence by increasing the flow rate on the other hand (Majumdar et al., 2005) - see Fig. 3.

Advantages of radically quick mixing and low sample flow rate were used immediately in protein folding research. The knowledge of three-dimensional structure of protein and its dynamics is crucial for life sciences. However, the main problem and limiting factor in the observation of such reactions is their time scale being of the order of microseconds. As proposed by Knight et al. (1998), the mixer offers a possibility to change the reaction environment in microseconds; it was adopted to trigger the protein folding due to rapid pH change removing the limiting time scale boundary (Dittrich et al., 2004; Hertzog et al., 2004; Pollack et al., 2001; Russell et al., 2002). The natural step forward for a mixer, integration with a reactor was done. The future development showed more flexibility and advantages of such continuous flow reactor design. Jahn et al., 2004 investigated the possibilities of hydrodynamic focusing application in generation of liposome vesicles. Liposomes, being a class of nanoparticles encapsulating the aqueous volume by phospholipid bilayer, play a crucial role in biotechnology and life sciences delivering drugs or genetic material into a cell. Jahn et al. (2004) took advantage of the laminar character of microfluidic flow, because the lack of temperature, shear stress and composition fluctuations in the reaction environment causes that the product can be characterized by high monodispersity compared to bulk produced liposomes - important progress bearing in mind that the vesicle size is one of the basic liposome characteristics determining the quantity of encapsulated material.

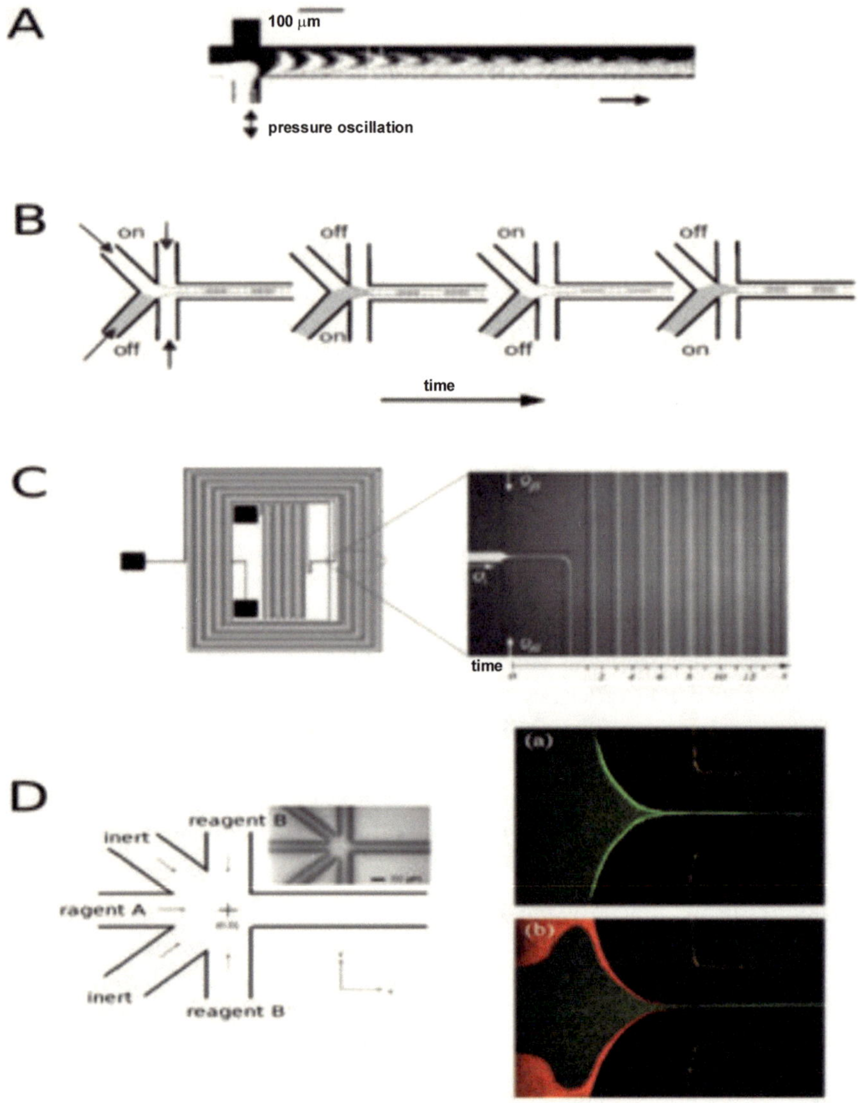

A - Tabeling et al., 2004; B - Nguyen and Huang, 2005; C - Park et al., 2006; D - Stiles et al., 2005

Fig. 3. Examples of mixing in microchannel

The use of hydrofocusing allows us to create well defined and predictable interphase surface maintaining the fully controlled environment. Such conditions, connected with low inertia typical of microfluidic systems caused by small volumes and laminar flow were used in polymer production. The developed method of continuous fabrication of polymeric microfibers consists in 'on the fly' photopolymerization of a hydrodynamically focused coaxial stream (Atencia and Beebe, 2005; Hyun et al., 2006; Jeong et al., 2004) - see Fig. 4.

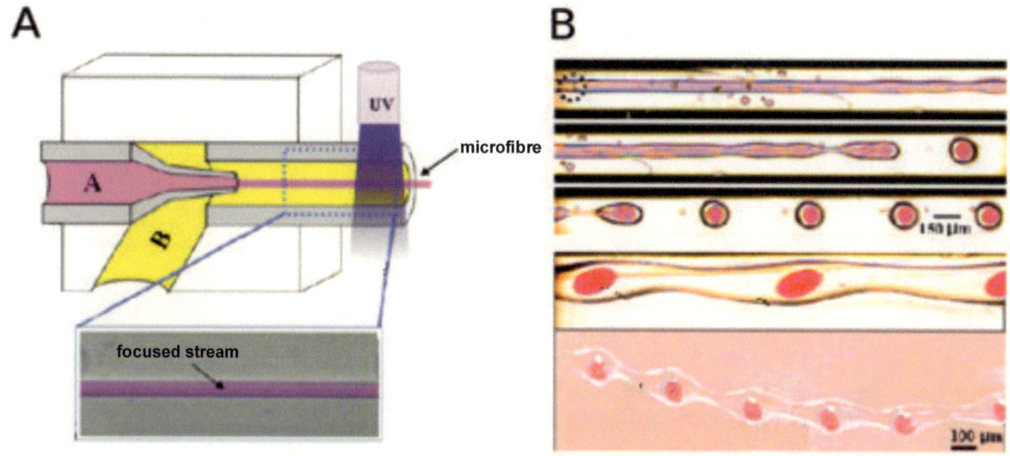

A - idea of photopolymerization, B - morphology of product

Fig. 4. Continuous fabrication of polymeric microfibres (Hyun et al., 2006; Jeong et al., 2004)

The main achievement in such technology is flexibility, as controlling the flowrates of sheath and core flow provides a tool to change the fiber diameter and morphology easily - by simple changing the flow conditions the same setup can be used for fabrication of polymeric microcapsules due to break-up of the liquid jet.

Hydrodynamic focusing microreactors can be used in microfabrication and patterning inside the capillaries as well (Kenis et al., 1999; Kenis et al., 2000; Takayama et al., 2001). The idea is to allow the reaction product to interact with the channel wall. Depending on the wall material and reaction product, a wide variety of structures and devices can be generated. Electrodes, wall etches, ridges or lines of crystals can be placed on the walls within the accuracy of 5 µm depending on flow volume rates control. Similarly, such an idea can be used in selective, precise and local treatment of biological cells. As demonstrated by Kam and Boxer (2003), Takayama et al. (2001), Takayama et al. (2003), it is possible to deliver reagents to a cell using multiple laminar streams with subcellular spatial resolution. Similarly, the discussed technique can be used in providing steady, controlled environment for cell population. That can mean equally distributed shear stress (Mohlenbrock et al., 2006) or stable in time, predictable and homogeneous chemical environment for lysis (Sethu et al., 2004). The applications of precisely controlled laminar fluid layers were also presented as a technique for fabrication of advanced membranes.

The characteristic features of hydrodynamic focusing can be used in rheology (Waigh, 2005; Wong et al., 2003). Diluted polymer particles delivered and focused precisely in the channel centre experience deformations due to shear forces and elasticity. Such an isolated molecule in the focused stream has a determined position in transverse axis and forced orientation parallel to flow direction. Labeling the endings of polymer chain by fluorescent probes allows one to observe a single molecule dynamics and its response to changes in flow conditions. Dynamics of such a single molecule can reveal complex rheological properties providing deeper insight into fundamental issues comparing to bulk rheological measurement.

The hydrofocusing geometry is an example how a channel modification, channel intersection can result in complication of physical phenomenon, implicating the possible applications. The geometry consisting of intersection of four rectangular cross section channel has found application also in two-phase flow. Many authors (Anna et al., 2003; Caubaud et al., 2005; Cristobal et al., 2006; Dreyfus et al., 2003; Joanicot and Ajdari, 2005; Garstecki et al., 2005a; Garstecki et al., 2005b; Raven et al., 2006; Seo et al., 2007; Utada et al., 2005; Ward et al., 2005; Xu and Nakajima, 2004;) used this geometry in a straight form or modified by a nozzle after the intersection to produce monodisperse two-phase systems (cf. Fig. 5).

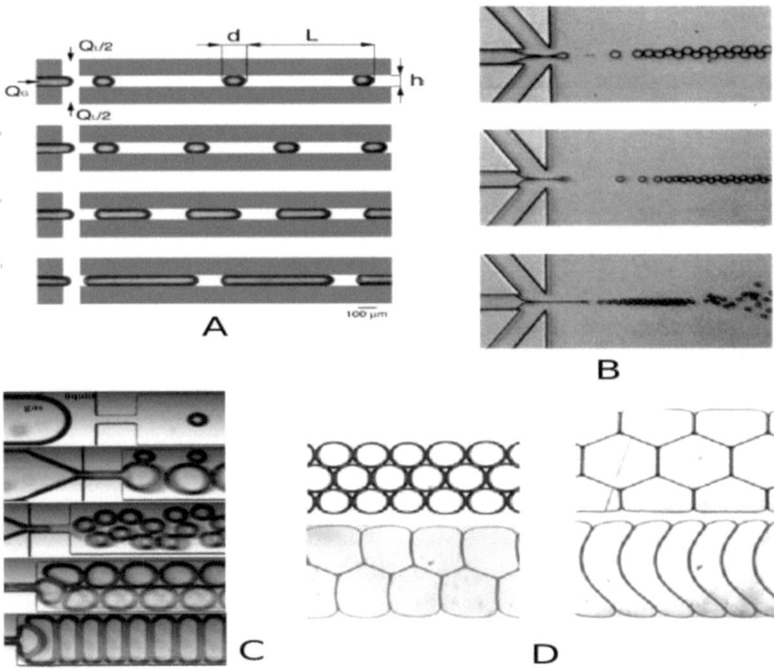

Fig. 5. Examples of generation of two-phase system in microchannels

Recently, hydrodynamic focusing has been applied in micro-PIV as a selective seeding technique (SeS-PIV) (Blonski et al., 2011; Domagalski et al., 2006; Domagalski et al., 2007; Domagalski, 2010; Domagalski, 2011; Mielnik and Saetran, 2006). Particle image velocimetry (PIV) is a flow visualization technique, where a flow velocity field is deducted from the displacement of tracer particles moving with investigated medium over time intervals. Simplifying, the flow field can be determined by correlating the tracer displacement on sequential frames. Due to small dimensions, in opposition to standard PIV technique where light is introduced in the form of a laser, in μ-PIV the flow is subject to bulk illumination to evade the technical problems with light alignment and light sheet generation. In SeS-PIV, a modification of μ-PIV, the tracer particles are introduced in the hydrofocused sheet making the width of the sheet responsible for the resolution (in opposition to the focal depth of optical system in the case of standard μ-PIV). That can de-bottleneck the system, making the measurement less dependent on the optical setup and permitting highly depth-resolved and instantaneous (time-resolved) velocity field measurements.

3. Three-dimensional aspect of hydrodynamic focusing

As was shown before, a lot of work concerning hydrodynamic focusing had been done and many examples of applications exist in the literature. However, little is known about this phenomenon on a basic level. So far, the papers on hydrodynamics of fluid focusing in microchannels present theoretical approach based on analytical solution of the flow in a rectangular cross-section channel (Chen et al., 2006; Solli et al., 2006; Wu and Nguyen, 2005 a,b,c).

The last publications, however, show the complexity of this phenomenon. A detailed investigation of the three-dimensional structure of hydrodynamic focusing performed by means of CLSM (confocal laser scanning microscopy) reveals two aspects of stream deformation (Blonski et al., 2011; Domagalski et al., 2007; Domagalski,2011; Domagalski and Dziubinski, 2010; Dziubinski and Domagalski, 2010). The first one consists in a non-uniform distribution of stream width and the second one relies on an additional curvature of the focused stream while pushing it away from the channel axis by non-symmetric side streams.

Experimental investigations indicated that in the case of symmetric side streams focused flow sheet was not necessarily uniform and its thickening close to the walls of a microchannel might reach undesirable values. A three-dimensional study of this effect confirmed that the focused sheet was not flat and with an increasing flow rate it exhibited nearly tripled thickness at both side walls, see Fig. 6.

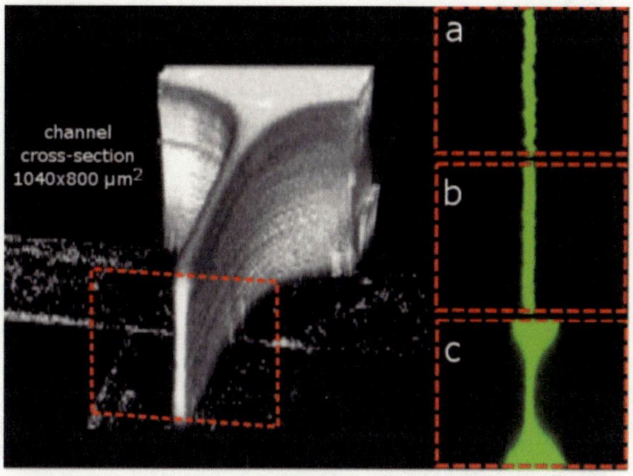

Fig. 6. The 3D projection of confocal microscopy image of hydrodynamic focusing with cross-section a,b,c - thickening of the focused plane close to side walls observed when increasing flow rate ($Q_A/Q_B = 1$; Re= 3.28; 6.46 and 12.92, respectively)

A detailed flow pattern analysis revealed possible regimes of focused stream shape: barrel-like shape, characterized by a decrease of width towards the top and bottom walls, Fig. 6a, flat uniform shape, Fig. 6b, and double concave shape, Fig. 6c.

Analyses of available experimental data reveal three regimes of the flow-focusing mechanism depending on the value of Reynolds number:

at $5<Re<8$ a nearly flat focused plane with constant width can be obtained,

$Re<5$ creates a slightly convex shape of the focused streams, at Re > 10 double concave shape is present, complete layering of the focused flow on the side walls takes place at the Reynolds number approaching 50.

Figure 7 show shapes of the focused stream obtained by CFD simulation for channels of rectangular cross section 300 × 400 μm. The geometry and dimensions of the CFD model were identical to the experimental device. The boundary conditions were set as mass flow at the inlets and pressure at the outlet of microchannel (Blonski et al., 2011).

Focusing ratio - the ratio of flow of the focused flow to the sum of focusing streams $Q_C/(Q_A+Q_B)$

A numerical simulation performed for four different focusing ratios indicates that thickness of the focused plane decreases with an increase of the focusing ratio, Fig. 7. However, a very strong effect on the focused sheet structure is observed by varying total flow rate (the flow Reynolds number), see Fig. 7 and 8. Increasing the Reynolds number above 10 practically destroys the flow focusing mechanism and for the Reynolds number above 50 the focused liquid is fully layered on the top and bottom walls, being absent from the channel center. Numerical calculation confirms very well the experimental data.

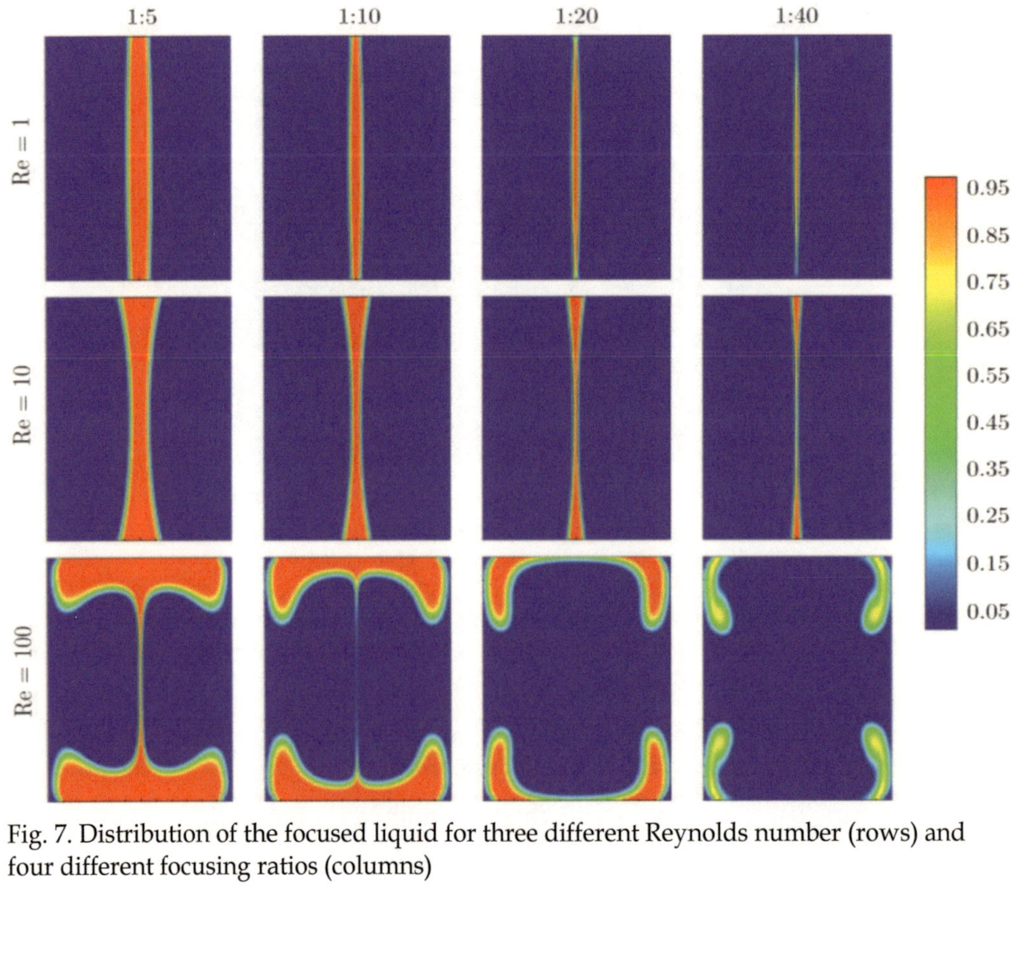

Fig. 7. Distribution of the focused liquid for three different Reynolds number (rows) and four different focusing ratios (columns)

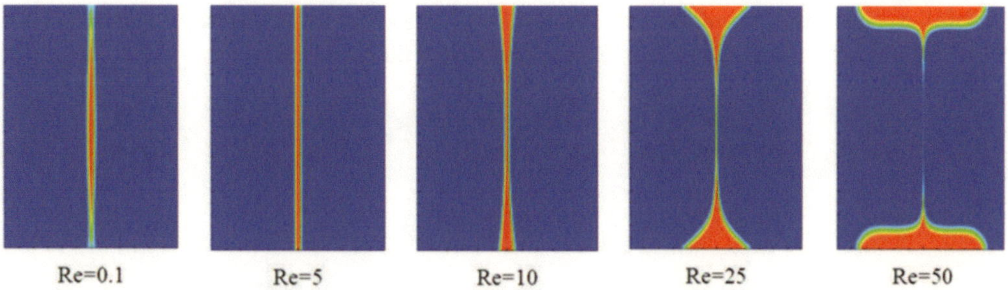

Fig. 8. Effect of Reynolds number on distributions of the focused liquid. Focusing ratio is 1:20 (Blonski et al.,2011)

Description of the shape of focused stream is complicated in the case when the focused stream is pushed away from the channel axis by non-symmetric side streams. Such behavior of the focused stream is shown in Fig. 9 which presents images of stream projection obtained by means of confocal microscopy with corresponding CFD simulations (Domagalski, 2011).

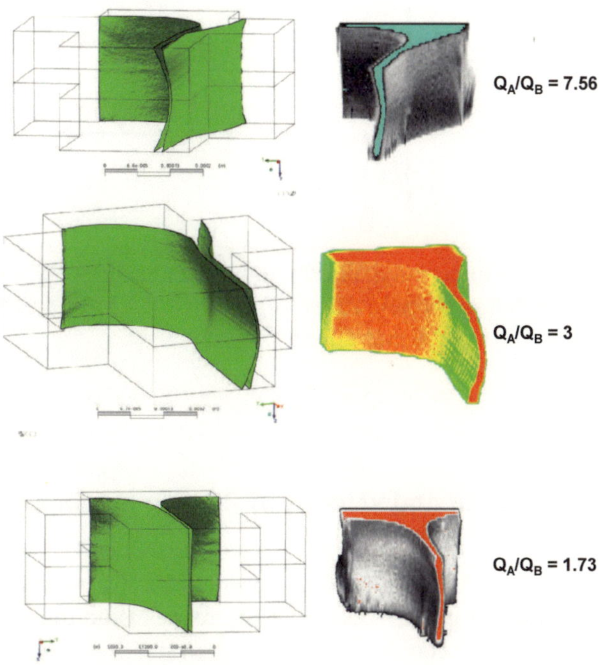

Fig. 9. The shapes of focused stream CFD results (left) and confocal microscope CLSM projections (right) for different ratios of side streams Q_A/Q_B. Channel rectangular cross section 1020 × 800 μm

When the stream is pushed away from the channel center, the previously described deformation in the form of uneven width of the stream is overlapped by the next deformation in the form of stream curvature perpendicular to flow direction. This behavior is confirmed by CFD simulations, as shown in Fig. 10 which presents a comparison of relevant cross sections obtained experimentally by the confocal microscopy with results of the CFD simulations for a channel cross section 1020 × 800 μm.

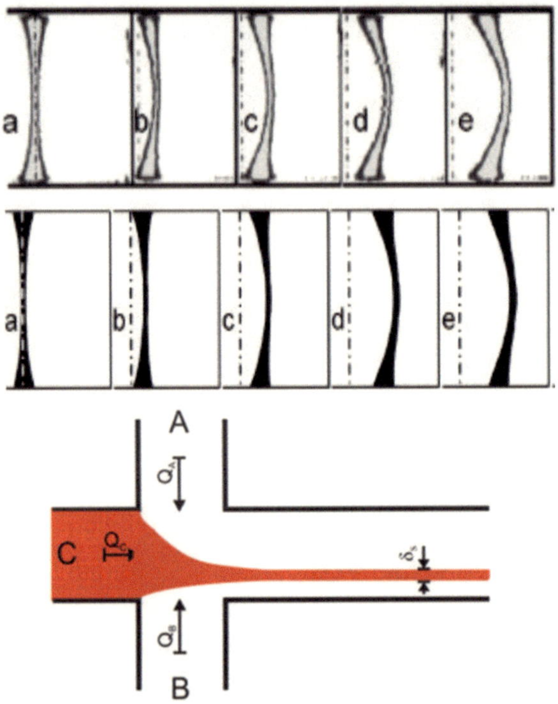

Fig. 10. The comparison of experimentally determined shape of focused stream (upper diagram) with CFD modeling (bottom diagram) for different values of side streams ratio Q_A/Q_B: a) 1.0, b) 1.73, c) 2.0, d) 3.0 and e) 7.56

This figure makes it possible to compare directly the shapes of deformed stream, showing good agreement of CFD simulation result and the observed system behavior.

4. The effect of properties and velocity of flowing media and channel size on the shape of focused stream

A very significant aspect of designing and operation of the systems based on hydrodynamic focusing is to determine the position of focused stream inside the outlet channel in given conditions of flow. As it has been stated earlier, when the flow is focused by identical side streams $Q_A=Q_B$, the focused stream leaves a microchannel flowing in the center of the outlet channel. In the case when the focusing streams are not symmetric, the focused stream is pushed away from the channel axis (Domagalski, 2007; Domagalski, 2008; Domagalski, 2011).

The basic geometric parameters that characterize the shape and position of stream in the microchannel were displacement of the stream from the center of channel axis z and its curving represented by distance c which is the difference of displacement of central part of the stream and its near-to-wall part, cf. Fig. 11.

For example, Figure 12 shows dependence of the position of focused stream in the outlet channel on the ratios of side flow rates for the system of channels with cross sections 1020 × 800μm and 260 x 200μm (Domagalski, 2011).

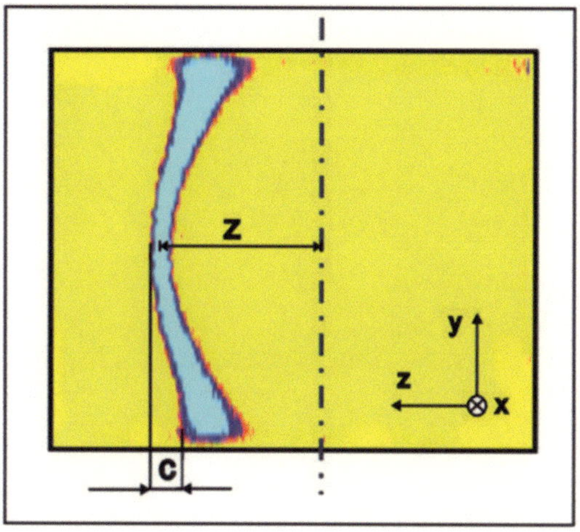

Fig. 11. The basic geometric parameters characterizing the shape and position of stream in a microchannel

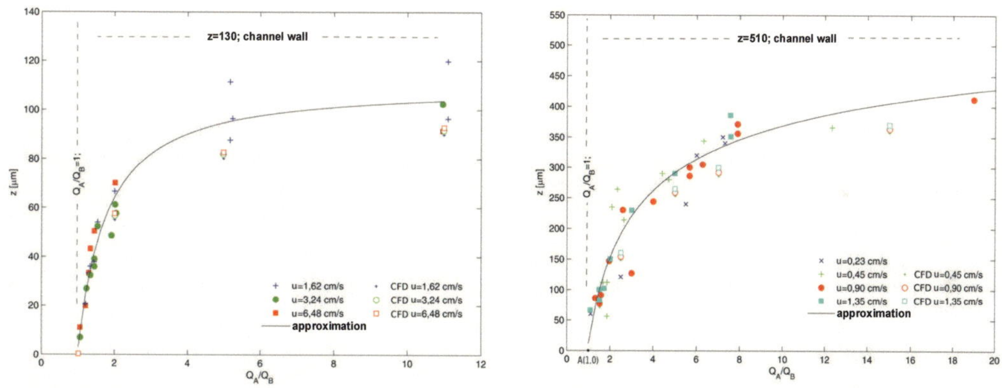

Channel cross section: 1020×800 μm (left-hand side diagram), 260×200 μm (right-side diagram)

Fig. 12. Displacement from the centre of channel axis z as a function of the side stream ratio Q_A/Q_B

These diagrams have a characteristic point with coordinates A(1,0) corresponding to the variant of symmetric focusing when volumetric flow rates of side flows are the same, and horizontal asymptote corresponds to a physical border in the form of the channel wall. It can easily be observed that pushing the focused stream away from the microchannel axis grows with an increase of the ratio of side streams.

Another aspect of deformation of the focused stream is the dependence of stream curvature on flow conditions. Figure 13 shows the dependence of curvature c of the focused stream on pushing the stream away from the channel axis z.

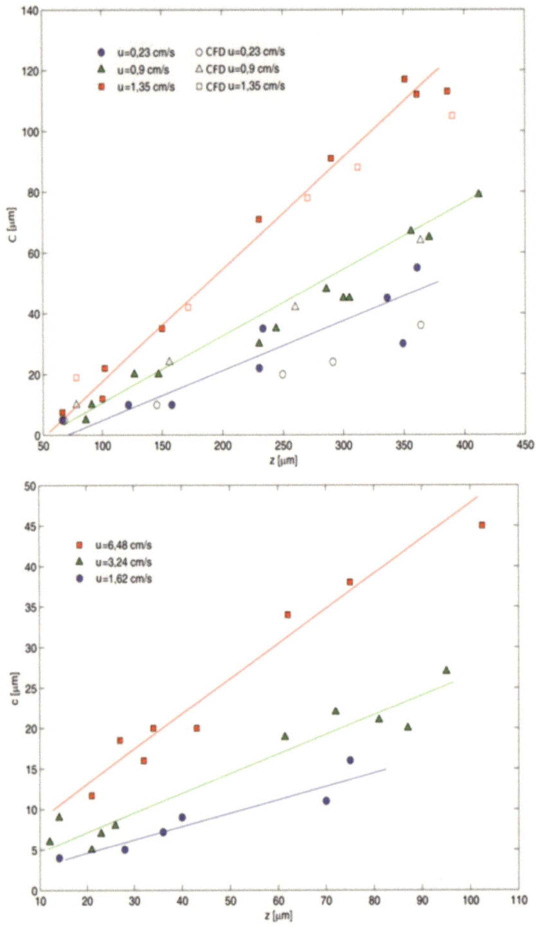

Fig. 13. Deformation of the focused stream as a function of pushing it away from the channel axis. Channel dimensions: 1020 × 800 µm (left-hand side diagram) and 260 × 200 µm (right-hand side diagram)

As follows from the figures, curving of the focused liquid stream pushed away from the microchannel axis increases with the process of pushing it away. Additionally, this effect is enhanced when the velocity of media flowing through the outlet channel grows.

While comparing diagrams shown in Fig. 12 and 13 one can observe that the relations have a similar character irrespective of the channel cross section. To investigate whether it is possible to exclude the effect of the channel cross section on the characteristics of hydrodynamic focusing, it was proposed to use dimensionless values. For this purpose the dimensionless pushing of the focused stream away from the channel center z' and dimensionless curving c' was used. These values are defined as follows:

$$z' = z/D_z \quad (1)$$

$$c' = c/D_z \quad (2)$$

where: D_z - equivalent channel diameter, z - pushing away from the channel axis, c - channel curvature.

In the case of a channel with rectangular cross section of dimensions a × b, the equivalent diameter has the form:

$$D_z = 4\,A/O = 2ab/(a+b) \tag{3}$$

where: A - cross section of the channel, O - channel perimeter flown by liquid, a and b - channel dimensions.

To investigate the effect of the channel cross section on hydrodynamic focusing, the cases of similar hydrodynamics were taken into account, and the criterion of similarity was the Reynolds number defined by the equation:

$$Re = u\,D_z\,\rho/\mu \tag{4}$$

where: u - liquid velocity, ρ - liquid density, μ - liquid viscosity

The effect of channel dimensions on pushing the focused stream away from its axis as a function of the ratio of side stream flow rates for different values of the Reynolds number is illustrated in Fig. 14, while the effect of the channel dimension on the dimensionless curvature c' is shown in Fig. 15.

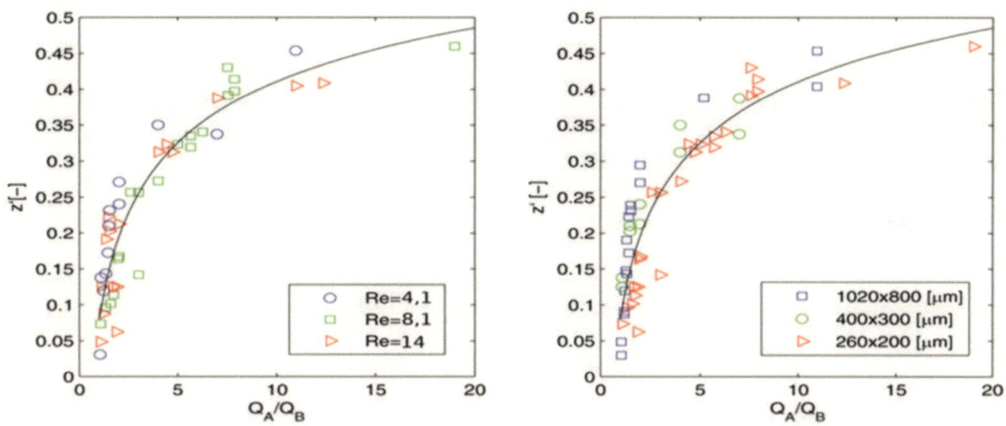

Fig. 14. Dimensionless pushing of the focused stream away from the channel axis as a function of the ratio of focused streams for different values of the Reynolds number (left-hand side diagram) and for channels of different dimensions (right-hand side diagram)

From the analyses of experimental data shown in Fig. 14 and 15, it follows that the dimensionless value of pushing the focused stream away does not depend on the channel dimensions, while the dimensionless curvature of the focused stream c' is dependent on the channel dimension. Curving of the focused stream is bigger in a smaller microchannel.

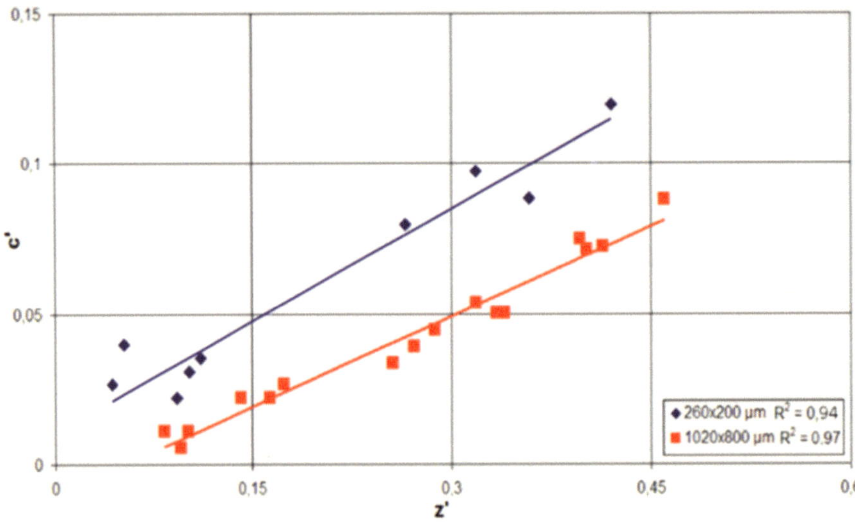

Fig. 15. Dimensionless deformation of focused stream c' as a function of dimensionless pushing off from the channel axis z' at constant Reynolds number Re=8.1

To estimate the position and shape of focused stream the following form of correlation equations was proposed:

$$z' = f(Q_A/Q_B) \tag{5}$$

$$c' = f\,(z', Re, D_z) \tag{6}$$

Preliminary knowledge of the shape of stream focused in given flow conditions is a basic information while designing and operating the devices in which hydrodynamic focusing is applied.

Based on available experimental data, the following form of Equations (5) and (6) is proposed

$$z'=-0.955\,(Q_A/Q_B)^{-0.184}+1.04 \tag{7}$$

$$c'= 0.027\,z'\,Re - 0.0393\,(D_z/D_{ref}) +0.0243 \tag{8}$$

A reference diameter D_{ref} is assumed to be 1000 µm which is the upper limit of dimension of the channel defined as a microchannel.

A comparison of experimental data with the values calculated using correlation equations (7) and (8) is shown in Fig. 16.

Using Equations (7) and (8) it is possible to determine the position and shape of focused stream with a maximum error reaching 25%. The equations are valid equally for channels of dimensions ranging from 260 × 200 to 1020 × 800 µm and for media with the following parameters: density 998 to 1097 kg/m^3, viscosity 0.997 to 12.5 mPas, surface tension 31 to 73 mN/m and for the range of liquid flow rate in the microchannel corresponding to the range of Reynolds number from 4.5 to 14.

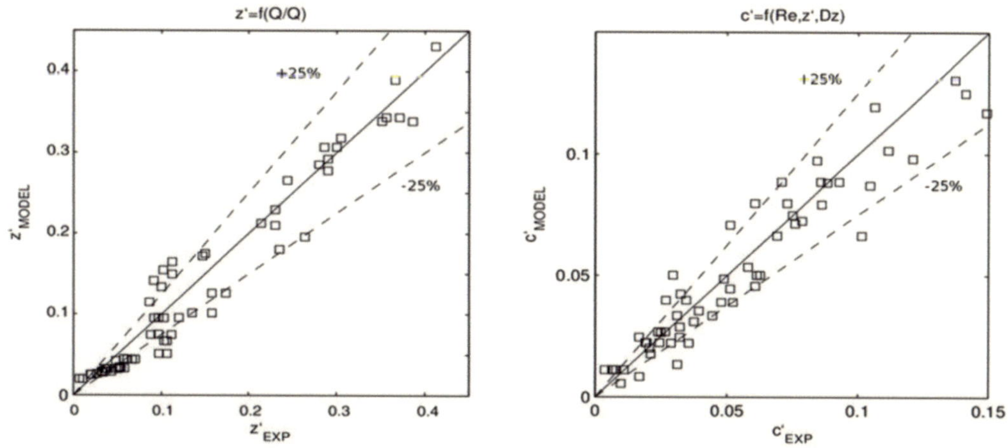

Fig. 16. The comparison of experimental value of z' and c' with the value calculated from eq. (7) and (8)

To investigate the effect of surface tension and liquid viscosity on the parameters of focused stream, measurements were taken for liquid with the surface tension 31 mN/m^2 (water with surfactant Triton X-100, Sigma-Aldrich) and liquid of viscosity higher than that of water amounting to 12.5 mPas. Results of exemplary measurements are given in Fig. 17 and 18.

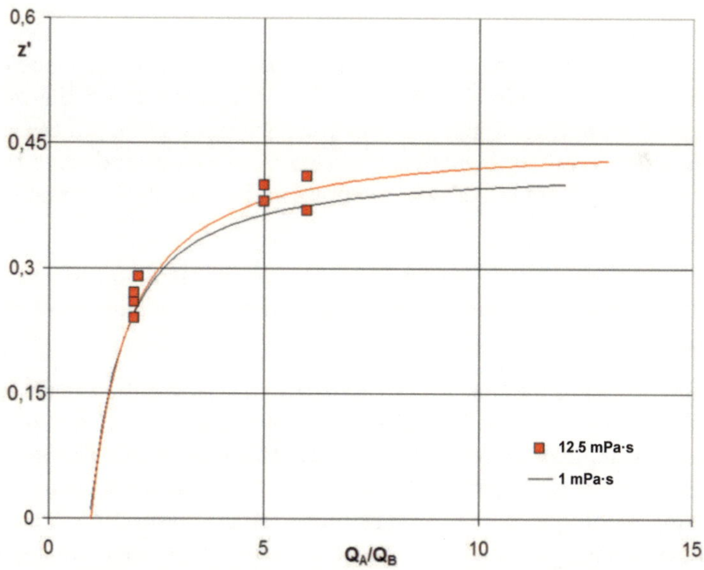

Fig. 17. The effect of liquid viscosity dimensionless pushing away of the focused stream from the channel axis for different values of the ratio of focusing streams

Based on the investigations it can be claimed that in the used range of measurements the effect of media properties on the character of a hydrodynamic phenomenon of stream focusing in the microchannel is negligibly small.

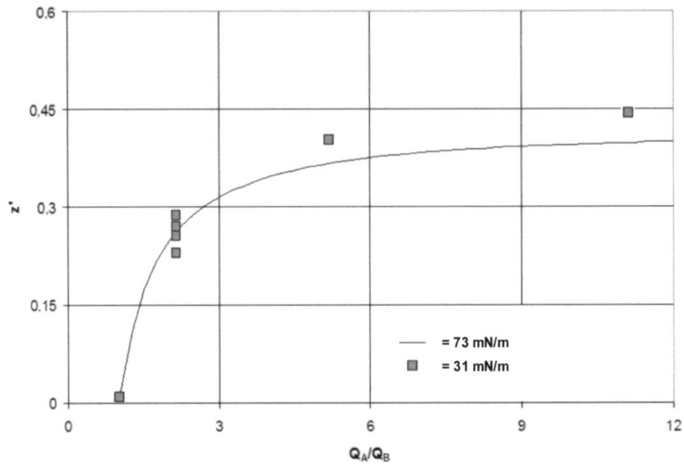

Fig. 18. The effect of surface tension of dimensionless stream curvature

5. Modification of micro-PIV flow visualization technique using hydrodynamic focusing phenomena

Micro Particle Image Velocimetry (micro-PIV) is a flow visualization technique for microfluidics (Raffel,2007), where a flow velocity field is constructed from the displacement of tracer particles moving with investigated medium over time intervals. Simplifying, the flow field can be determined by correlating the tracer's displacement on sequential frames. Due to small length scales of the observed phenomena, the flow is a subject of volume illumination, meaning the whole channel volume is illuminated and the measurement is based on focal depth of optical system, as only tracers within the depth of focus are clearly visible. That caused several problems as particles from below and over the focal plane participate in image brightness as background noise and evidently depreciate the evaluation accuracy.

To overcome this drawback, low concentration of tracers is usually used and the performed averaged correlation procedure averages the results over a large number of pairs of images, this however costs time.

Recently, the flow focusing method was proposed to introduce the tracers as a thin layer instead of whole volume seeding (SeS-PIV Selective Seeding PIV) (Mielnik and Saetran, 2006; Domagalski et al., 2008; Blonski et al., 2011; Domagalski, 2011). Such layer can be obtained in a rectangular cross-section channel via hydrodynamic focusing, which is shown in Fig. 19. Limiting seeding to a thin layer improves spatial resolution of the velocity field evaluation and permits to apply higher tracers concentration, hence allowing for acquisition of shorter sequences of images.

As visible, the stream containing tracers is squeezed by tracer-free side streams, which makes it possible to create the confined, narrow layer of tracers. Now, two conditions are necessary to take the advantage of such a setup. First, the flow has to be laminar, so the focused stream will not be perturbed, second, the diffusion effect has to be negligible. The first condition is generally fulfilled in a micro-area – with sub-millimeter characteristic length and liquid flow that is usually true. The second condition is dependent on tracer particle diameter, as the diffusion speed is proportional (Einstein-Stokes formula) to particle diameter.

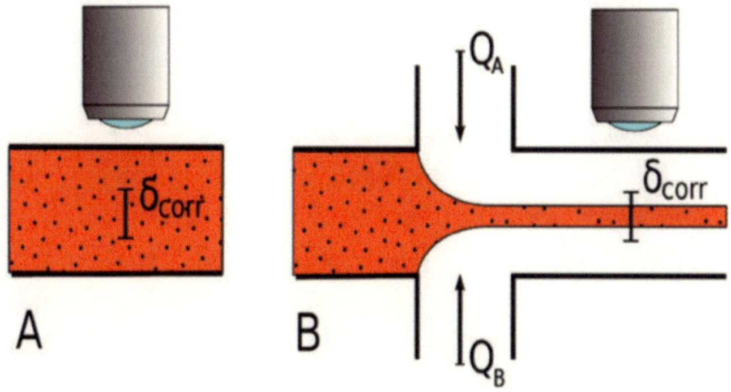

Fig. 19. The classical micro-PIV (A) and the idea of its modification as SeS-PIV (B)

Assuming that the focused stream is narrower than the depth of focus of a microscope, the measurement becomes independent of optical parameters. Due to limiting the source of fluorescent light to well-defined thin surface, the signal to noise ratio is strongly improved - see Fig. 20.

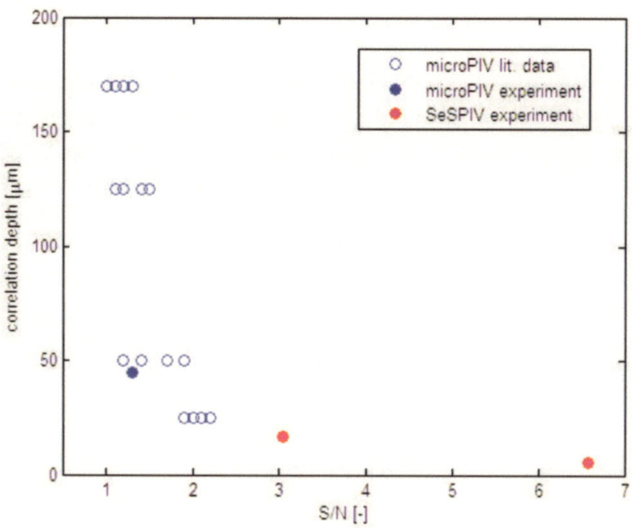

Fig. 20. Signal to noise ratio S/N as a function of test section depth (correlation depth)

What is more important, such a modification eliminates the seeding concentration limit, allowing the tracer layer to seed densely, resulting in the possibility of analyzing flow field on the basis of reduced (compared to standard, low concentration, volume illuminated micro-PIV) number of image pairs. The measurement can take less time, allowing for the measurement of non steady flows.

The whole depth section of the flow contributing to the measured velocity field is called the depth of correlation. Physically it is the depth of focus of the microscope, extended by the effect of diffraction and tracer particles geometry (Meinhart et al., 2000).

At the first step of analysis it is useful to compare the raw images of micro-PIV and its modification SeS-PIV. The comparison shown in Fig. 21 reveals a visible quality increase in the case of the SeS-PIV method.

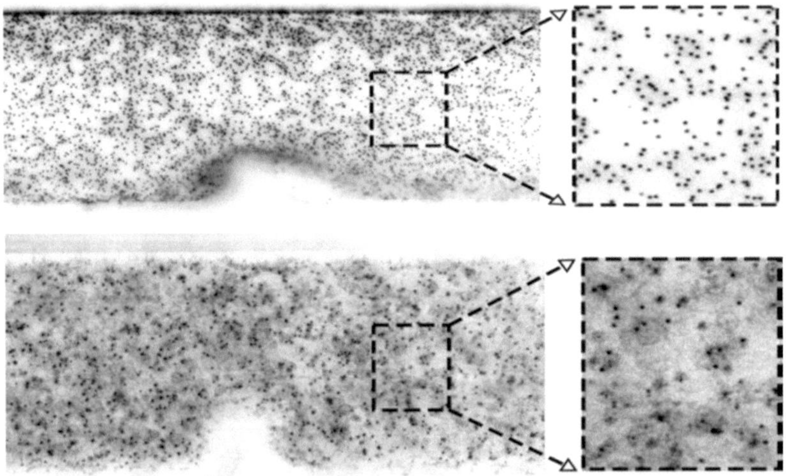

Fig. 21. Comparison of SeS-PIV image (upper) and micro-PIV raw images (bottom)

One may find better contrast of the image with focused seeding. This is caused by the lack of background, as no tracers are present outside focused, controlled tracer streams of known geometry. Moreover, the tracers are flowing in a thin layer, thinner than the depth of focus, so their images lack the diffraction rings as opposed to out-of-focus particles present in the micro-PIV picture. The visible blurred area near the obstacle in SeS-PIV picture is due to the three-dimensional deformation of focused stream. During flow over the ridge casing the tracers are coming out of the depth of focus. However, this effect was observed only at the highest tested velocities.

The velocity profiles determined by micro-PIV and SeS-PIV methods are presented in Fig. 22 (Blonski et al., 2011).

5.1 Applicability of the novel SeS-PIV technique

Applicability of the presented measuring technique is related directly to focusing hydrodynamics, the shape of cross section of the focused stream and diffusion of tracers used in the measurements. This applicability is limited by deformations of the focused stream and tracer diffusion rate.

The limit of acceptable deformation of the stream is determined by its curvature not exceeding the depth of correlation. The main mechanisms deforming the focused stream are Dean vortices, Moffat vortices and diffusion (Dascopoulos and Lenhoff, 1989; Domagalski, 2009; Ismagilov et al., 2000; Kamholtz et al., 1999; Kamholtz and Yager, 2001; Kamholtz and

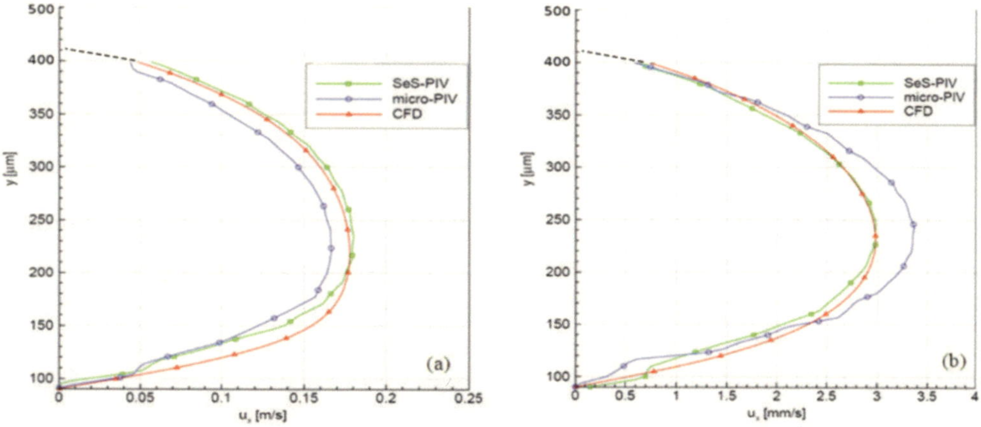

Fig. 22. Comparison of the numerical (CFD) and experimental results (SeS-PIV and micro-PIV) of vertical profile of velocity in the microchannel for Re=25 (left) and Re=0.4 (right)

Yager, 2002; Munson et al., 2005). Dean vortices formed as a result of unbalanced centrifugal force during motion along the curved line play a key role in deformation of the stream, inducing an increase of the cross section of the focused stream in near-to-wall regions. Intensity of the Dean vortices is characterized by the Dean number which is directly proportional to the Reynolds number (Munson et al., 2002). The effect of Dean vortices decreases with a decrease of the Reynolds number which is illustrated by rectangular cross section of the stream at Re=5. Transition into the region Re<1 (particularly in the creeping flow regime Re « 1) totally neutralizes the effect of Dean vortices on flow in the microchannel but causes generation of Moffat vortices (Mercer, 2004; Moffat, 1964). These are the structures formed due to wall action in the immediate vicinity of stagnation points. However, they do not have such a destructive effect on the stream shape as the Dean vortices have.

The effect of diffusion depends strongly on the particle size of tracers used in measurements. For a typical tracer size d=2 µm, diffusion coefficient is of the order D=2.2 10^{-13} m^2/s. In this case the path of diffusion is of the order of several micrometers along the whole length of a typical microchannel. Since the width of focused stream is much bigger, the effect of diffusion on the accuracy of measurements is negligible. However, in the literature there are examples of researches carried out with the use of quantum dots of size 20 nm, which much enhance the process of diffusion and do not allow us to abandon its effect on the shape of focused stream.

Hydrodynamic focusing of a liquid stream provides an opportunity to control the position of focused stream in the outlet channel. This technique used to illustrate velocity fields, enables mapping of the velocity field in subsequent channel cross sections (at different positions of the stream in the outflow channel), and consequently allows velocity to be measured in the entire liquid volume. The limit of applicability of this technique is determined by curvature of the focused stream which cannot be bigger than the depth of correlation of micro-PIV. Deformations limiting in this way applicability of the hydrodynamic focusing to a modification of the micro-PIV technique are primarily the functions of the Reynolds number, and for the non-symmetric variant of focusing also the ratio of velocities of the flowing media.

Concluding, it should be stated that the applicability of the SeS-PIV method is determined by three conditions:

1. The Reynolds number for the outlet stream is below Re=10.
2. The diffusion path is much smaller than the width of focused stream.
3. The width of focused stream is smaller than the depth of correlation of the microscope.

The proposed correlation equations (7) and (8) are used to estimate preliminarily the shape and position of focused stream inside the channel which enables determination of the region in which condition 3 is satisfied, and consequently enables quick identification of the applicability of the SeS-PIV method.

6. Conclusions

This chapter presents a review of applications of hydrodynamic focusing and the latest research in this area. Hydrodynamic focusing being a well established technique in microfluidic area has found many applications. Due to specific features it has been successfully involved in several microfluidic applications ranging from ultra-fast mixers and microreactors via flow addressed in Lab-on-a-Chip applications and cytometry, two-phase systems generators, rheometry and flow visualization to microfabrication. Chemical synthesis in microscale is faster, small volumes and high area-to-volume ratios reduce risks and can improve economics, short diffusion lengths allow for fast mixing, generally showing a way for process intensification.

The latest researches, however show precisely a new complicated three-dimensional aspect of this phenomenon indicating novel promising possibilities of future applications and development. A detailed investigation of the three-dimensional structure of hydrodynamic focusing performed by means of CLSM (confocal laser scanning microscopy) reveals two aspects of stream deformation. The first one consists in a non-uniform distribution of stream width and the second one relies on an additional curvature of the focused stream while pushing it away from the channel axis by non-symmetric side streams. The influence of properties and velocity of flowing media and channel size on the shape and position of focused stream in the microchannel has been presented.

A modification of the micro-PIV technique by introducing the tracers in hydrodynamically focused thin layer instead of volume seeding was proposed. Such modification known as SeS-PIV improves the raw image quality by removing the background noise ratio and permits higher seeding concentration. These features drastically improve the analysis of raw images - comparing to micro-PIV technique, making SeS-PIV techniques a valuable tool for microfluidic flow visualization.

7. Acknowledgment

This work was supported by National Science Center of Poland as research project No. N N 209 764 640.

8. References

Anna, S.L.; Bontoux, N. & Stone, H.A. (2005). Formation of dispersions using "flow focusing" in microchannels, *Applied Physics Letters,* Vol. 82, No. 3 pp. 364-366.

Atencia, J. & Beebe, D.J. (2005). Controlled microfluidic interfaces, *Nature,* Vol. 437, No. 29 pp. 648-655.

Bang, H.; Chung, C.; Kim, J.K.; Kim, S.H.; Chung, S.; Park, J.; Lee, W.G.; Yun, H.; Lee, J.; Cho, K.; Han, D-C. & Chang, J.K. (2006). Microfabricated fluorescence-activated cell sorter through hydrodynamic flow manipulation, *Microsystem Technologies,* Vol. 12, No. 8, pp. 746-753.

Blonski, S.; Domagalski, P.; Dziubinski, M. & Kowalewski, T. (2011). Selective Seeding in microPIV., *Archives of Mechanics,* Vol. 63, No. 2, pp. 163-182.

Brody, J.P.; Yager, P.; Goldstein, R.E. & Austin, R.H. (1996). Biotechnology at low Reynolds number, *Biophysical Journal,* Vol. 71, pp. 3430-3441.

Caubaud, T.; Tatineni ,M.; Zhong, X. & Ho C.-M. (2005). Bubble dispenser in microfluidic devices, *Phys. Rev.* E 72 037302.

Chein, R. & Tsai, S.H. (2004). Microfluidic Flow Switching Design Using Volume of Fluid Model, *Biomed. Microdev.,* Vol. 6, No. 1, pp. 81-90.

Chen, J.M., Horng, T.-L. & Tan, W.Y. (2006). Analysis and measurements of mixing in pressure-driven microchannel flow. *Microfluid Nanofluid,* Vol. 2, No. 6, pp. 455–469.

Chung, S.; Park, S.J.; Kim, J.K.; Chung, C.; Han, D.C. & Chang, J.K. (2003). Plastic microchip flow cytometer based on 2- and 3-dimensional hydrodynamic flow focusing, *Microsystem Technologies,* Vol. 9, pp. 525–533.

Cristobal, G.; Arbouet, L.; Sarrazin, F.; Talaga, D.; Brunnel, J.-L.; Joanicot, M. & Servant, L. (2006). On-line laser Raman spectroscopic probing of droplets engineered in microfluidic devices, *Lab Chip,* Vol. 6, pp. 1140-1146.

Daskopoulos, P. & Lenhoff, A.M. (1989). Flow in curved ducts - Bifurcation, *Journal of Fluid structure for stationary ducts, Mechanics,* Vol. 203, pp. 125-148.

Dittrich, P.S. & Schwille, P. (2003). An Integrated Microfluidic System for Reaction, High-Sensitivity Detection, and Sorting of Fluorescent Cells and Particles, *Anal. Chem.,* Vol. 75, pp. 5767-5774.

Dittrich, P.S.; Muller, B. & Schwille, P. (2004). Studying reaction kinetics by simultaneous FRET and cross-correlation analysis in a miniaturized continuous flow reactor, *Phys. Chem. Chem. Phys.,* Vol. 6, pp. 4416-4420.

Domagalski, P.M.; Mielnik, M.M.; Lunde I. & Saetran, L.R. (2006). Characteristics of Hydrodynamically Focused Streams for Use in Microscale Particle Image Velocimetry (Micro-PIV), *Proceedings of ASME 4th International Conference on Nanochannels, Microchannels and Minichannels,* CD-ROM Proceedings ISBN 0-7918-3778-5, June, Limmerick, Ireland.

Domagalski, P.M.; Dziubinski, M.; Budzynski, P.; Mielnik, M.M. & Saetran L.R. (2007). Width variations of hydrodynamically focused streams in low to moderate Reynolds number, *Proceedings of European Conference of Chemical Engineering (ECCE-6),* Copenhagen, Denmark, September, CD-ROM Proceedings ISBN 978-87-91435-57-9.

Domagalski, P.M.; Mielnik, M.M.; Lunde I. & Saetran, L.R. (2008). Characteristics of Hydrodynamically Focused Streams for Use in Microscale Particle Image Velocimetry (Micro-PIV), *International Journal of Heat Transfer Engineering*, Vol. 28, No. 8, pp. 680-688.

Domagalski, P.M.; Bardow A. & Ottens, M. (2009). Netherlands Rapid microfluidic-based measurement of diffusion coefficients, *Process Technology Symposium*, NPS-9, Veldhoven, the Netherlands, 26-28 October.

Domagalski, P.M. & Dziubiński, M. (2010). The 3D characteristics of hydrodynamic focusing in rectangular microchannels, *19th International Congress of Chemical and Process Engineering CHISA* 2010, Prague, Czech Republic.

Domagalski, P.M. (2011). Hydrodynamic focusing of liquid in intersection of microchannels, *Ph.D. thesis*, Lodz Technical University, Lodz, Poland.

Dongeun, H.; Wei, G.; Kamotani, Y.; Grotberg, J.B.; & Shuichi Takayama. (2005). Microfluidics for flow cytometric analysis of cells and particles, *Physiol. Meas.*, Vol. 26, pp. 73-98.

Dreyfus, R.; Tabeling, P. & Willaime, H. (2005). Ordered and Disordered Patterns in Two-Phase Flows in Microchannels, *Phys. Rev. Lett.*, Vol. 90, No. 14, pp. 144505.

Dziubinski, M. & Domagalski, P.M. (2007). Hydrodynamic focusing inside rectangular microchannels, *Chemical and Process Engineering*, Vol. 28, No. *3*, pp. 567-577.

Dziubiński, M. & Domagalski, P. (2010). Structure of hydrodynamically focused stream in an intersection of michrochannels, *Paper S8-16, VIII Euromech Fluid Mechanics Conference EFMC8,* Bad Reichenhall, Germany.

Garstecki, P.; Stone, H.A. & Whitesides, G.M. (2005a). Mechanism for Flow-Rate Controlled Breakup in Confined Geometries: A Route to Monodisperse Emulsions, *Phys. Rev. Lett.*, Vol. 94, No. 16, pp. 164501.

Garstecki, P.; Ganan-Calvo, A.M. & Whitesides, G.M. (2005b). Formation of bubbles and droplets in microfuidic systems, *Bulletin of the Polish Academy of Sciences,* Vol. 53, No. 4, pp. 361-372.

Givan, A.L. (2011). Flow Cytometry: First Principles, *Wiley-Liss.*

Goranovic, G.; Perch-Nielsen, I.R.; Larsen, U.D.; Wolff, A.; Kutter, J.P. & Telleman, P. (2001). Three-Dimensional Single Step Flow Sheathing in Micro Cell Sorters, *Proceedings of MSM Conference.*

Hertzog, D.E.; Michalet, X.; Jäger, M.; Kong, X.; Santiago, J.G.; Weiss, S. & Bakajin, F. O. (2004). Femtomole Mixer for Microsecond Kinetic Studies of Protein Folding, *Anal Chem.*, Vol. 76, No. 24, pp. 7169-7178.

Hyun, J.O; So, H.K.; Ju, Y.B.; Gi H.S. & Sang, H.L. (2006). Hydrodynamic micro-encapsulation of Aqueous fluids and cells via 'on the fly' photopolymerization, *J. Micromech. Microeng.*, Vol. 16, pp. 285-291.

Hyunwoo ,B.; Chung, C.; Kim, J.K.; Kim, S.H.; Chung, S.; Park, J.; Lee, W.G.; Yun, H.; Lee, J.; Cho, K.; Han, D-C. & Chang, J.K. (2006). Microfabricated fluorescence-activated cell sorter through hydrodynamic flow manipulation, *Microsyst. Technol.*, Vol. 12, pp. 746-753.

Ismagilov, R.F.; Stroock, A.D.; Kenis, P.J.A.; Whitesides, G. & Stone, H.A. (2000). Experimental and theoretical scaling laws for transverse diffusive broadening in two-phase laminar flows in microchannels. *Applied Physics Letters,* Vol. 76, No. 17, pp. 2376-2378.

Jahn, A.; Vreeland, W.N.; Gaitan, M. & Locascio, L.E. (2004). Controlled Vesicle Self-Assembly in Microfluidic Channels with Hydrodynamic Focusing, *J. Am. Chem. Soc.*, Vol. 126, No. 9, pp. 2674-2675.

Jeong, W.; Kim J.; Kim S.; Lee, S.; Mensing, G. & Beebe, D.J. (2004). Hydrodynamic microfabrication via "on the fly" Photopolymerization of microscale fibers and tubes, *LabChip*, Vol. 4, pp. 576-580.

Jiang, F.; Drese, K.S.; Hardt, S.; Kupper, M. & Schonfeld, F. (2004). Helical Flows and Chaotic Mixing in Curved Micro Channels, *AIChE Journal*, Vol. 50, No. 9, pp. 2297-2305.

Joanicot, M. & Ajdari, A. (2005). Droplet Control for Microfluidics, *Science,* Vol. 309, No. 5736, pp. 887-888.

Kam, L. & Boxer, S.G. (2003). Spatially Selective Manipulation of Supported Lipid Bilayers by Laminar Flow: Steps Toward Biomembrane Microfluidics, *Langmuir,* Vol. 19, pp. 1624-1631.

Kamholz, A.E.; Weigh, I.B.H.; Finlayson, B.A. & Yager, P. (1999). Quantitative analysis of molecular interaction in a microfluidic channel: the T-sensor, *Anal. Chem.*, Vol. 71, pp. 5340-5347.

Kamholz, A.E. & Yager, P. (2001). Theoretical Analysis of Molecular Diffusion in Pressure-Driven Laminar Flow in Microfluidic Channels, *Biophysical Journal,* Vol. 80, pp. 155-160.

Kamholz, A.E. & Yager, P. (2002). Molecular diffusive scaling laws in pressure-driven microfluidic channels: deviation from one-dimensional Einstein approximations, *Sensors and Actuators B*, Vol. 82, No. 1, pp. 117-121.

Kenis, P.J.; Ismagilov, R.F. & Whitesides, G.M. (1999). Microfabrication Inside Capillaries Using Multiphase Laminar Flow Patterning, *Science,* Vol. 285, No. 5424, pp. 83-85.

Kenis, P.J.A.; Ismagilov, R.F.; Takayama, S. & Whitesides, G.M. (2000). Fabrication inside Microchannels Using Fluid Flow, *Acc. Chem. Res.*, Vol. 33, No. 12, pp. 841-847.

Klank, H.; Goranovic, G.; Kutter, J.P.; Gjelstrup, H.; Michelsen J. & Westergaard, C.H. (2002). PIV measurements in a microfluidic 3D-sheathing structure with three-dimensional flow behaviour, *J. Micromech. Microeng.*, Vol. 12, pp. 1503-1506.

Knight, J.B.; Vishwanath, A.; Brody, J.P. & Austin, R.H. (1998). Hydrodynamic Focusing on a Silicon Chip: Mixing Nanoliters in Microseconds,. *Phys Rev. Lett.*, Vol. 80, pp. 3863-3866.

Kruger, J.; Singh, K.; O'Neill, A.; Jackson, C.; Morrison, A. & O'Brien, P. (2002). Development of a microfluidic device for fluorescence activated cell sorting, *J. Micromech. Microeng.*, Vol. 12, pp. 486-494.

Lee, G.B. Hwei B.-H. & Huang, G.-R. (2001a). Micromachined pre-focused MxN flow switches for continuous multi-sample injection, *J. Micromech. Microeng.*, Vol. 11, pp. 654-661.

Lee, G.B.; Lin, C.H. & Chang, S.C. (2005b). Micromachine-based multi-channel flow cytometers for cell/particle counting and sorting, *J. Micromech. Microeng.*, Vol. 15, pp. 447-454.

Majumdar, Z.K.; Sutin, J.D.B. & Clegg, R.M. (2005). Microfabricated continuous-flow, turbulent, microsecond mixer, *Rev. Sci. Instrum.*, Vol. 76, 125103-125103-11.

Meinhart, C.D.; Wereley S.T. & Santiago, J.G. (2000). Volume illumination for two-dimensional particle image velocimetry. *Meas. Sci. Technol.*, Vol. 11, pp. 809-814.

Mercer, A. McD. (2004). Moffatt eddies in viscous flow through a curved tube of square cross section, *AIChE Journal*, Vol. 32, No. 1, pp. 159-162.

Mielnik, M.M. & Saetran, L.R. (2006). Selective Seeding for micro-PIV, *Exp. Fluids*, Vol. 41, pp. 155-159.

Moffat, H.K. (1964). Viscous and resistive eddies near a sharp corner, *Journal of Fluid Mechanics*, Vol. 18, pp. 1-18.

Mohlenbrock, M.J.; Price, A.K. & Martin, R.S. (2006). Use of microchip-based hydrodynamic focusing to measure the deformation-inducted release of ATP from erythrocytes, *Analyst*, Vol. 131, pp. 930-937.

Munson, B.R.; Young, D.F. & Okiishi, T.H. (2002). Fundamentals of Fluid Mechanics, *Wiley*

Munson, M.S.; Hawkins, K.R.; Hasenbank M.S. & Yager, P. (2005). Diffusion based analysis in a sheath flow microchannel: the sheath flow T-sensor, *Lab. Chip*, Vol. 5, pp. 856-862.

Nguyen, N.T. & Huang, X. (2005). Mixing in microchannels based on hydrodynamic focusing and time-interleaved segmentation: modelling and experiment, *Lab Chip*, Vol. 5, pp. 1320-1326.

Nieuwenhuis J.H., Bastemeijer J., Sarro P.M., & Vellekoop M.J. (2003). Integrated flow-cells for novel adjustable sheath flows, *Lab Chip*, Vol. 3, pp. 56-61.

Park, H.Y.; Qiu, X.; Rhoades, E.; Korlach, J.; Kwok, l.W.; Zipfel, W.R.; Webb, W.W. & Pollack, l. (2006). Achieving Uniform Mixing in a Microfluidic Device: Hydrodynamic Focusing Prior to Mixing, *Anal. Chem.*, Vol. 78, No. 13, pp. 4465-4473.

Pollack, L.; Tate, M.W.; Finnefrock, F.C.; Kalidas, C.; Trotter, S.; Darnton, N.C.; Lurio, L.; Austin, R.H.; Batt, C.A.; Gruner, S.M. & Mochrie, S.G.J. (2001). Time Resolved Collapse of a Folding Protein Observed with Small Angle X-Ray Scattering, *Phys. Rev. Lett.*, Vol. 86, No. 21, pp. 4962-4965.

Raffel, M.; Willert, C.E.; Wereley, S.T. & Kompenhans, J. (2007). Particle Image Velocimetry: A Practical Guide, *Springer*

Raven, J.P.; Marmottant, P. & Graner, F. (2006). Dry microfoams: formation and flow in a confined channel, *The European Physical Journal B - Condensed Matter and Complex Systems*, Vol. 50, No. 1, pp. 137-143.

Russell, R.; Millett, I.S.; Tate, M.W.; Kwok, I.W.; Nakatani, B.; Gruner, S.M.; Mochrie, S.G.J.; Pande, V.; Doniach, S.; Herschlag, D. & Pollack, I. (2002). Rapid compaction during RNA folding, *PNAS* Vol.99, No7, pp. 4266–4271.

Ryo, M.; Hiroshi, O.; Isao, Y. & Ryohei, Y. (1991). A Development of Micro Sheath Flow Chamber, *Micro Electro Mechanical Systems An Investigation of Micro Structures, Sensors, Actuators, Machines and Robots.* IEEE: 265-270.

Schrum, D.P.; Culbertson, C.T.; Jacobson, S.C. & Ramsey, J.M. (1999). Microchip Flow Cytometry Using Electrokinetic Focusing, *Anal. Chem.*, Vol. 71, pp. 4173-4177.

Seo, M.; Paquet, C.; Nie, Z.; Xua, S. & Kumacheva, E. (2007). Microfluidic consecutive flow-focusing droplet generators, *Soft Matter*, Vol. 3, pp. 986-992.

Sethu, P.; Anahtar, M.; Moldawer, L.L.; Tompkins, R.G. & Toner, M. (2004). Continuous Flow Microfluidic Device for Rapid Erythrocyte Lysis, *Anal. Chem.*, Vol. 76, pp. 6247-6253.

Shapiro, H. (2003). Practical Flow Cytometry, *Wiley-Liss*.

Solli, L.A.; Saetran, L.R., & Mielnik, M.M. (2006). Numerical Modeling of Fluid Layers in Hydrodynamic Focusing, *Proceedings of Second International Conference on Transport Phenomena in Micro and Nanodevices*, Barga, Italy.

Stiles, T.; Fallon, R.; Vestad, T.; Oakey, J. & Marr, D.W.M. (2005). Hydrodynamic focusing for vacuum-pumped microfluidics, *Microfluidics and Nanofluidics*, Vol. 1, No .3, pp. 280-283.

Sundararajan, N.; Pio, M.S.; Lee, L.P. & Berlin, A.A. (2004). Three-dimensional hydrodynamic focusing in polydimethylsiloxane (PDMS) microchannels, *Journal of Microelectromechanical Sy*stems, Vol. 13, No. 4, pp. 559-567.

Tabeling, A.; Chabert, M.; Dodge A.; Jullien C. & Okkels, F. (2004). Chaotic mixing in cross-channel micromixers, *Phil. Trans. R. Soc. Lond. A.*, Vol. 362, No. 1818, pp. 987-1000.

Takayama, S.; Ostuni, E.; Leduc, P.; Naruse, K.; Ingber, D.E. & Whitesides, G.M. (2001). Subcellular positioning of small molecules, *Nature* Vol. 411, pp. 10-16.

Takayama, S.; Ostuni, E.; Leduc, P.; Naruse, K.; Ingber, D.E. & Whitesides G.M. (2003). Selective Chemical Treatment of Cellular Microdomains Using Multiple Laminar Streams, *Chemistry & Biology*, Vol.,10, No.,2, pp. 123-130.

Takayama, S.; Ostuni, E.; Qian, X., Mcdonald, J.C.; Jiang, X.; Leduc, P.; Wu, M-H.; Ingber, D.E. & Whitesides, G.M. (2001). Topographical Micropatterning of Poly(dimethylsiloxane) Using Laminar Flows of Liquids in Capillaries, *Adv. Mater*, Vol. 13, No. 8, pp. 570-574.

Utada, A.S.; Lorenceau, E.; Kaplan, P.D.; Stone, H.A. & Weitz, A. (2005). Monodisperse Double Emulsions Generated from a Microcapillary Device, *Science*, Vol. 308, No. 5721, pp. 537-541.

Vestad, T.; Marr, D. W. M. & Oakey, J. (2004). Flow control for capillary-pumped microfluidic systems, *J. Micromech. Microeng.*, Vol. 14, pp. 1503-1506.

Waigh, T.A.; (2005). Microrheology of complex fluids, *Rep. Prog. Phys.*, Vol. 68, pp. 685–742.

Ward, T.; Faivre, M.; Abkarian, M. & Stone, H.A. (2005). Microfluidic flow focusing: Drop size and scaling in pressure versus flow-rate-driven pumping, *Electrophoresis*, Vol. 26, pp. 3716-3724.

Wong, P.K.; Lee, Y.K. & Ho, C.M. (2003). Deformation of DNA molecules by Hydrodynamic focusing, *J. Fluid Mech.* Vol. 497, pp. 55-65.

Wu, Z. & Nguyen N.-T. (2005a). Hydrodynamic focusing in microchannels under consideration of diffusive dispersion: theories and experiments, *Sensors and Actuators B: Chemical*, Vol. 107, No. 2, pp. 965-974.

Wu, Z. & Nguyen, N.-T. (2005b). Rapid Mixing Using Two-Phase Hydraulic Focusing in Microchannels, *Biomedical Microdevices*, Vol. 7, No. 1, pp. 13-20.

Wu, Z. & Nguyen, N-T. (2005c). Convective–diffusive transport in parallel lamination micromixers, *Microfluidics and Nanofluidics*, Vol. 1, No. 3, pp. 208-217.

Xu, Q. & Nakajima, M. (2004). The generation of highly monodisperse droplets through the breakup of hydrodynamically focused microthread in a microfluidic device, *Appl. Phys. Lett.*, Vol. 85, No. 17, pp. 3726-3728.

Yamada, M. & Seki, M. (2005). Hydrodynamic filtration for on-chip particle concentration and classification utilizing microfluidics, *Lab Chip*, Vol. 5, pp. 1233-1239.

3

Analysis of a Coupled-Mass Microrheometer

David Cheneler
University of Birmingham
United Kingdom

1. Introduction

Many large companies, especially those involved in chemical synthesis, for instance those that manufacture personal care products or pharmaceuticals, are increasingly reliant on high throughput screening techniques to develop their next generation products. This may be to facilitate product optimisation or quality control of existing processes. Many of these products are viscoelastic in nature and require sophisticated rheological techniques to determine their properties (Hansen & Quake, 2003). However, current rheological techniques are not well suited to high throughput characterisation of the generally small volumes of liquid available (Kumble, 2003). Recently a number of devices, known as microrheometers, have been fabricated with the ability to determine the properties of small samples of viscoelastic fluids (Cheneler, 2011, Crecea et al., 2009, Christopher et al., 2010). These devices are generally designed to dynamically manipulate isolated drops of fluid, usually contained in liquid bridges (Cheneler, 2011, Christopher et al., 2010), and are not in a form that is easily integrated into a commercial automated process, precluding high-throughput analysis.

One of the limitations to the development of microrheometers and their subsequent integration into commercial processes is the difficulty in modelling microfluidic systems. The issue is that only a few of the possible fluid geometries that can be realised in useful experimental set-ups lend themselves to analytical mathematical investigation. As such this causes, in the design of microrheometers, the analysis of the fluid mechanics to be highly idealised or restricted to simple cases (Christopher et al., 2010). Most microfluidic systems can be modelled in a numerical fashion, whether it is using finite element analysis, molecular dynamics or computational fluid dynamics (Sujatha et al. 2008). This would suffice if the microfluidic system was in isolation, but in a microrheometer, the system is integrated into a micro electro-mechanical system (MEMS). In such devices the mechanical dynamics, the electrical circuit analysis and fluid mechanics are all coupled together. These three facets require different techniques to model them and in general these techniques are incompatible. Therefore in order to get accurate quantitative data from a microrheometer, each aspect of the analysis has to be performed in a more consistent form. This requires there to be an analytical solution to the equations describing the microfluidics and for the solution to be in a form that can be coupled into the analysis of the rest of the system. This will allow the complete deterministic response of the microrheometer to be known.

The purpose of this chapter is to describe the complete system analysis of a new kind of coupled-mass microrheometer and show how such a device can be integrated into a real production process. It will be shown how the fluid mechanics can be fully analysed, starting from the basic governing equations, as a sinusoidal viscoelastic squeeze flow problem. The solution of which is coupled with the dynamical and electrical analysis of the microrheometer and a deterministic and quantitative method of measuring the viscoelastic fluid properties is given.

2. Principles of the coupled-mass microrheometer

The microrheometer to be described here is an oscillating coupled-mass device driven by a comb drive actuator (see Fig. 1). Its response is measured capacitively using a structure akin to the comb drive actuator. It incorporates a microfluidic channel through which the sample is delivered and removed, allowing simple integration with production lines and automated systems. At present this microrheometer has not been fabricated, but to do so would require only minor modifications to the standard techniques used to presently fabricate silicon-based comb-drive actuated oscillators as discussed by Lee (Lee, 2011).

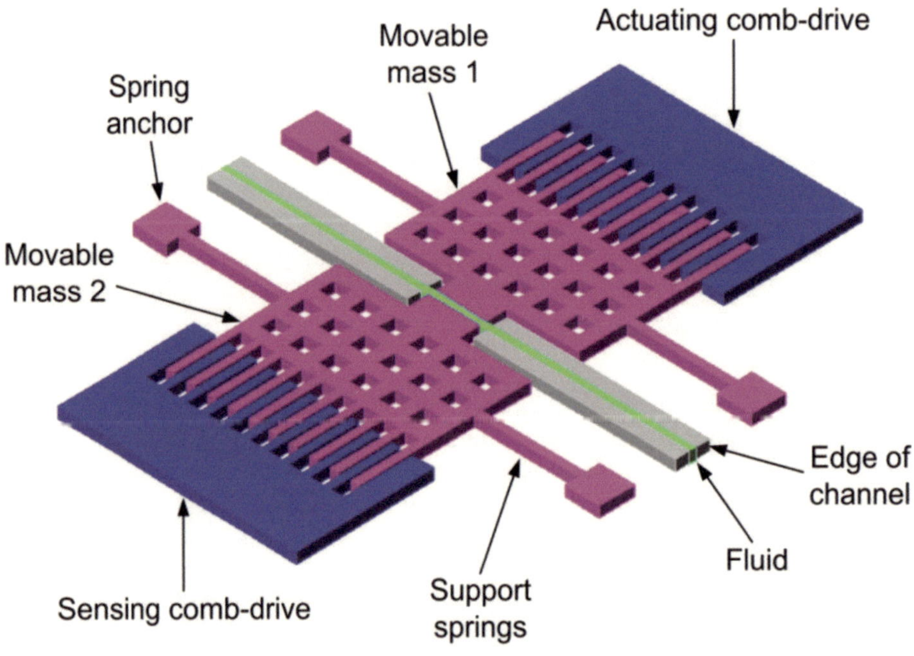

Fig. 1. A schematic of the coupled-mass microrheometer

The premise is that when a signal is applied to the actuating comb drive, it exerts a force on mass one (see Fig. 2) causing it to oscillate at a prescribed frequency. This squeezes between two parallel plates the fluid in the channel, which is formed from a compliant membrane. The fluid in turn applies a force to the second mass which then moves at the same frequency as the first. This uses the phenomena whereby micro-oscillators on the same chip suffer from unintentional cross-talk and frequency locking (Wei, 2009). Whilst normally this would be a problem, here it is beneficial. The movement of the second mass is detected by measuring the change in capacitance of the sensing comb drive. Knowing the force applied to the first mass and the response of the second mass allows for the determination of the fluids viscoelastic properties.

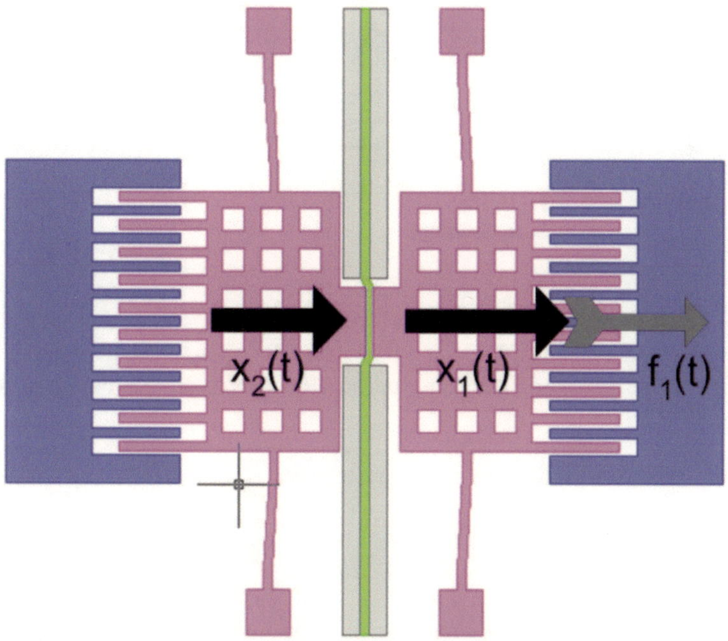

Fig. 2. A schematic showing the dynamics of the microrheometer. Here x(t) denotes the displacement of each of the masses respectively and $f_1(t)$ denotes the force applied by the actuating comb drive

3. Theoretical aspects

As stated, the microrheometer described here is an oscillating coupled-mass device (see Fig. 3) and as such lends itself to be modelled using a lumped-mass analysis (Wei, 2009). This method assumes that the dynamics of the device is identical to that of an equivalent system of masses, springs and dashpots which represent the inertial, restorative and dissipative components of the device respectively. In order for this analysis to represent the entire system however, the fluid mechanics and electrical circuitry will also have to be represented as spring, dashpots and external forces in a consistent fashion. This will be discussed in more fully in the subsequent sections.

3.1 Dynamics of a coupled-mass microrheometer

In lumped-mass analysis, it is assumed that the device can be represented by a system of masses, springs and dashpots. By this it is meant that the parameters such as stiffness and damping factors are constants so that forces are linearly dependent on velocity or displacement for instance. This is generally the case when displacements are small (Rao, 2010) and nonlinearities in the motion are negligible. In this instance, it is assumed that all the parameters are defined as constants as shown Fig. 3:

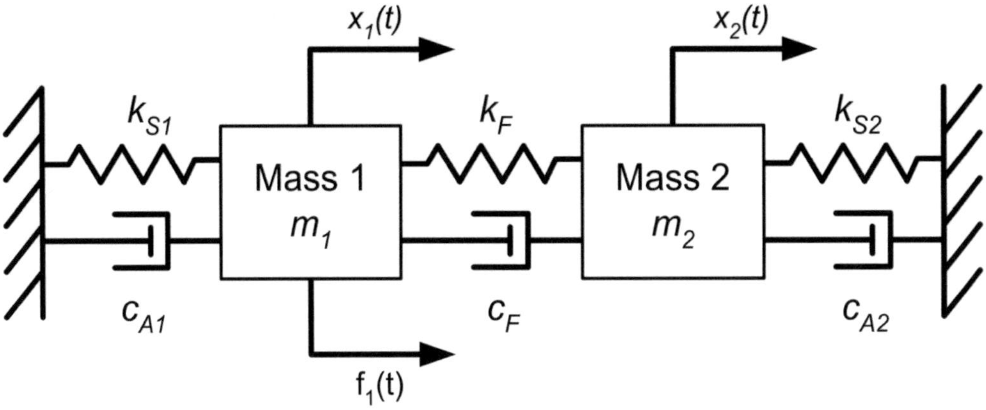

Fig. 3. A schematic of the equivalent lumped-mass system. k_S and c_A denote the stiffness of the support springs and the air damping associated with each of the masses, m. k_F and c_F denote the effective stiffness and damping parameters which are a function of the fluids viscoelastic properties

The coupled equations of motion for such a system is:

$$\begin{aligned} m_1\ddot{x}_1 + (c_{A1}+c_F)\dot{x}_1 - c_F\dot{x}_2 + (k_{S1}+k_F)x_1 - k_F x_2 = f_1(t) \\ m_2\ddot{x}_2 - c_F\dot{x}_1 + (c_F + c_{A2})\dot{x}_2 - k_F x_1 + (k_F + k_{S2})x_2 = 0 \end{aligned} \tag{1}$$

where one dot signifies velocity and two dots denote acceleration. These equations need to be solved to give x_2. This is the instantaneous position of mass 2 which is the value measured by the sensing electrode as shown in Fig. 1. The fluid properties can therefore be calculated from this value. As the force applied to mass 1 is sinusoidal and has the form $f_1 = Fe^{i\omega t}$ where F is the amplitude of the applied force and ω is the angular frequency, the responses of mass 1 and 2 can be assumed to have the form $x_1 = X_1 e^{i\omega t}$ and $x_2 = X_2 e^{i\omega t}$ where X_1 and X_2 are the amplitudes of the displacements of mass 1 and 2 respectively. Differentiating and substituting these factors into eq. 1 allows the equations to be represented in matrix form:

$$\begin{bmatrix} -\omega^2 m_1 + i\omega(c_{A1}+c_F) + (k_{S1}+k_F) & -i\omega c_F - k_F \\ -i\omega c_F - k_F & -\omega^2 m_2 + i\omega(c_F + c_{A2}) + (k_F + k_{S2}) \end{bmatrix} \begin{bmatrix} X_1 \\ X_2 \end{bmatrix} = \begin{bmatrix} F \\ 0 \end{bmatrix} \tag{2}$$

From eq. 2 the following factors can be defined:

$$Z_1 = -\omega^2 m_1 + k_{S1} + i\omega c_{A1} \tag{3}$$

$$Z_2 = -\omega^2 m_2 + k_{S2} + i\omega c_{A2} \tag{4}$$

$$Z_F = k_F + i\omega c_F \tag{5}$$

These are the mechanical impedances (Rao, 2010) related to mass 1, mass 2 and the fluid respectively. Note that all the coefficients are known in eqs. 3 and 4 and the coefficients in eq. 5 are those related to the dynamic properties of the fluid and are hence the coefficients to be measured. Substituting these impedances into eq. 2 allows for the following simplification:

$$\begin{bmatrix} Z_1 + Z_F & -Z_F \\ -Z_F & Z_2 + Z_F \end{bmatrix} \begin{bmatrix} X_1 \\ X_2 \end{bmatrix} = \begin{bmatrix} F \\ 0 \end{bmatrix} \tag{6}$$

Multiplying both sides of eq. 6 by the inverse of the impedance matrix gives the amplitudes X_1 and X_2. X_2 can be shown to be:

$$X_2 = \frac{Z_F F}{(Z_1 + Z_F)(Z_2 + Z_F) - Z_F^2} \tag{7}$$

This can be simplified to:

$$X_2 = \frac{Z_F F}{Z_1 Z_2 + (Z_1 + Z_2) Z_F} \tag{8}$$

Therefore the unknown fluid impedance can be given as:

$$Z_F = k_F + i\omega c_F = \frac{X_2 Z_1 Z_2}{F - X_2(Z_1 + Z_2)} \tag{9}$$

As evidenced in eqs. 3, 4 and 7, Z_1, Z_2 and X_2 are complex. This means the displacement of mass 2 can be given as:

$$x_2 = |X_2|(\cos\phi + i\sin\phi)e^{i\omega t} \tag{10}$$

where ϕ is the phase between the displacement of mass 2 and the force (and related voltage) applied to mass 1 and $|X_2|$ is the absolute (measured) amplitude of the displacement of mass 2. $X_2 = |X_2|(\cos\phi + i\sin\phi)$ is the form that needs to be substituted into eq. 9 as all the values are measured and therefore known. The effective stiffness and damping parameters due to the fluids viscoelastic properties are given by the real and imaginary components of eq. 9, thus:

$$k_F = \mathrm{Re}\left(\frac{X_2 Z_1 Z_2}{F - X_2(Z_1 + Z_2)}\right) \tag{11}$$

$$c_F = \frac{1}{\omega}\mathrm{Im}\left(\frac{X_2 Z_1 Z_2}{F - X_2(Z_1 + Z_2)}\right) \tag{12}$$

If the air damping in eq. 1 is negligibly small, eqs. 11 and 12 become:

$$k_F = \frac{|X_2|(k_{S1} - \omega^2 m_1)(k_{S2} - \omega^2 m_2)\left[F\cos\phi - |X_2|\left[(k_{S1} + k_{S2}) - \omega^2(m_1 + m_2)\right]\cos(2\phi)\right]}{F^2 - 2|X_2|F\left[(k_{S1} + k_{S2}) - \omega^2(m_1 + m_2)\right] + \left(|X_2|\left[(k_{S1} + k_{S2}) - \omega^2(m_1 + m_2)\right]\right)^2} \tag{13}$$

$$c_F = \frac{|X_2|(k_{S1} - \omega^2 m_1)(k_{S2} - \omega^2 m_2)\left[F\sin\phi - |X_2|\left[(k_{S1} + k_{S2}) - \omega^2(m_1 + m_2)\right]\sin(2\phi)\right]}{F^2 - 2|X_2|F\left[(k_{S1} + k_{S2}) - \omega^2(m_1 + m_2)\right] + \left(|X_2|\left[(k_{S1} + k_{S2}) - \omega^2(m_1 + m_2)\right]\right)^2} \quad (14)$$

It should be noted that eq. 10 can be expanded to give:

$$x_2 = |X_2|\left[\cos\phi\cos\omega t - \sin\phi\sin\omega t + i(\cos\phi\sin\omega t + \sin\phi\cos\omega t)\right] \quad (15)$$

This simplifies to:

$$x_2 = |X_2|\left(\cos(\omega t + \phi) + i\sin(\omega t + \phi)\right) \quad (16)$$

As the force applied to mass 1 will actually be the real component of f_1, i.e. $f_1 = \mathrm{Re}(Fe^{i\omega t}) = F\cos(\omega t)$, the actual response of mass 2 that is measured is the real component of eq. 16:

$$x_2 = |X_2|\cos(\omega t + \phi) \quad (17)$$

Therefore given that x_2 is measured, f_1 is what was applied and everything in Z_1 and Z_2 is known, k_F and c_F, and hence the fluids viscoelastic properties, can be explicitly calculated.

3.2 Derivation of the viscoelastic fluid dynamics

The squeezing of a Newtonian fluid between two plates was first detailed analytically by Reynolds (Reynolds, 1886) although it wasn't solved at that time for viscoelastic materials or for fluids being squeezed between rectangular plates. There have been many subsequent studies into viscoelastic squeeze flow due to its pertinence to lubrication (Bell et al., 2005). These studies have mostly concentrated on axisymmetric squeeze flow (Engmann et al., 2005) or constant load/velocity squeezing when considering rectangular plates and has generally been limited to plane strain cases (Denn & Marrucci, 1999). Therefore here a detailed derivation of the sinusoidal squeezing of a viscoelastic fluid between two parallel rectangular plates will be given. This will be achieved with a particular view of producing the relevant coefficients needed for the analysis in §3.1. To solve the squeeze flow problem consider the inertialess Navier-Stokes and continuity equations for Newtonian liquids in Cartesian coordinates (Reynolds, 1886):

$$\frac{dp}{dx} = \mu\left(\frac{d^2u}{dx^2} + \frac{d^2u}{dy^2} + \frac{d^2u}{dz^2}\right) \quad (18)$$

$$\frac{dp}{dy} = \mu\left(\frac{d^2v}{dx^2} + \frac{d^2v}{dy^2} + \frac{d^2v}{dz^2}\right) \quad (19)$$

$$\frac{dp}{dz} = \mu\left(\frac{d^2w}{dx^2} + \frac{d^2w}{dy^2} + \frac{d^2w}{dz^2}\right) \quad (20)$$

$$0 = \frac{du}{dx} + \frac{dv}{dy} + \frac{dw}{dz} \tag{21}$$

where z is on one of the surfaces of the channel in the direction of relative motion, y is on the same surface in the direction perpendicular to relative motion and x is mutually perpendicular (see Fig. 4). u, v and w are the velocity components in the x, y and z directions respectively. p is the pressure distribution in the channel.

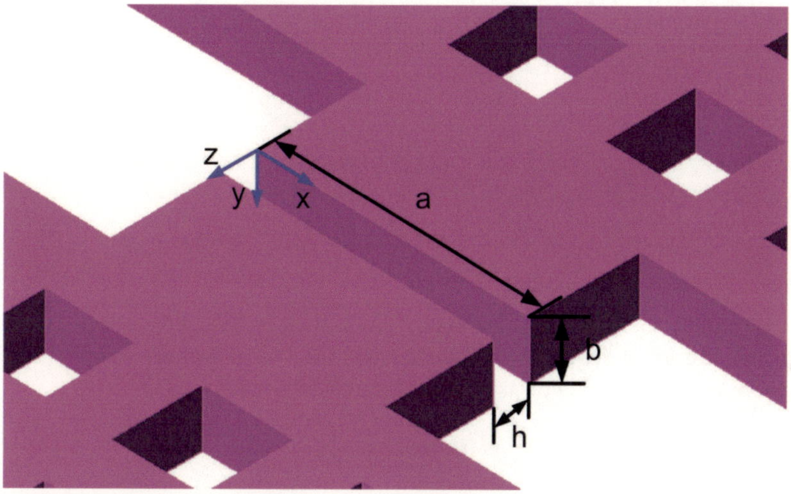

Fig. 4. A close up of the geometry between the plates squeezing the fluid. Fluid and channel have been omitted for clarity. The blue arrows denote the coordinate system used

As the channel width is small, w will be small compared to u and v, and the variations of u and v in the directions x and y are small compared with their variations in the z direction. The equations for the interior of the film then become:

$$\frac{dp}{dx} = \mu \frac{d^2u}{dz^2} \tag{22}$$

$$\frac{dp}{dy} = \mu \frac{d^2v}{dz^2} \tag{23}$$

$$\frac{dp}{dz} = 0 \tag{24}$$

These equations are subject to the following boundary conditions:

$$\begin{aligned} z = 0 &\Rightarrow u = 0, \ v = 0, \ w = 0 \\ z = h &\Rightarrow u = 0, \ v = 0, \ w = V \end{aligned} \tag{25}$$

Where h is the instantaneous width of the channel and V is the relative velocity of the sides. As eq. 24 shows the pressure to be independent of z, eqs. 22 and 23 are directly integrable. Integrating gives:

$$u = \frac{1}{2\mu}\frac{dp}{dx}(z-h)z \tag{26}$$

$$v = \frac{1}{2\mu}\frac{dp}{dy}(z-h)z \tag{27}$$

Differentiating these equations with respect to x and y respectively and substituting into the continuity equation (eq. 21) gives:

$$\frac{dw}{dz} = -\frac{1}{2\mu}\left[\frac{d}{dx}\left\{\frac{dp}{dx}(z-h)z\right\} + \frac{d}{dy}\left\{\frac{dp}{dy}(z-h)z\right\}\right] \tag{28}$$

Integrating from z = 0 to z = h and using the boundary conditions in eq. 25 gives:

$$\frac{d^2p}{dx^2} + \frac{d^2p}{dy^2} = \frac{12\mu V}{h^3} \tag{29}$$

This is the equation that needs to be solved to calculate the pressure distribution in the channel so that the force required to squeeze the fluid at a certain velocity can be found. This is possible because this is Poisson's equation of the form:

$$\nabla^2 p = -q(x,y) \tag{30}$$

Therefore we can assume it to have a solution of the form (Strauss, 2007):

$$p(x,y) = \sum_{m=1}^{\infty}\sum_{n=1}^{\infty} C_{mn}\phi_{nm}(x,y) \tag{31}$$

where $\phi_{nm}(x,y)$ are the eigenfunctions of the related Helmholtz equation:

$$\nabla^2\phi + \lambda\phi = 0 \tag{32}$$

Substituting eq. 31 into eq. 30 gives:

$$\sum_{m=1}^{\infty}\sum_{n=1}^{\infty} C_{mn}\nabla^2\phi_{nm}(x,y) = -q(x,y) \tag{33}$$

From eq. 32 we get:

$$\nabla^2\phi_{mn} = -\lambda\phi_{mn} \tag{34}$$

This can be solved using separation of variables by assuming a solution of the form:

$$\phi(x,y) = X(x)Y(y) \tag{35}$$

Substituting this into eq. 34 gives:

$$X''(x)Y(y) + X(x)Y''(y) + \lambda X(x)Y(y) = 0 \tag{36}$$

Upon rearranging, this leads to:

$$\frac{X''(x)}{X(x)} = -\frac{Y''(y)}{Y(y)} - \lambda = -\mu \tag{37}$$

This results in two ODEs and their associated boundary conditions:

$$X''(x) + \mu X(x) = 0, \quad X(0) = 0, \quad X(a) = 0 \tag{38}$$

$$Y''(y) + (\lambda - \mu)Y(y) = 0, \quad Y(0) = 0, \quad Y(b) = 0 \tag{39}$$

Where a is the width of the silicon plates and b is the depth (see Fig. 4). The boundary conditions are equivalent to stating the pressure is zero outside the gap between the plates. Eq. 38 has the solution:

$$X_m(x) = \sin\left(\frac{m\pi x}{a}\right), \quad m = 1,2,3... \tag{40}$$

So that:

$$\mu = \frac{m^2\pi^2}{a^2} \tag{41}$$

Similarly:

$$Y_n(y) = \sin\left(\frac{n\pi y}{b}\right), \quad n = 1,2,3... \tag{42}$$

And:

$$\lambda - \mu = \frac{n^2\pi^2}{b^2} \tag{43}$$

Combining eq. 41 and eq. 43 gives:

$$\lambda_{nm} = \pi^2\left(\frac{m^2}{a^2} + \frac{n^2}{b^2}\right), \quad m,n = 1,2,3... \tag{44}$$

Substituting eq. 40 and eq. 42 into eq. 35 gives:

$$\phi_{mn} = \sin\left(\frac{m\pi x}{a}\right)\sin\left(\frac{n\pi y}{b}\right), \quad m,n = 1,2,3... \tag{45}$$

Combining eqs. 33 and 34 give:

$$q(x,y) = \sum_{m=1}^{\infty}\sum_{n=1}^{\infty} C_{mn}\lambda_{mn}\phi_{mn}(x,y) \tag{46}$$

Everything in eq. 46 is known except for the coefficients C_{mn}. These can be found using the generalised Fourier series in two variables (Strauss, 2007) thus:

$$C_{mn} = \frac{12}{ab\lambda_{mn}} \int_0^b \int_0^a q(x,y) \sin\left(\frac{m\pi x}{a}\right) \sin\left(\frac{n\pi y}{b}\right) dx dy, \quad m,n = 1,2,3 \tag{47}$$

As $q(x,y) = -\frac{12\mu V}{h^3}$ (from comparing eqs. 29 and 30) this becomes:

$$C_{mn} = \frac{48\mu V}{mn\pi^2 h^3 \lambda_{mn}} \left[\cos(m\pi)\cos(n\pi) - \cos(m\pi) - \cos(n\pi) + 1\right], \quad m,n = 1,2,3 \tag{48}$$

Substituting this into eq. 31 gives:

$$p(x,y) = \sum_{m=1}^{\infty} \sum_{n=1}^{\infty} \frac{48\mu V}{mn\pi^2 h^3 \lambda_{mn}} \left[\cos(m\pi)\cos(n\pi) - \cos(m\pi) - \cos(n\pi) + 1\right] \times \ldots$$
$$\ldots \sin\left(\frac{m\pi x}{a}\right) \sin\left(\frac{n\pi y}{b}\right), \quad m,n = 1,2,3\ldots \tag{49}$$

The force is found by integrating this pressure over the top surface thus:

$$F_F = \sum_{m=1}^{\infty} \sum_{n=1}^{\infty} \frac{48\mu V}{mn\pi^2 h^3 \lambda_{mn}} \left[\cos(m\pi)\cos(n\pi) - \cos(m\pi) - \cos(n\pi) + 1\right] \times \ldots$$
$$\ldots \int_0^b \int_0^a \sin\left(\frac{m\pi x}{a}\right) \sin\left(\frac{n\pi y}{b}\right) dx dy, \quad m,n = 1,2,3\ldots \tag{50}$$

This equals:

$$F_F = \sum_{m=1}^{\infty} \sum_{n=1}^{\infty} \frac{48ab\mu V}{mn\pi^4 h^3 \lambda_{mn}} \left[\cos(m\pi)\cos(n\pi) - \cos(m\pi) - \cos(n\pi) + 1\right]^2, \quad m,n = 1,2,3 \tag{51}$$

This is the force required to squeeze a Newtonian liquid between two parallel rectangular plates at a known velocity. To see if this is a sensible result, compare it to the well-known force required to squeeze the fluid between two parallel circular plates, known as the Stefan-Reynolds equation (Bell et al. 2005):

$$F_{FC} = \frac{3\pi\mu R^4 V}{2h^3} \tag{52}$$

In eqs. 51 and 52, is can be seen that the solutions for the rectangular and for the circular plates are both linearly dependent on the viscosity and the velocity and are both inversely proportional to the cube of the gap between the plates. Furthermore, if it assumed that the dimensions of the rectangular plates make them square and that the radius of the circular plates are such that the surface area of the square and circular plates are equal, as can be seen in Fig. 5, where the two plates are far apart the solution for the circular and rectangular plates are nearly identical as is expected.

The damping coefficient for a Newtonian liquid squeezed between two rectangular plates can then given simply as:

$$c = \sum_{m=1}^{\infty}\sum_{n=1}^{\infty}\frac{48ab\mu}{mn\pi^4 h^3 \lambda_{mn}}\left[\cos(m\pi)\cos(n\pi) - \cos(m\pi) - \cos(n\pi) + 1\right]^2, \quad m,n = 1,2,3 \tag{53}$$

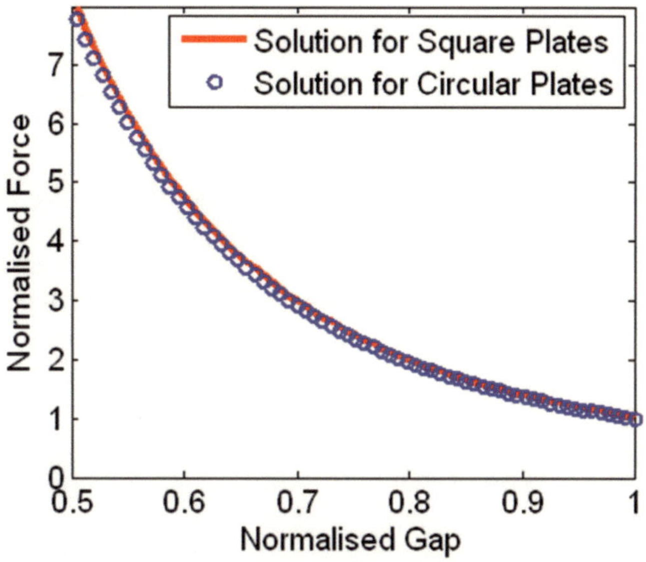

Fig. 5. Comparison between the solution for squeeze flow between rectangular plates, eq. 51, and the classical Stefan-Reynolds equation, eq. 52, for circular plates. For the purposes for comparison it is assumed the plates are square with the same surface area as the circular plates

For a viscoelastic liquid it is necessary to take into account the memory of the liquid. Therefore the force must take on the form (Bird et al., 1987):

$$F_F = \sum_{m=1}^{\infty}\sum_{n=1}^{\infty}\frac{48ab}{mn\pi^4 h^3 \lambda_{mn}}\int_{-\infty}^{t} G(t-t')\frac{dh}{dt'}dt'\left[\cos(m\pi)\cos(n\pi) - \cos(m\pi) - \cos(n\pi) + 1\right]^2 \tag{54}$$
$$m,n = 1,2,3...$$

If the boundaries of the channels are oscillating, the channel width has the form (Bell et al., 2005):

$$h(t) = \bar{h} + \varepsilon e^{i\omega t} \tag{55}$$

where $\bar{h}$ is the average gap, ω is the angular frequency and ε is the amplitude of oscillation. And by substituting:

$$t - t' = \xi \tag{56}$$

into the integral found in eq. 54, we obtain:

$$\int_{-\infty}^{t} G(t-t')\frac{dh(t')}{dt'}dt' = i\omega\varepsilon e^{i\omega t}\int_{0}^{\infty} G(\xi)e^{-i\omega\xi}d\xi \tag{57}$$

$G(\xi)$ represents the memory function of the material. If the fluid inside the channel is assumed to be a Maxwell liquid, this is defined as (Macosko, 1994):

$$G(t) = G_0 e^{-\frac{t}{\tau}} \tag{58}$$

where G_0 is the shear modulus and τ is the relaxation time. Substituting into eq. 57 and integrating gives:

$$i\omega\varepsilon e^{i\omega t}\int_{0}^{\infty} G_0 e^{-\frac{\xi}{\tau}} e^{-i\omega\xi}d\xi = G_0\frac{1}{1/\tau + i\omega}\frac{dh}{dt} = \left(\eta' - i\frac{G'}{\omega}\right)\frac{dh}{dt} \tag{59}$$

where η' is the dynamic viscosity and G' is the storage modulus. Substituting this result back into eq. 54 gives the force required to squeeze a viscoelastic fluid sinusoidally between two rectangular plates as:

$$F_F = \varepsilon e^{i\omega t}(iG'' + G')\sum_{m=1}^{\infty}\sum_{n=1}^{\infty}\frac{48ab}{mn\pi^4\bar{h}^3\lambda_{mn}}\left[\cos(m\pi)\cos(n\pi) - \cos(m\pi) - \cos(n\pi) + 1\right]^2 \tag{60}$$

$$m,n = 1,2,3...$$

G'' is the loss modulus of the fluid and is defined as $G'' = \omega\eta'$. It is the storage and loss modulus, i.e. G' and G'' that is of rheological importance (Macosko, 1994) and are therefore the most commonly measured dynamic viscoelastic properties. Separating the components of eq. 60 that are in-phase with the displacement and velocity respectively gives the final stiffness, k_F, and damping, c_F, coefficients for the liquid that are to be used in §3.1:

$$k_F = \sum_{m=1}^{\infty}\sum_{n=1}^{\infty}\frac{48abG'}{mn\pi^4\bar{h}^3\lambda_{mn}}\left[\cos(m\pi)\cos(n\pi) - \cos(m\pi) - \cos(n\pi) + 1\right]^2 \tag{61}$$

$$c_F = \sum_{m=1}^{\infty}\sum_{n=1}^{\infty}\frac{48abG''}{mn\pi^4\bar{h}^3\lambda_{mn}}\left[\cos(m\pi)\cos(n\pi) - \cos(m\pi) - \cos(n\pi) + 1\right]^2 \tag{62}$$

3.3 Electrical considerations

As shown in Fig. 1, the system is actuated using a capacitive lateral comb-drive. The force caused by applying a voltage to the comb-drive actuator is proportional to the change in capacitance and to the square of the applied voltage. Specifically for a comb-drive actuator this becomes (Lee, 2001):

$$F_{cd}(t) = \frac{1}{2}\frac{\partial C}{\partial x}v(t)^2 = n\frac{\varepsilon t}{g}v(t)^2 \tag{63}$$

where C is the capacitance of the comb-drive, ε is the permittivity of the medium between the teeth, t is the thickness of a tooth, g is the gap between opposing teeth and n is the number of teeth. $v(t)$ is the applied voltage. As the force is proportional to the square of the voltage, if the voltage was purely a sinusoidal signal, the force will have double the frequency of the voltage. Therefore it is common practice (Lee, 2011) to add a DC bias voltage to the applied voltage signal, i.e.:

$$v(t) = V_{DC} + V_{AC}\cos(\omega t) \tag{64}$$

where V_{DC} is the amplitude of the DC bias voltage and V_{AC} is the amplitude of the sinusoidal voltage. An example of the circuitry that could be used in this instance is shown in Fig. 6:

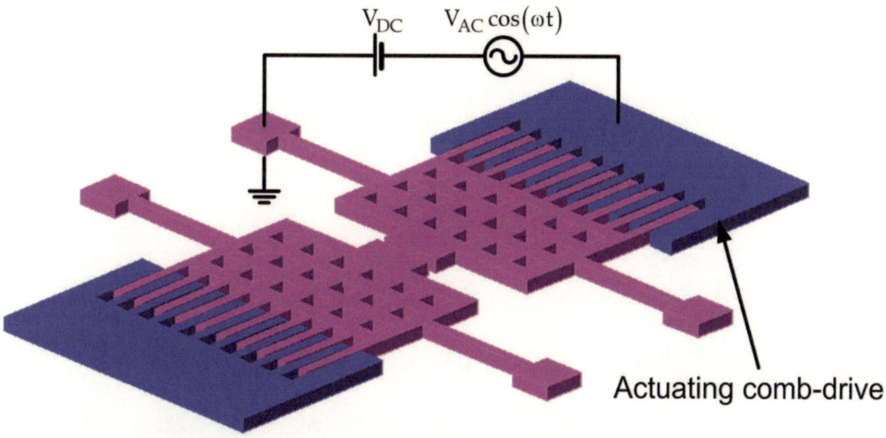

Fig. 6. A schematic of the circuitry for the actuating comb-drive. Note the moving masses are grounded through the anchors on the springs

The force therefore becomes:

$$F_{cd}(t) = n\frac{\varepsilon t}{g}\left(V_{DC} + V_{AC}\cos(\omega t)\right)^2 = n\frac{\varepsilon t}{g}\left(V_{DC}^{\ 2} + 2V_{DC}V_{AC}\cos(\omega t) + V_{AC}^{\ 2}\cos^2(\omega t)\right) \tag{65}$$

When $V_{DC} >> V_{AC}$ eq. 65 can be approximated as:

$$F_{cd}(t) = F_C + F\cos(\omega t) \tag{66}$$

where F_C is a constant force and F is the amplitude of the applied force as used in eqs. 11 and 12. These constants can be shown to be:

$$F_C = n\frac{\varepsilon t}{g}\left(V_{DC}^{\ 2} + \frac{V_{AC}^{\ 2}}{2}\right) \tag{67}$$

$$F = 2n\frac{\varepsilon t}{g}V_{DC}V_{AC} \tag{68}$$

The additional constant force naturally has an effect which wasn't considered in §3.1. Plugging a constant force into eq. 2 gives the static displacement of mass 1 and 2 as:

$$X_{1S} = \frac{k_{S2} + k_F}{k_{S1}k_{S2} + k_F(k_{S1} + k_{S2})} F_C \tag{69}$$

$$X_{2S} = \frac{k_F}{k_{S1}k_{S2} + k_F(k_{S1} + k_{S2})} F_C \tag{70}$$

However as the fluid in the channel is defined as a Maxwell liquid, and k_F is proportional to the storage modulus as shown in eq. 60, k_F becomes zero as the force is constant. This is because the storage modulus varies as (Macosko, 1994):

$$G'(\omega) = \frac{G_0 \tau^2 \omega^2}{\tau^2 \omega^2 + 1} \tag{71}$$

where G_0 is the shear modulus introduced in eq. 57 and τ is the characteristic relaxation time of the liquid. Therefore:

$$X_{1S} = \frac{F_C}{k_{S1}} \tag{72}$$

$$X_{2S} = 0 \tag{73}$$

It is important to note is that the average gap between the plates, $\bar{h}$, is therefore smaller by a factor X_{1S}.

4. Discussion

For the sake of simplifying the discussion of the theory, it is assumed that the channel is filled with a simple Maxwell liquid whose relaxation modulus is described by the data given in Fig. 7.

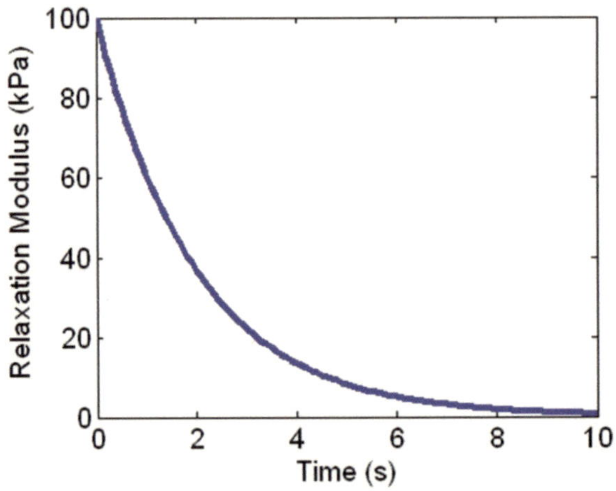

Fig. 7. Typical relaxation data for a simple Maxwell fluid

Fitting a Prony series to this data, such as that given by eq. 57, shows that the behaviour of this liquid can be described by a single relaxation time constant and a single shear modulus. For this hypothetical liquid these happen to be τ = 2 s and G_0 = 100 kPa. These constants suggest that the storage and loss moduli of the liquid should take the form of that shown in Fig. 8:

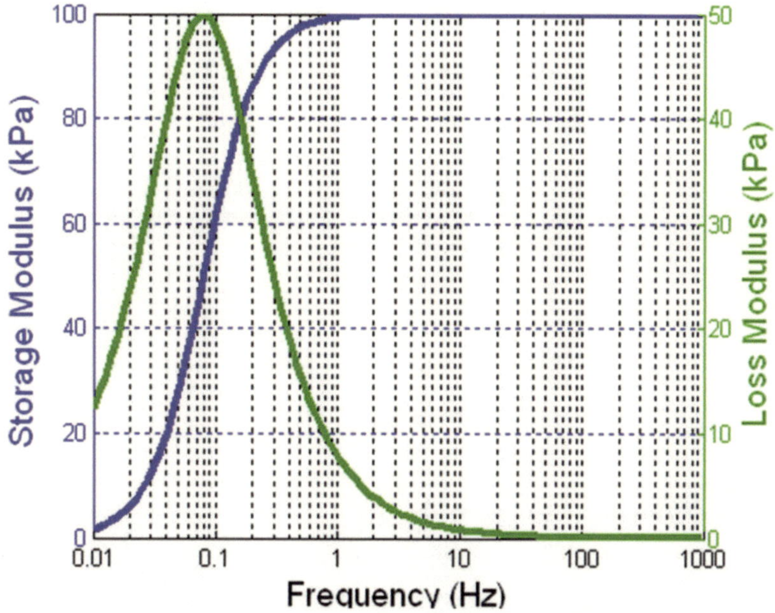

Fig. 8. The resultant dynamic properties of the simple Maxwell Fluid

Here the storage modulus is defined as in eq. 71 and the loss modulus is defined as (Macosko, 1994):

$$G''(\omega) = \frac{G_0 \tau \omega}{\tau^2 \omega^2 + 1} \tag{74}$$

The mechanical parameters of the coupled-mass microrheometer should also by known by design. Here, for simplicity it shall be assumed that $m_1 = m_2 = 4\text{x}10^{-8}$ kg, $k_{S1} = k_{S2} = 600$ N/m and $c_{A1} = c_{A2} = 0.1$ kg/s. Also for the electrical parameters it will be assumed that $n = 100$, $t = 50$ µm and $g = 0.1$ µm. The applied voltages are $V_{DC} = 100$ V and $V_{AC} = 1$ V. This means the sinusoidal force given by eq. 68 is 9 µN and the constant force given by eq. 67 is 443 µN. Therefore if the original channel width was 10µm, $\bar{h} = 10.75$ µm. As the measured material properties are calculated from eqs. 61 and 62, which are inversely proportional to the cube of the average channel width, ignoring the effects of the static deflection given by eq. 72 could lead to substantial errors. In this case the error in the measured storage and loss moduli would have been *c.a.* 24%.

Given these values, the displacement of mass 2 - the displacement that will be measured, can be predicted. The results are shown in Fig. 9.

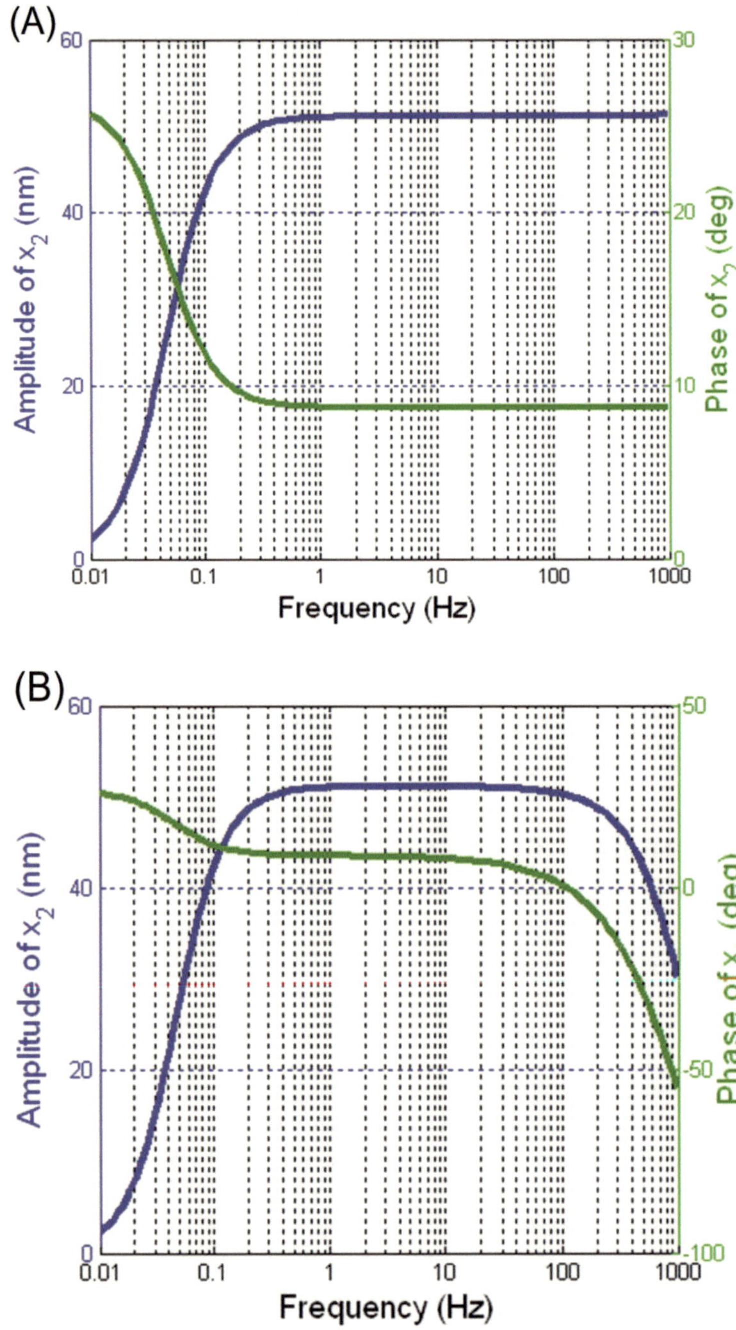

Fig. 9. The dynamic response of mass 2 for the case (A) no air damping and for (B) significant air damping

As can be seen the maximum displacement is only 52 nm. This is despite the relatively large voltage that was applied to the actuating comb-drive. The advantage of this is that it is very likely that any non-linear effects are negligibly small and that the strains in the liquid are sufficiently low that linear viscoelastic theory is valid (Bird et al., 1987). The issue is that it may be too small to be measured. Ignoring parasitic capacitances, the change in the capacitance of the sensing comb-drive is:

$$dC(t) = n\frac{\varepsilon t}{g}x_2(t) \tag{75}$$

In this instance, the capacitance is nominally c.a. 39.8 pF. The sensitivity of the comb-drive, given by eq. 76, can be shown to be 0.44 pF/µm, which is low given that displacements are in the nanometre regime.

$$\frac{\partial C}{\partial x} \approx n\frac{\varepsilon t}{g} \tag{76}$$

This is one of the main problems with this sort of device: you need to apply large voltages to achieve reasonable displacements and it still results in small changes in capacitances which may be difficult to measure, as can be seen in Fig. 10. Fortunately, new capacitance measurement chips with accuracies as good as 4 aF/√Hz (Irvine Sensors, n.d.) are commercially available, making these devices viable.

Another issue is the effect of air damping. Unless the comb-drives are sealed in a vacuum, it is likely that air-damping will be significant and will have to be included in the analysis using eqs. 11 and 12 rather than 13 and 14 (see Fig. 9). Unfortunately, in cases of complicated geometries such as this, damping is difficult to calculate. Even numerical analysis using CFD or FEA is unlikely to be accurate due to the number of assumptions and simplifications needed to get convergence (Ye et al., 2003). However, due to the comprehensive modelling, it is quite a simple task to curve fit to experimental data like that shown in Fig. 9, given a model fluid in the channel whose rheometry is well known, and ascertain the mechanical parameters more accurately. In practice it may be necessary to calibrate the rheometer in this manner anyway due to the variability of MEMS fabrication technologies in order to establish real values for the masses and stiffnesses.

The purpose of this chapter was to describe the complete system analysis of a new kind of coupled-mass microrheometer and show how such a device can be integrated into a real production process. The main issue is that in the installation of a sensor into a commercial process there should be no disruption or chance of contamination of the main process flow line. Naturally, issues such as space constraints and cleaning cycles need to be addressed as well. A possible lay-out is shown in Fig. 11. Here the microrheometer is connected externally to the main process flow line and to cleaning lines via electrically actuated three-way valves. Both the fluid to be tested and cleaning fluid are likely to need to be pumped into the microfluidic channel as the pressure required to push the fluid through the device will possibly be higher than that available in the main lines due to the constrictions in the channel. It can be imagined that at certain intervals of time, process fluid will be pumped into the channel, a frequency sweep performed and the channel washed clean while the data is analysed. Overall, as the device is channel-based rather than based on a liquid bridge, integration into automated commercial process lines is simple and allows for real-time, high-throughput analysis that can be used for quality control and product optimisation.

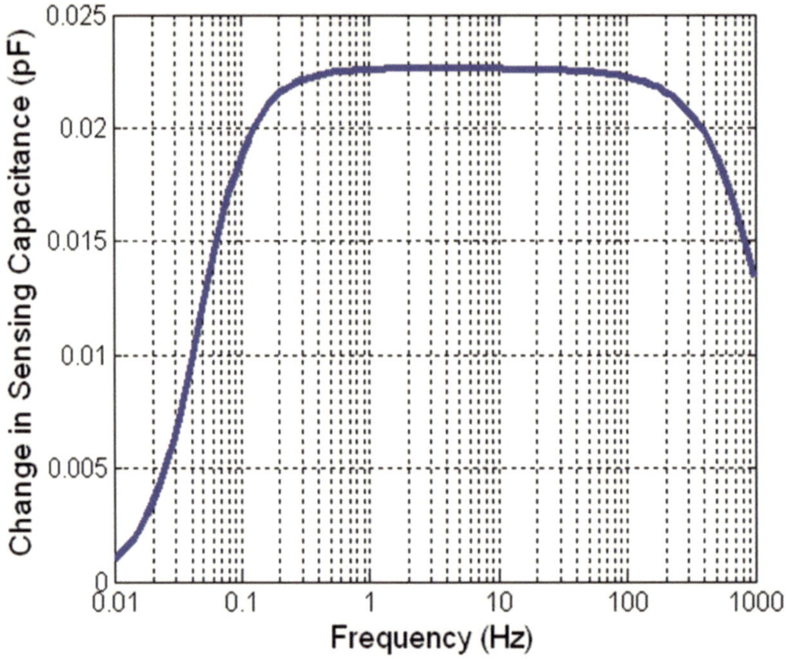

Fig. 10. Resultant change in amplitude of capacitance of sensing comb-drive

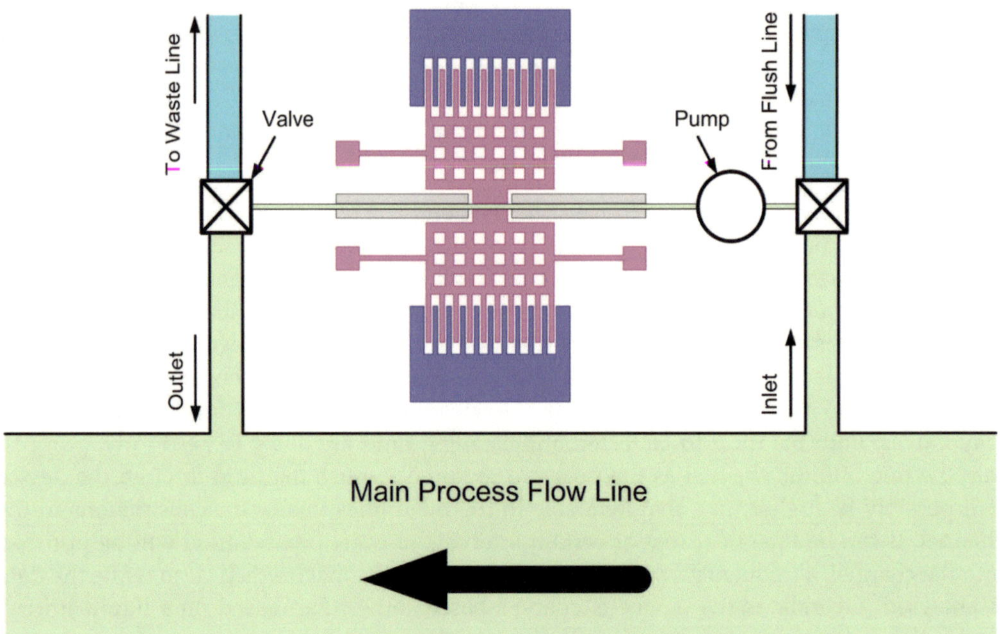

Fig. 11. Potential layout for integration of sensor into a commercial process flow line

5. Conclusion

A deterministic method of measuring viscoelastic fluid properties using a new kind of coupled-mass microrheometer has been shown. The fluid mechanics has been fully analysed as a viscoelastic squeeze flow problem and coupled with dynamical and electrical analysis. It has also been shown how such a device can be integrated into a real production process.

6. Acknowledgments

The author would like to thank Bethany Hanson for proofreading this chapter.

7. References

Bell, D., Binding, D. M. & Walters, K. (2005). The Oscillatory Squeeze Flow Rheometer - Comprehensive Theory and a New Experimental Facility, *Rheologica Acta*, Vol.46, No.1, (May 2006), pp. 111-121, ISSN 0035-4511

Bird, R. B., Armstrong, R. C. & Hassager, O. (1987). *Dynamics of Polymeric Liquids, Fluid Mechanics Vol. 1*, 2nd Ed., Wiley-Interscience, ISBN 047180245X, New York

Cheneler, D., Bowen, J., Ward, M. C. L. & Adams, M. J. (2011). Principles of a Micro Squeeze Flow Rheometer for the Analysis of Extremely Small Volumes of Liquid, *Journal of Micromechanics and Microengineering*, Vol.21, No.4, (April 2011), 045030, ISSN 0960-1317

Christopher, G. F., Yoo, J. M., Dagalakis, N., Hudson, S. D. & Migler, K. B. (2010). Development of a MEMS based Dynamic Rheometer, *Lab on a Chip*, Vol.20, (August 2010), pp. 2749-2757, ISSN 1473-0197

Crecea, V., Oldenburg, A. L., Liang, X., Ralston, T. S. & Boppart, S. A. (2009). Magnetomotive Nanoparticle Transducers for Optical Rheology of Viscoelastic Materials, *Optical Express*, Vol.17, No.25, (December 2009) pp. 23114-23122, ISSN 1094-4087

Denn M. M. & Marrucci, G. (1999). Squeeze Flow between Finite Plates, *Journal of Non-Newtonian Fluid Mechanics*. Vol. 87, pp. 175–178, ISSN 0377-0257

Engmann, J., Servais, C. & Burbidge, A. S. (2005). Squeeze Flow Theory and Applications to Rheometry: A Review, *Journal of Non-Newtonian Fluid Mechanics*, Vol. 132, No. 1-3, (December 2005), pp. 1-27, ISSN 0377-0257

Hansen, C. & Quake, S. R. (2003). Microfluidics in Structural Biology: Smaller Faster . . . Better. *Current Opinion in Structural Biology*, Vol.13, (October 2003), pp. 538–454, ISSN 0959-440X

Irvine Sensors (n.d.). MS3110 Universal Capacitive Readout™ IC, 23.08.2011, Available from http://www.irvine-sensors.com/pdf/MS3110%20Datasheet%20USE.pdf

Kumble, K. D. (2003). Protein Microarrays: New Tools for Pharmaceutical Development, *Analytical and bioanalytical chemistry*, Vol.377, (July 2003), pp. 812–819, ISSN 1618-2642

Lee, K. B. (2011). *Principles of Microelectromechanical Systems*, Wiley-IEEE, ISBN 0470466340, Singapore

Macosko, C. W. (1994). *Rheology: Principles, Measurements, and Applications,* Wiley-VCH, ISBN 0471185752, New York

Rao, S. S. (2010). *Mechanical Vibrations,* Prentice Hall, ISBN 0132128195, New York

Reynolds, O. (1886). On the Theory of Lubrication and Its Application to Mr. Beauchamp Tower's Experiments, Including an Experimental Determination of the Viscosity of Olive Oil, *Philosophical Transactions of the Royal Society of London,* Vol. 177, (February 1886), pp. 157-234, ISSN: 02610523

Strauss, W. A. (2007). *Partial Differential Equations,* Wiley, ISBN 0470054565, New York

Sujatha, K. S., Matallah, H., Babaai, M. J. & Webster, M. F. (2008). Modelling Step-Strain Filament-stretching CaBER-type using ALE Techniques, *Journal of Non-Newtonian Fluid Mechanics.,* Vol.148, (January 2008), pp. 109-121, ISSN 0377-0257

Walters, K. (1975). *Rheometry,* Chapman & Hall, ISBN 0412120909, London

Wei, X., Anthony, C., Lowe, D. & Ward, M. C. L. (2009). Design and Fabrication of a Nonlinear Micro Impact Oscillator, *Procedia Chemistry,* Vol.1, (September 2009), pp. 855–858, ISSN 1876-6196

Ye, W., Wang, X., Hemmert, W., Freeman, D. & White, J. (2003). Air Damping in Laterally Oscillating Microresonators: A Numerical and Experimental Study, *Journal of* Microelectromechanical Systems, Vol. 12, No. 5, (October 2003), pp. 557-566, ISSN 1057-7157

Part 2

Technology

4

Droplet-Based Microfluidic Scheme for Complex Chemical Reactions

Venkatachalam Chokkalingam[1], Ralf Seemann[2,3],
Boris Weidenhof[3] and Wilhelm F. Maier[3]
[1]*Experimental Physics, Saarland University, Saarbrücken*
[2]*Max Planck Institute for Dynamics and Self-Organisation (MPIDS), Göttingen*
[3]*Technical Chemistry, Saarland University, Saarbrücken*
Germany

1. Introduction

Liquid flow in microfluidic channels typically occurs at low Reynolds numbers, such that the flow is purely laminar [Solomon et al. 2003; Oddy et al. 2001]. As a consequence, two or more streams of miscible fluids flowing side-by-side only mix by diffusion. Such a mixing is very slow and it would be difficult to perform chemical reactions within the typical length scales respectively time scales used in standard microfluidic settings [Bruus et al. 2007; Stone et el., 2004; Teh et al., 2008]. Furthermore, the axial dispersion of concentration gradients parallel to the flow direction (Taylor dispersion [Taylor, 1954]) is a problem associated with the laminar flow in confined geometries due to the parabolic flow profile within microfluidic channels.

To overcome these microfluidic obstacles in mixing and to avoid axial dispersion, there has been considerable development towards the use of multiphase flows within microchannels. Continuous phase liquids can e.g. be segmented into allotments by injecting bubbles or droplets of a second immiscible fluid phase [Günther et al. 2004; Khan et al., 2004]. In this case, axial dispersion is eliminated because each liquid allotment is restricted between two bubbles and fast mixing is achieved by a friction induced circulating flow inside each allotment [Song et al., 2003 ; Li et al., 2007]. But as the liquid of each allotment is directly in contact with the microfluidic channel walls, cross-contaminations cannot be fully avoided and sticking of samples or reagents to the microchannel walls is unavoidable in this type of segmented flow microfluidics provided certain chemicals are processed.

Alternatively to the segmentation of a continuous flow by bubbles, reagents can be compartmented in emulsion droplets in a surrounding carrier phase for microfluidic processing. The latter approach is termed "droplet-based" microfluidic systems [Atencia & Beebe, 2005; Tabling, 2006]. The (emulsion-) droplets never get in contact with the channel walls and the sticking of reagents which are contained inside the droplets to the microfluidic channel walls is avoided. Mixing of a droplet content is also fast due to the friction induced convective flow pattern emerging inside droplets similar to the flow pattern in the individual allotments of segmented flow microfluidics [Song et al., 2003 ; Li et al., 2007]. But even that the transport and the processing of liquids in the form of individual droplets has many benefits, the development of new techniques and approaches were required to make droplet based microfluidics a useful tool to perform complex chemical reactions.

The first crucial step is to dispense different reagents into individual droplets with good volume control. In initial approaches using droplets as micro-reactors, the different reactants were commonly dispensed into droplets from a single inlet channel [Song et al., 2003; Chen et al., 2005; Li et al., 2007]. Using this method, the reactants are already partly mixed and reactive at the channel inlet even before being dispersed into droplets. This method of common dispensing will thus result in channel clogging for fast chemical reactions where a gel or a precipitate is formed within milliseconds of contact.

The problem of channel clogging can be avoided by initiating the reaction inside droplets by either adding a reagent from a nozzle directly into passing droplets [Abate et al., 2010] or by combining two individual droplets containing the different reactants. The microfluidic reaction scheme of combining droplets on demand is feasible to process also reactive fluids, which might precipitate, solidify or stick to channel walls [Evans et al., 2009] and which cannot be processed in single phase microfluidics approaches. This reaction scheme enables a microfluidic device to function for a long operation time with stable reaction conditions. To perform droplet based chemical reactions in the above mentioned manner several issues have to be addressed: First, the volume of the chemicals needs to be known precisely for full control over the chemical reaction. Thus, primary droplet sizes have to be produced with excellent monodispersity. Second, at least two kinds of the primary droplets containing the different reactants have to be generated in a pair like manner. Thus either two trains of droplets have to be synchronized subsequent to droplet production or the droplet production has to be synchronized directly to form strictly alternating pairs of droplets containing the different chemicals. Third, we have to reliably combine the formed primary droplet pairs and mix the content of the subsequently formed secondary droplets to induce the desired chemical reaction. And last, we either have to observe and analyze the results *in-situ* in the microfluidic device or we have to collect and post-process the reactants outside of the microfluidic device.

In this chapter, we discuss a droplet-based microfluidic scheme to perform fast chemical reactions without channel clogging and with precise volume and process control. This microfluidic scheme enables a process to function for long operation times with stable conditions. For that every microfluidic processing step is considered and adapted to meet the experimental requirements. We discuss this scheme explicitly for the production of mesoporous silica particles and platinum doped silica particles from a sol-gel synthesis route for heterogeneous catalysis. In the first sections the emulsion system is introduced, the device fabrication, and its operation is explained followed by the basics of sol-gel chemistry and how to evaluate the results. In the subsequent sections the individual microfluidic techniques and their optimization for the considered sol-gel reaction are explained. The explanation of the individual techniques is followed by a discussion of the entire microfluidic scheme which is used for the production of silica and Pt-doped silica microspheres from a sol-gel synthesis route and finalized by the physical and chemical analysis of the microfluidically produced microspheres.

2. Background, materials and experimental set-up

2.1 Emulsion systems

Water-in-oil (W/O) emulsions are good candidates for applications where aqueous phase chemical reactions are required as the two liquid phases are immiscible and the droplets can be stabilized easily and also their volume is maintained. If additional organic phases like methanol are contained in the aqueous phase particular oils, like perfluorinated oils, have to be chosen as carrier fluids where neither of the compounds of the droplet phase is soluble. Throughout this chapter the following emulsion systems is applied: A water-methanol mixture containing reagents are dispensed in perfluorodecalin (PFD) ($C_{10}F_{18}$, ABCR GmbH, Germany), as the continuous oil phase. The emulsion system is stabilized by the fluorinated surfactant molecules penta decaflouoro-1-octanol ($C_8H_3F_{15}O$, ABCR GmbH & Co. KG) which are added to the PFD phase. To approximate the amount of surfactant needed to stabilize the oil/aq. phase - interface, the interfacial tension of water/PFD and water-methanol/PFD (with the vol. ratio of 60/40 of methanol/water) was determined as function of the surfactant concentration using the pendant drop method (OCA 20, Data Physics GmbH, Germany), c.f. fig. 1.

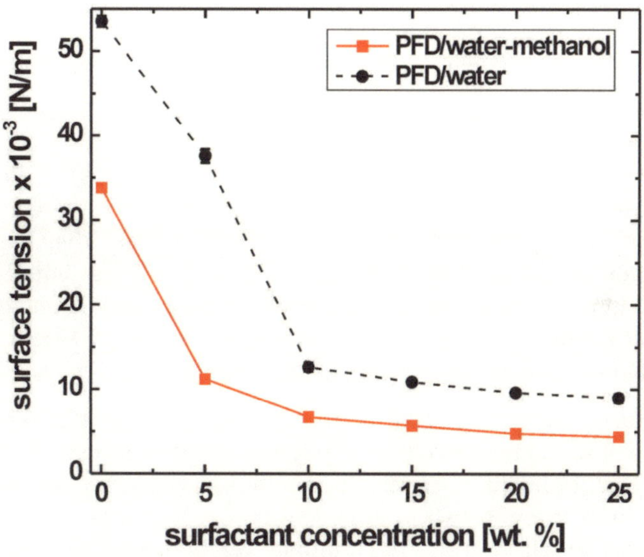

Fig. 1. Surface tension of perfluorodecalin/water and perfluorodecalin/methanol-water (60/40 v/v) interfaces as function of penta decaflouoro-1-octanol concentration.

I turns out that a surprisingly large amount of penta decaflouoro-1-octanol is needed to lower the interfacial tension to its plateau value. The addition of methanol (60 vol%) lowers the surface tension compared to pure water and increases the stability of the emulsion droplets significantly whereas the critical surfactant concentration needed to reach the plateau value is hardly affected by the addition of methanol. For the experiments discussed in this chapter 20 wt.% of penta decaflouoro-1-octanol is added to perfluorodecalin to lower the oil/water surface tension sufficiently to guarantee for stable droplet generation. However, penta decaflouoro-1-octanol stabilizes the droplets only weakly compared to standard emulsion systems and allows for easy droplet coalescence techniques, as will be discussed later.

2.2 Microfluidic device fabrication

The fabrication of microfluidic devices has become one of the specialities of micro-electro-mechanical systems (MEMS). Silicon microfabrication was well-established already in integrated circuit (IC) processes and therefore first microfluidic systems applied silicon as a device material. On the other hand glass has been the traditional material for chemical vessels for analysis and reactions and was also a natural choice for first microchips. However, microfabrication in glass and silicon substrates is technical demanding due to several required fabrication steps like etching and wafer bonding. To accelerate device fabrication times and to increase fabrication volumes the typically much easier to handle polymer microfabrication techniques made their way. Today a wide selection of polymeric materials and fabrication techniques provide suitable materials and methods for most applications. For droplet-based microfluidics it is particularly important that the used device material has a suitable chemical resistance to the used organic carrier phase to avoid or minimize swelling and dissolution of the polymeric materials and to avoid diffusion of the organic carrier phase through the device material. Additionally, the wettability of the device material is very important in droplet based microfluidics as the carrier phase must preferentially wet the channel material to guarantee stable droplet generation.

The microfluidic devices used in this study are made of a commercially available Poly(dimethylsiloxane) (PDMS) elastomer kit, 'Sylgard 184' (SG184, Dow Corning), and were fabricated by soft lithographic techniques [Xia & Whitesides, 1998; Wong & Ho, 2009]. PDMS has a unique combination of properties resulting from the presence of an inorganic siloxane backbone and organic methyl groups attached to silicon. They have very low glass transition temperatures and are liquid at room temperature. These liquid materials can be readily converted into solid elastomers by a cross-linking reaction. A photo-lithographically prepared SU-8 structure on a silicon wafer was used as a mould with inverted pattern structure. The main steps of the device fabrication are sketched in fig. 2a.

The Sylgard 184 base and curing agent were mixed in a ratio of 10:1 w/w, degassed and decanted onto the SU-8 mould. Once mixed, poured over the master and heated to elevated temperatures, the liquid mixture becomes a cross-linked elastomer within a few hours via the hydrosilylation reaction between vinyl ($SiCH=CH_2$) groups and hydrosilane (SiH) groups [Whitesides et al., 2001]. The multiple reaction sites on both the base and crosslinking oligomers allow for three-dimensional crosslinking. After thermal curing overnight at 65 °C, the Sylgard 184 facilitates easy peeling of the cross linked silicon elastomer structure from the SU-8 mold. In order to use these structures as microfluidic device, inlet holes were punched and the channels were sealed by a glass cover. To bond the glass cover covalently to the PDMS device the PDMS surface was activated by an oxygen plasma (Diener electronic GmbH, Germany) [Whitesides et al., 2001; Eddings et al., 2008] rendering the PDMS surface hydrophilic. To finally produce stable aqueous drops in the surrounding perfluorinated oil phase within the PDMS microchannels, the surfaces of the bonded devices were rendered back to hydrophobic nature by heating the devices to 60 °C for at least 6 hours. Teflon tubing was connected to the punched holes (fig. 2b) and the other ends of the tubing were connected to glass syringes.

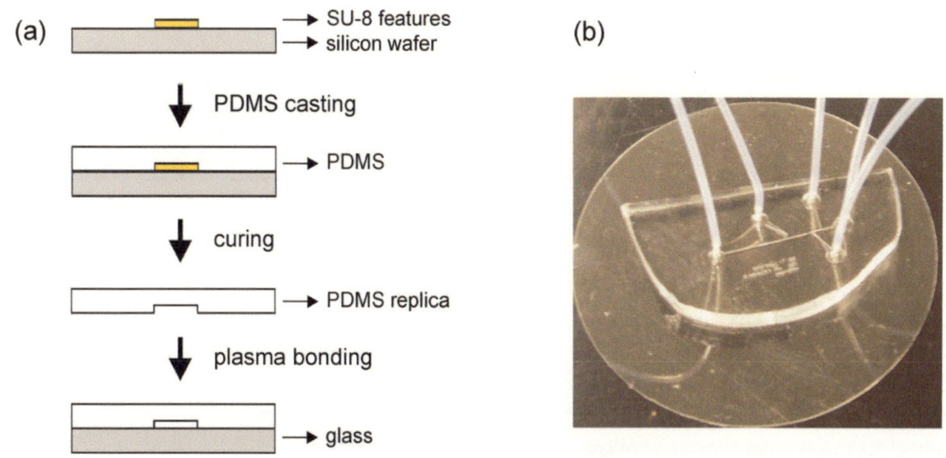

Fig. 2. (a) Schematic of PDMS Device Fabrication. (b) Ready to use PDMS device with tubes connected; the diameter of the cover glass slide: 5 cm.

2.3 Experimental set-up

A Leica Z16 APO optical microscope with a Leica L5 FL light source (fig. 3a), was typically used to monitor the microfluidic processes. Images and movies were recorded using a high resolution CCD camera (PCO 1600 hs, PCO AG, Germany); whenever needed, a high speed CMOS camera (PCO 1200 hs, PCO AG, Germany) was used which is able to capture more than 600 frames per second. Typical exposure times ranged from 500 µs to 50 ms, depending on flow velocity. The flow rates were adjusted precisely using custom-made syringe pumps (fig. 3b) driven by dc motors with decoder which were computer controlled using LabView programs (National Instruments Corporation). Using 1 ml gas tight precision glass syringes (Hamilton Bonaduz AG, Switzerland) volumetric flow rates between 10 µl/h and 1 ml/h were typically applied. Image analysis was done using the software: Image-Pro Plus 6.3 (Media Cybernetics, Inc., USA) and ImageJ 1.42 (National Institute for Health, USA).

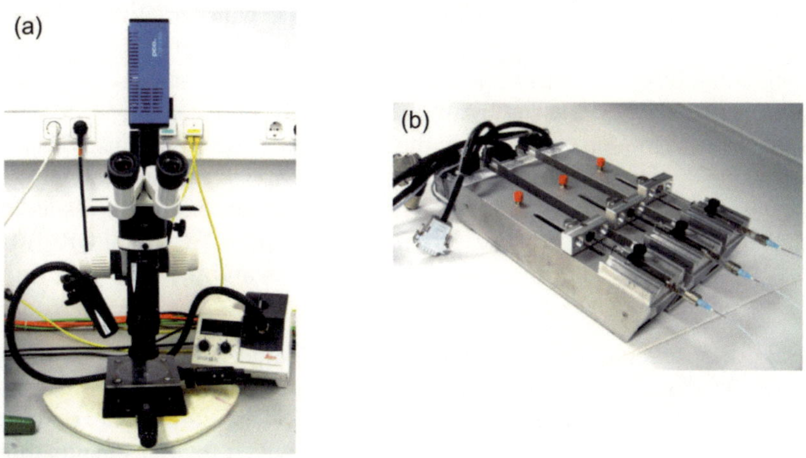

Fig. 3. a) Leica macroscope Z16 APO with a Leica L5 FL light source and camera. b) Custom made syringe pumps with glass syringes (Hamilton).

2.4 Sol-gel chemistry

Mesoporous materials with pores of 2 - 50 nm possess a potential combination of high surface areas and accessible pore sizes that make them ideal candidates for e.g. catalysts, catalyst supports and for the separation of complex molecules. A widely used material with such merits is mesoporous silica which is primarily produced through sol-gel processing. As we employ such a sol-gel process in our microfluidic approach, the reaction scheme will be explained in the following.

The most common starting point of the sol-gel process is the mixing of alkoxides (any organic compound derived from an alcohol by replacement of a hydrogen atom with a metal or other species) in a hydrolyzing environment (a chemical reaction of a compound with water). Gelation of the alkoxide solution occurs via a two-step process. First, alkoxides react readily with water. This reaction is called hydrolysis, because a hydroxyl ion becomes attached to the silicon atom as follows:

$$Si(OR)_4 + H_2O \rightarrow HO\text{-}Si(OR)_3 + R\text{-}OH$$

where R typically represents CH_3, C_2H_5, or C_3H_7. Depending on the amount of water and catalyst present, hydrolysis may proceed completely, so that all of the OR groups are replaced by OH groups, as follows:

$$Si(OR)_4 + 4\ H_2O \rightarrow Si(OH)_4 + 4\ R\text{-}OH$$

Any intermediate species $[(OR)_2\text{–}Si\text{-}(OH)_2]$ or $[(OR)_3\text{–}Si\text{-}(OH)]$ would be considered the result of partial hydrolysis. Two partially hydrolyzed molecules can link together in a condensation reaction to form a siloxane [Si–O–Si] bond:

$$(OR)_3\text{–}Si\text{-}OH + HO\text{–}Si\text{-}(OR)_3 \rightarrow [(OR)_3Si\text{–}O\text{–}Si(OR)_3] + H\text{-}O\text{-}H$$

or

$$(OR)_3\text{–}Si\text{-}OR + HO\text{–}Si\text{-}(OR)_3 \rightarrow [(OR)_3Si\text{–}O\text{–}Si(OR)_3] + R\text{-}OH$$

The condensation reaction liberates small molecule such as water or alcohol. These reactions can continue to build larger and larger silicon-containing molecules. The hydrolysis of the alkoxide can be catalyzed by both acids and bases, but the condensation step occurs much slower under acidic conditions. The hydrolysis and polycondensation reactions initiate at various sites within the precursor and the water as mixing occurs. When sufficient interconnected Si-O-Si bonds are formed in a region, they respond cooperatively as colloidal (submicrometer) particles or a sol. The size of the sol particles and the cross-linking within the particles depend upon the pH and the molar ratio r ($r = [H_2O]/[Si(OR)_4]$) [Brinker & Scherer, 1990]. As a function of time, the colloidal particles and the condensed silica species link together to a three-dimensional network.

The physical characteristics of the gel network depend largely upon the size of the formed particles and the degree of cross-linking prior to gel formation. At gel formation, the viscosity increases sharply and results in a gel. The gel is allowed to condense further during the following aging step (also called syneresis, which involves maintaining the gel for a period of time; hours to days) in order to increase the degree of cross-linking in the metal oxide network, and thereby to increase the strength and the structural stability of the gel. After aging, the gel still contains excess water not used in hydrolysis of the alkoxide, alcohol produced during hydrolysis, and other additives present in the starting mixture. Accordingly, the gel must be dried under atmospheric conditions to remove these solvents. After drying, gels are calcined at a temperature of 550 °C in order remove the left-over organics and surface absorbed species. After this processing, the physical properties like surface morphology, pore size, and pore volume of the synthesized silica can be analyzed using the techniques which will be explained in the following sections.

The specific surface area and the pore structure of silica synthesized under strong acidic ($pH < 2$) or strong basic ($pH > 12$) conditions usually lead to silica with specific surface areas below 600 m^2/g. Silica prepared under acidic conditions is microporous and has a fairly high surface area whereas the silica obtained under strong basic conditions (Stöber conditions [Stöber et al., 1968]) has a low surface area and is nonporous (pseudo mesoporous) [Tilgner et al., 1995]. To achieve a mesoporous silica with high surface area we will adopt a base catalyzed condensation at pH between 6 and 7 in our microfluidic scheme.

2.5 Techniques to evaluate the produced particles

In this section, the experimental techniques are shortly introduced which were used to quantify the properties of the produced silica (particles) with respect to physical and chemical properties like pore structure and composition.

Thermo-Gravimetric Analysis (TGA)

In our microfluidic approach silica gel particles were produced by means of droplet based microfluidics and collected outside of the microfluidic device. These silica gel particles have to be calcined to silica by a temperature treatment. To determine the ideal calcination temperature a thermo-gravimetric analysis (TGA) was performed using a TGA/DSC1 (Mettler-Toledo GmbH, Germany). The catalyst particles were filled into an alumina cup and heated at a ramp rate of 10 °C/min from ambient temperature to 800 °C. An empty alumina cup was used as the reference material. A flow of synthetic air (mixture of O_2 and N_2) of 40 ml/min, was maintained during the experiment. Evolving gases were monitored online with a Balzers ThermoStar GSD 300 T quadrupole mass spectrometer, see also fig. 9.

Specific surface area by physisorption

The porosity of the produced silica can be analyzed by 'physisorption'. The term 'physisorption' or 'physical adsorption' refers to the phenomenon of gas molecules adhering to a surface at a pressure less than their vapour pressure. The attraction between the adsorbed molecules and the surface are relatively weak. The energy of adsorption is approximately 20 kJ/mol, i.e. much lower than the typical strength of chemical bonds and therefore the adsorbed molecules retain their identities. An adsorption isotherm is recorded by carefully varying the partial pressure of the gas at constant temperature while the amount of the adsorbed gas is measured. The most widely used analysis of adsorption isotherms considers multilayer adsorption and was derived by Stephen **B**runauer, Paul **E**mmet and Edward **T**eller. Such a multilayer adsorption isotherm is called **BET** isotherm [Anderson et al., 1998; Brunauer et al., 2002]. Here, the initially adsorbed layer on a solid surface can act as a substrate for further adsorption and the adsorption isotherm is expected to rise infinitely, in contrast to a monolayer isotherm which saturates at high pressures (Langmuir isotherm [Langmuir, 1917]).

In our experiments, the specific surface area (m^2/g) and pore volumes (cm^3/g) of the porous silica was measured by the Brunauer-Emmet-Teller (BET) multipoint technique using an automated gas adsorption analyzer (Carlo Erba Sorptomatic 1990). The produced samples (at least 200 mg to achieve a good signal to noise ratio) were loaded and degassed at 200 °C for 2 h in vacuum, followed by the analysis of the surface area at the temperature of liquid nitrogen (77 K) with N_2 as adsorbate gas (the molecular area of the adsorbate N_2 is taken to be 15.8 $Å^2$), see also Figs. 9 and 11.

Electron Microscopy and X-ray fluorescence

Hitachi S-4500 Scanning Electron Microscopy (SEM) was used in the presented work to image the surface structure of the produced catalyst particles with high magnification. For that, the catalyst particles were fixed to an electrically conducting carbon bed stuck to the SEM sample holder and then coated with thin gold layer to avoid charging effects. The accelerating voltage used for the measurements was 10 kV, see also figs. 10 and 12. To characterize the metal phase, JEOL JEM 2011 (JEOL GmbH, Germany) transmission electron microscopy (TEM) was used. An EDAX-Eagle II X-ray fluorescence (XRF) analyzer was used to analyze the platinum content in the microfluidically produced catalyst particles, see also fig. 14.

3. Microfluidic techniques

In this section the individual techniques will be explained in some detail, which are the basis of the microfluidic reaction scheme used in section 4 for the production of silica particles.

3.1 Synchronized droplet production

To perform chemical reactions by merging droplets inside a microfluidic device it is not just enough to produce one kind of monodisperse droplets. We rather have to produce two types of monodisperse droplets (with appropriate reagents) in a strictly alternating manner. Step emulsification geometry was demonstrated to produce droplets with excellent monodispersity (coefficient of variance in diameter < 1.2%) [Priest et al. (2006-1)]. The step geometry facilitates a triggered Rayleigh-Plateau type instability of a quasi 2d stream of an aqueous phase surrounded by an oily phase at a topographic step. For the synchronized production of two kinds of droplets having different chemical content, we combine two of these individual step-emulsification units into one double step-emulsification device (fig. 4A) [Chokkalingam et al., 2008]. A double-step-emulsification device consists of two single step-emulsification units combined to a common main channel, which is fed by the continuous oil phase.

A time series of the droplet formation process captured with the high speed camera is shown in fig. 4b: When a droplet is forming in one step-emulsification unit (e.g. bottom unit in fig. 4 b-a), the Laplace pressure of the forming drop decreases as its interface curvature decreases. This leads to a pressure drop redirecting the continuous phase flow towards this step-emulsification unit (bottom channel in Figure 4 b-b,c). During this droplet formation the volume flow of the continuous phase is hindered and the stream of dispersed phase coming from the upper step-emulsification unit retracts. When the dark droplet pinched-off into the main channel, the Laplace pressure changes rapidly, the stream of the dispersed phase coming from the bottom unit retracts, and the continuous phase can flow more freely into the main channel (fig. 4 b-d). Subsequently, the dispersed phase coming from the top step-emulsification unit enters the main channel (fig. 4 b-e) and the process loops again. This pressure cross talk between the individual step-emulsification units synchronizes the drop production and leads to the defined formation of droplet pairs which might contain different reactants.

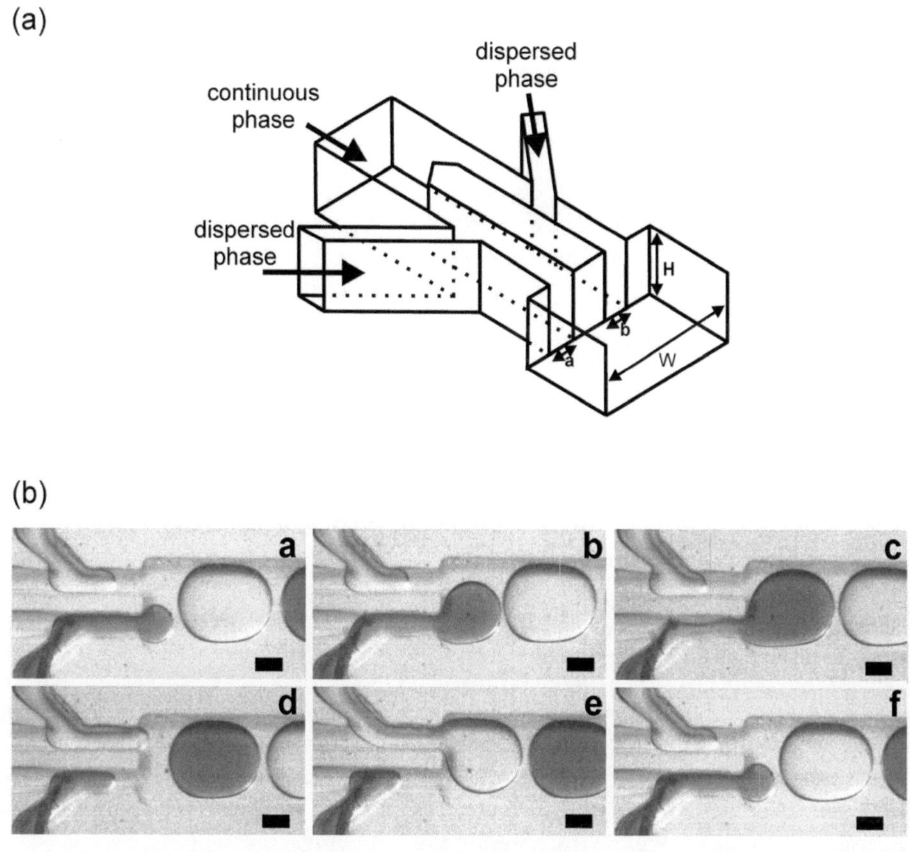

Fig. 4. (a) Schematic of the double step-emulsification device. The typical device dimensions were a = b = 35 μm, H = 120 μm and W = 160 μm. (b) Time series of optical images showing the droplet formation using a double step-emulsification device. For clarity, the aqueous phase injected from bottom is dyed with nile blue. Scale bar: 50 μm.

In case the volumetric flow rates of both step-emulsification units differ, the double step emulsification can produce droplet pairs having not only different chemical contents but also different volumes. In either case the volume control of each droplet type remains on the level of a single step emulsification device, i.e. having a variance in drop volume < 4%, see also fig. 5. For a better judgment of the dispersing precision for the reactants, we denote the droplet monodispersity in terms of droplet volumes rather than in droplet radius. The droplet volumes are calculated as follows: If the radius R of the projected droplet area is smaller or equal than the height of the microfluidic channel, the droplet is spherical and its volume can be simply calculated by $V = 4/3\ \pi R^3$. If the size of the droplets exceeds the height of the channel h, as is our case, the droplets are assumed as disk shaped droplets and their volumes are calculated by $(\pi/12)\cdot[2D^3 - (D - h)^2 (2D + h)]$ where D is the diameter of the disk.

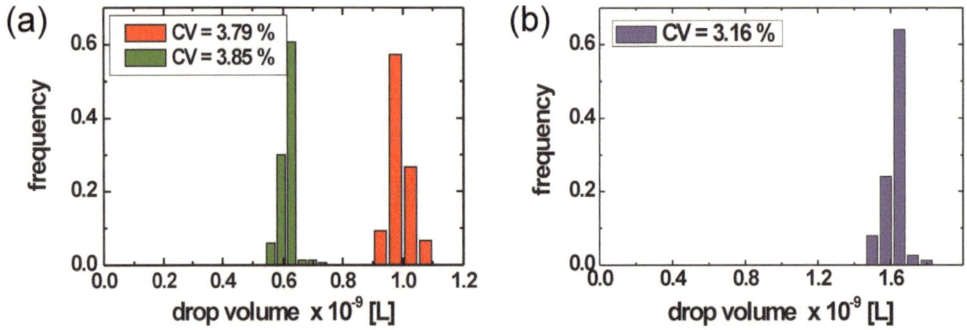

Fig. 5. (a) Histogram of the droplet volumes *A* and *B* that were generated with the self-synchronizing double step-emulsification device. (b) Histogram of the volumes of the merged droplets *C*.

The alternating droplet production is very robust, even up to about 300 droplets per second for the droplet size used here, and works up to a dispersed phase volume fraction of about 96 %. All these characteristics make the double step-emulsification device ideal for applications in complex microfluidic reactions.

3.2 Droplet merging and mixing

In the subsection before, it was explained how pairs of droplets can be produced in a strictly alternating manner. If the two droplets forming a pair contain different reactants, a chemical reaction can be initiated by merging and mixing the individual droplet pairs. In fig. 6, the controlled merging of two droplets, *A* and *B*, forming a pair and the subsequent droplet mixing in the combined droplet *C* is shown. The droplet merging can be achieved via a geometrical constriction as will be discussed in the following.

To keep the technical requirements for the drop coalescence as low as possible and still account for all experimental needs, we apply a technique where the aqueous droplet pairs are merged when they pass through a geometrical constriction. Such a merging technique works very effective for an emulsion system not being stabilized with surfactants or being stabilized with less effective surfactants, i.e. surfactants that do not stabilize the emulsion very well (c.f. section 2.1). When the droplet pairs reach a geometrical constriction, the merging of the two droplets is induced by slowing down the leading droplet at the constriction while the subsequently arriving droplet is pushed towards the first one (see fig. 6). If the constriction size is chosen appropriately, the droplets will pass the constriction during or right after the droplet coalescence. This technique has the lowest demands on the device fabrication and works very reliably also for droplet pairs arriving at the constriction with several hundred hertz. The precondition for the reliable droplet merging is that the droplet pairs arrive at the constriction with a certain separation from the previous and the following droplet pairs. This can be achieved easily by producing two kinds of droplets having slightly different volumes: both kinds of droplets do not touch the side wall and experience the approximately parabolic flow profile in the microfluidic channel. Due to the parabolic flow profile, the smaller droplets move to the center of the channel where they experience a larger mean flow velocity and travel faster than the larger droplets. Accordingly, at a small distance downstream from the production unit, we find clearly separated droplet pairs where the larger droplet is leading and the smaller droplet is trailing behind (see fig. 6). The coalescence occurs at the constriction as explained earlier. Figure 5b shows the histogram of the volume of the merged gel drops C.

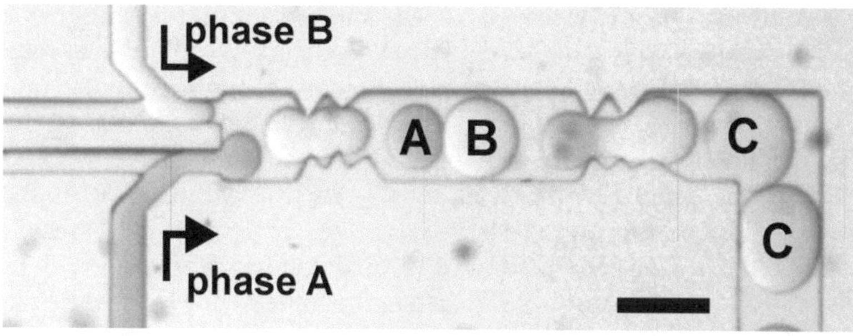

Fig. 6. Optical micrograph showing the production of droplets *A* and *B* having different volumes, the formation of separated droplet pairs lead by the bigger droplet and the subsequent coalescence into droplets *C* at a geometrical constriction. The aqueous phase injected from the bottom channel is colored with nile blue for clarity. Scale bar: 200 µm.

Downstream mixing of the reagents inside the combined droplet proceeds very efficiently due to the twisty flow pattern inside the droplets which is generated by the flow induced friction of the surfactant lamellae with the channel walls. Due to the dimensions typically used in microfluidic devices the circulating flow pattern emerging inside the travelling droplets is mirror symmetric with respect to a plane through the middle of the droplets parallel to the flow direction. Accordingly mixing occurs primarily within the 'top' and 'bottom' half of the droplet, whereas the mixing between the top and bottom half is slow which is a major obstacle if a chemical gradient is perpendicular to the flow direction. In this case the mixing can be accelerated by repeatedly altering the symmetry of the flow pattern using e.g. back fold channel geometries [Song et al, 2003]. However, in the situation considered here, where two droplets are combined that are flowing behind each other the resulting chemical gradient is parallel to the flow direction, c.f. the combined droplet in the constriction in Fig 6, and the symmetry of the concentration gradient does not need to be repeatedly varied to achieve fast mixing. With a concentration gradient parallel to the flow direction the elongation of the merged droplet at the constriction and the followed expansion in the wider channel additionally promotes fast mixing. In our experiments complete mixing is achieved within a travel distance of about 500 µm, i.e. about twice the droplet length, or about 100 ms at typical flow velocities of about 8 mm/s.

4. Microfluidic synthesis and analysis of silica particles

To produce silica particles by microfluidic routes, we optimized the individual microfluidic steps for this reaction, as explained in the previous section 3, and also optimized the sol-gel recipe to adapt the gelation time to typical microfluidic processing time of a few seconds, cf. section 2. In this section we will now present the microfluidic reaction scheme for the fabrication of silica microspheres and discuss the achieved results.

The applied microfluidic sol-gel synthesis scheme is sketched in Scheme 1. One type of droplets, *A*, contains a 1.57 M acidified solution of the silica precursor tetramethoxysilane, TMOS ($Si(OCH_3)_4$, ABCR GmbH & Co. KG) in a mixture of methanol (Merck KGaA, Darmstadt, Germany) and water (Millipore™) at a volumetric ratio of 60/40 . The second type of droplets, *B*, contains ammonia (Merck KGaA, Darmstadt, Germany) based on the same water/methanol mixture. These two kinds of monodisperse droplets were produced with different volumes (fig. 5a) for the sake of the subsequent droplet merging as explained in section 3. A rapid acid catalyzed TMOS hydrolysis occurs in the aqueous solution *A* even before dispensing them into droplets at pH 1-2. Because of the high molar ratio of Si:H_2O ($r > 10$) and the nearly nonexistent retarding effect of the methoxide group on the hydrolysis rate, a rapid and nearly quantitative hydrolysis of the precursor to silicic acid can be assumed within a few minutes [Pouxviel et al., 1987]. The formation of the silica gel, however, is very slow at this pH. To accelerate the condensation of silicic acid in the microfluidic environment, we combined the TMOS containing droplets *A* with the ammonia containing droplets *B* to form droplets *C* and thereby adjust the pH to be between 6 and 7, above the iso-electric point of silica (pH > 3), where the gelation time is minimum [Stöber et al., 1968].

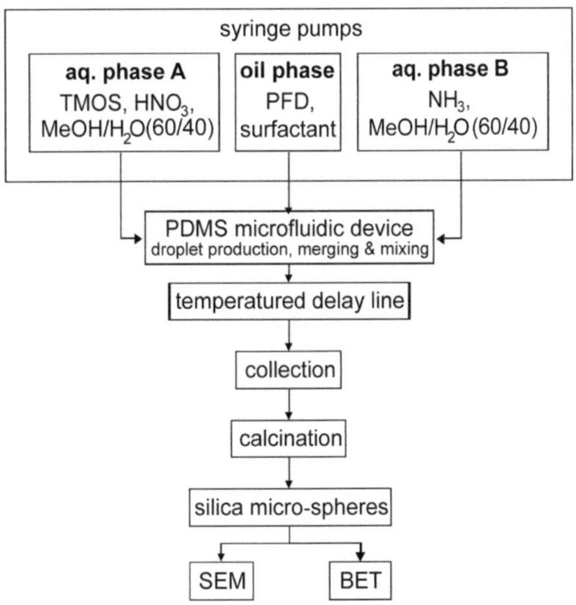

Scheme 1. Microfluidic synthesis route including the analysis of the produced silica spheres.

The final molar ratio in the mixed solution in droplet *C* was TMOS/methanol/HNO_3/ NH_3/H_2O = 1/17.8/0.13/0.33/34.3. The condensation reaction starts with the deprotonation of silanols due to the reaction with hydroxyl ions. This condensation along with the aggregation of the condensed species leads to a continuous growth of the formed gel particles. This continues until the silica particles increasingly fuse and fill the complete drop volume (drop volume determines particle size). Apart from the condensation, the dissolution of silica due to the reverse reactions of condensation also exhibits a strong pH-dependence which increases by more than three orders of magnitude between pH 3 and 8 in aqueous solution [Brinker & Scherer, 2000]. Due to the large methanol concentration in the merged droplet *C*, a low dissolution rate could be assumed as the solubility of silica in methanol is much lower than in water. During merging and mixing of droplets inside the microfluidic device, none of the reactive mixture gets in contact with the microfluidic channel which avoids any precipitation or sticking of the silica gel to the microfluidic channels and allows for long operation times.

Because of the large concentration of fluoroalkyl groups present in the surfactant being slightly soluble in the aqueous phase and the large surfactant concentration in the continuous phase, it seems possible that the pH of the aqueous solution is influenced. To guarantee that the pH of the merged droplets C is in the desired range for the sol-gel reaction, as adjusted by the reactant concentrations in droplets *A* and *B* we conducted two control experiments using the same microfluidic scheme: We added subsequently the pH indicators methyl red and bromothymol blue to one of the droplets [Methyl red: red at pH below 4.4, yellow at pH over 6.2, and orange in between (fig. 7a). Bromothymol blue: yellow at pH below 6, blue at pH over 7.6 and dark green in between (fig. 7b)]. With these reactions we could confirm that the pH inside the combined droplets *C* is in the range between 6.2 and 7.6 which is desired for fast gelation and we can exclude a significant pH change by the large concentration of surfactant.

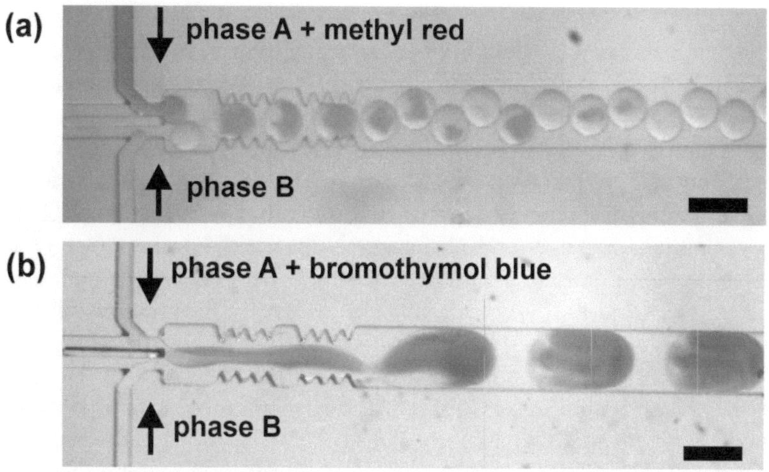

Fig. 7. Control experiments using pH sensitive dyes to determine the pH of the combined droplets C. (a) pH indicator: methyl red. The color change indicates a pH > 6.2. (b) pH indicator: bromothymol blue. The color change indicates a pH < 7.6. Scale bar: 150 µm.

To complete the microfluidic reaction scheme, the merged and mixed droplets *C* are subsequently given some time to develop the gel network. To avoid the formation of silica gel clumps at the rear side of the droplets [Evans et al., 2009], we guide the droplets into Teflon tubing with increased cross sectional area where the flow velocity is reduced by more than one order of magnitude. As the droplets do not touch the side walls of the Teflon tubing the convective flow pattern inside the droplets effectively vanishes. To guarantee that the gel droplets stay separated in the Teflon tubing and do not start to clump together, the distance between them is slightly increased by adding a small percentage of continuous phase after the back fold channels (fig. 8), just before entering the Teflon tubing. The Teflon tubing is about 1 m in length and placed in a temperature controlled unit at 65 °C. The elevated temperature accelerates the gel formation and the removal of the solvent from the droplets.

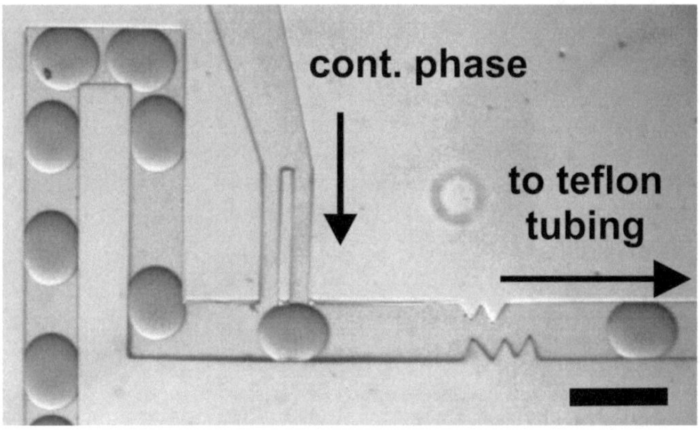

Fig. 8. Optical micrograph showing the end of the mixing line. The flow is from left to right. Right after the mixing line additional continuous oil phase is injected via the side channel to increasing the distance between the merged droplets before they enter the Teflon tubing. Scale bar: 200 µm.

The monodisperse silica particles are allowed to travel within the Teflon tube for about 20 min. Subsequently, they are collected outside of the microfluidic device in a beaker containing the same perfluorinated oil at the same temperature of 65 °C. In this beaker, the gel particles remain under continuous stirring for another 120 min, to ensure proper network formation and gentle solvent evaporation. The collected gel particles are removed from the beaker and subsequently stored at room temperature for about 48 hours. To remove the remaining organics and the surface absorbed species, the gel particles are calcined at elevated temperature. To identify the right temperature being sufficiently large to remove all organics and as low as possible not to densify the forming silica network, the evolving gases and the mass loss were monitored online as function of temperature, see fig. 9. With the help of the mass spectrometer signal, the mass loss of the samples can be divided into two main temperature regions: (1) between ambient temperature and 120 °C where water is evolved and (2) between 120 and 550 °C. Within the second region the feature between 120 °C to 250 °C denotes the temperature range where the PFD passes out whereas the plateau between 300 °C to 420 °C denotes the temperature range where PFD is combusted leading to CO_2 desorption. As a result, a minimum temperature of about 500 °C is required for calcination to expel all volatile organic components completely from the silica gel particles. Accordingly, all the collected gel particles that were used for the following analysis were calcined at a temperature of 550 °C.

The surface structure of the calcined silica particles are imaged by Scanning Electron Microscopy; images are displayed in fig. 10. The silica particles are almost perfectly round rarely showing small raised surface corrugations. The surface morphology is cloudy and reminiscent of a crumble topping.

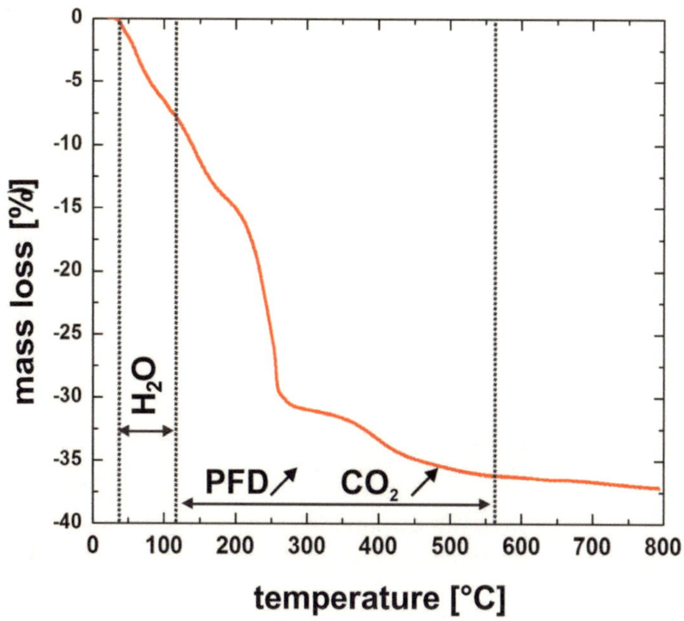

Fig. 9. Thermo-Gravimetric Analysis curve of the produced silica particles.

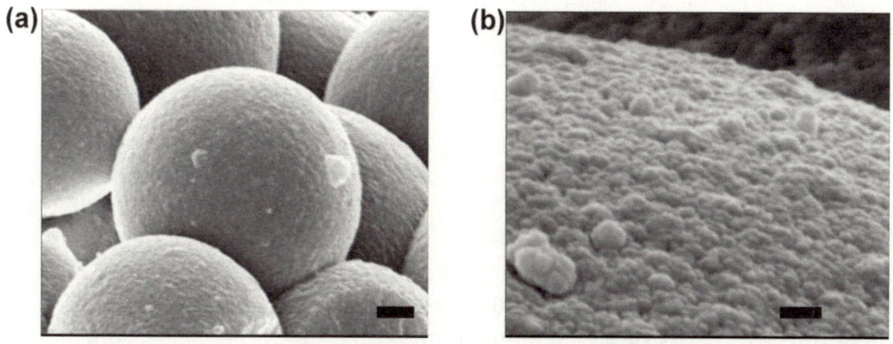

Fig. 10. Scanning electron micrographs of silica spheres after calcinations. (a) Scale bar: 1 μm. (b) Close up of the surface structure. Scale bar: 50 nm.

The porosity and the pore volume of the produced silica particles were investigated by a BET analysis, c.f. section 2.5. A typical nitrogen adsorption-desorption isotherm recorded for the microfluidically synthesized silica particles is shown in fig. 11a. It shows a steep rise in the low-pressure region at a normalized pressure of about $p/p_0 < 0.05$ which indicates the presence of micropores in our analyzed silica samples. At larger partial pressure a characteristic hysteresis loop appears, between a relative pressure of $p/p_0 = 0.40$ and 0.85. According to IUPAC this type of isotherm can be classified as type IV and indicates the presence of mesopores (pore size 2 – 50 nm).Moreover, the shape of hysteresis indicates cylindrical pores with bimodal pore openings.. Based on the adsorption isotherm in fig. 11a the pore size distribution is calculated and displayed in fig. 11b. The pore size distribution (black solid curve in fig. 11b) is narrow having a distribution maximum at a pore radius of 2.4 nm.

The specific surface area of the produced silica particles averaged over several production runs and for various measurements was determined to 820 m^2/g (± 20 m^2/g) with a cumulative pore volume of 0.93 cm^3/g (± 0.02 cm^3/g).

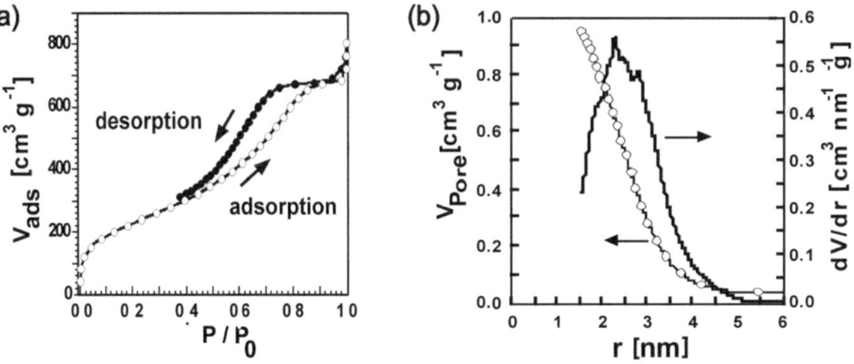

Fig. 11. (a) BET nitrogen adsorption-desorption isotherm of the produced silica spheres. V_{ads} = adsorbed volume, p/p_0 = reduced pressure (b) Pore size distribution. V_{pore} = pore volume, r = pore radius.

The specific surface area of the microfluidically produced silica spheres reveal significant differences compared to bulk synthesized silica using the same chemical recipe. For the microfluidically synthesized material we achieved a surface area which is about 40 % larger than the corresponding bulk value which is below 600 m^2/g. The surface area of the microspheres itself is just about 1 m^2/g and negligible with respect to the total surface area.

In order to further evaluate the superior results of our microfluidic reaction scheme, we compare the properties of our microfluidically produced silica particles to other microfluidic synthesis routes by Carroll *et al.* [Carroll et al., 2008], Lee *et al.* [Lee et al., 2008], Chen *et al.* [Chen et al., 2008]. However, the formation of mesoporous silica particles differ from our chemical scheme as the silica formation in these references was supported by surfactant templates as the solvents evaporate into the continuous carrier phase and cannot be compared directly. Furthermore, the microfluidically produced silica spheres in these references were not analyzed explicitly in terms of surface area and pore structure and rather results from bulk processing using the same chemical recipes were adopted. This might be a consequence of the limited operation time of the microfluidics device due to channel clogging and the presumably small amount of produced material which was not sufficient for a detailed analysis of the pore structure. Comparing our results to the adopted bulk results from Carroll et al., and Chen et al., we achieved a 50 % larger specific surface area even without surfactant templating due to the optimized droplet-based microfluidic reaction scheme.

5. Microfluidic synthesis of precious metal catalysts

Due to their enormous internal surface area and their homogeneous pore size distribution the microfluidically produced silica particles seem to be ideal supports for catalytically active elements. Platinum (Pt) or palladium (Pd) doped silica catalysts [Yazawa et el., 2002] are e.g. highly active for catalytic combustion of hydrocarbon and other organic exhausts.

Yao et al. [Yao et al., 1980] investigated the propane oxidation reactions and found Pt doped materials to be better catalysts than Palladium and Rhodium doped materials. In the following we modify the chemical recipe and use the above presented microfluidic reaction scheme to produce Pt doped silica spheres.

Platinum compounds ($Pt(NO_3)_2$) were readily dissolved in an aqueous phase and can thus be applied in a straight forward manner to the production of platinum doped silica spheres using the same microfluidic scheme as discussed above. Thus $Pt(NO_3)_2$ was added (up to about 7 mol.% (46 wt.%) of Pt) to the silica precursor tetramethoxysilane (TMOS) in a mixture of methanol and water at a volumetric ratio of 60/40 as dispersed phase *A*. The dispersed phase *B* contains again ammonia based on the same water/methanol mixture. Following the above described microfluidic protocol, platinum doped silica particles were produced, post-processed and analyis. The surface morphology of these particles depends on their Pt concentration as is evident from scanning electron micrographs. The SEM images displayed in fig. 12 reveal that the surface morphology of catalyst particles get rougher and cloudier as the Pt concentration increases.

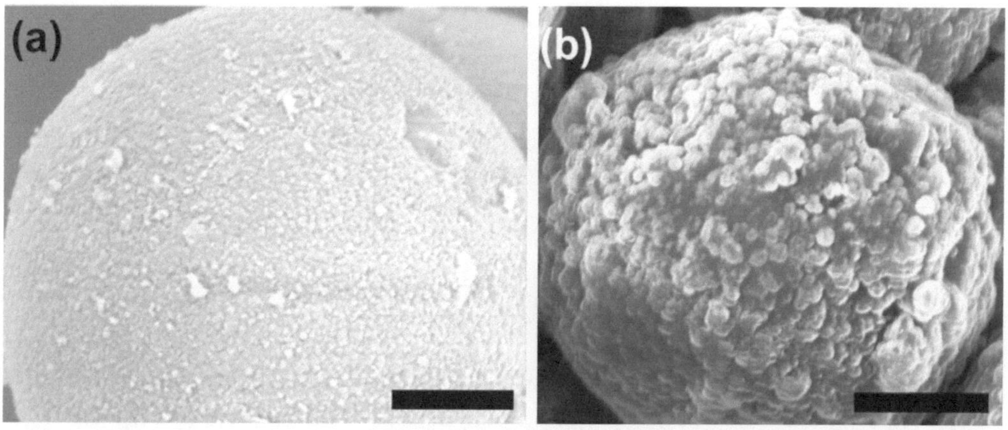

Fig. 12. SEM images of Pt doped silica particles: (a) 3 mol.%, (b) 7 mol.%. Scale bar: 1 μm.

We analyzed again the surface area and pore structure of the microfluidically synthesized platinum doped silica particles by a BET analysis. The specific surface area of the microfluidic catalyst particles with 4.91 mol.% platinum concentration averaged over several production runs and for various measurements was determined to be around 590 m^2/g (± 20 m^2/g) with a pore size distribution maximum at 1.9 nm (± 0.1 nm) pore radius and a cumulative pore volume of 0.54 cm^3/g (± 0.02 cm^3/g). Platinum concentration was also varied to 0.15 mol.% and the results are displayed in Table 1. The BET isotherm for 4.91 mol.% Pt concentration is shown in fig. 13.

Transmission electron micrographs indicate that the microfluidically produced silica particles contain metallic platinum (face centered cubic lattice) arranged in nanoscopic dots of about 2 nm size, see fig. 14. Based on the pore structure, the large surface area and the distribution of metallic platinum the microfluidically synthesized Pt doped silica particles are expected to be excellent catalysts for heterogeneous catalysis in general and for propane oxidation in particular.

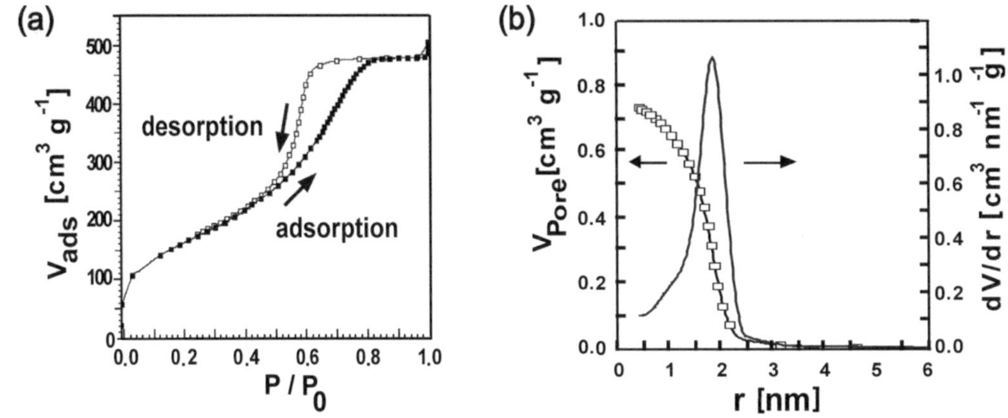

Fig. 13. BET nitrogen adsorption-desorption isotherm of the platinum doped silica catalyst microspheres produced through our microfluidic reaction scheme. V_{ads} = adsorbed volume, p/p_0 = reduced pressure.

Fig. 14. TEM measurements of microfluidically synthesized platinum doped silica. a) TEM micrograph; scale bar: 100 nm.' b) electron diffraction pattern of the Pt nanodots. The face centered cubic lattice indicates metallic Pt; scale bar 2 nm^{-1}.

Material	Specific Surface Area (m^2/g)	Pore radius (nm)	Cumulative pore volume (cm^3/g)
SiO_2 particles	820 (± 20)	2.4	0.61 (± 0.02)
0.15 mol.% Pt/SiO_2 particles	592 (± 20)	1.961	0.58 (± 0.02)
4.91 mol.% Pt/SiO_2 particles	590 (± 20)	1.994	0.54 (± 0.02)

Table 1. Characterization results from the BET isotherms and XRF Analysis.

6. Conclusion

We discussed a droplet-based microfluidic reaction scheme to perform chemical reactions with precise volume and process control that allows for fast reactions that might even form gels or precipitates. The important step for the microfluidic scheme is the handling of the reactive chemical mixture that never touches the channel walls and allows for long and stable operation conditions. We applied this microfluidic scheme to perform sol-gel reactions and produced silica particles and platinum doped silica particles with superior properties.

To achieve the synchronized production of droplet pairs containing different reactants a double step-emulsification device was used. Each kind of droplets containing the different reactants is produced with precise volume control and adjustable production frequency. The synchronization mechanism is very robust and makes the step-emulsification devices ideal for complex chemical reactions. The gel formation inside the droplets is started by a controlled merging of the droplet pairs at a geometric constriction forming droplets with a concentration gradient parallel to the flow direction. Mixing of the reactants inside the merged droplets is very fast due to the particular concentration gradient and a friction induced convective flow pattern. To optimize the protocol for the considered sol-gel reaction, an additional delay line and a thermal post processing line are included into the microfluidic device.

The silica particles produced by this microfluidic reaction scheme were mesoporous and had a superior surface area of 820 m^2/g (± 20 m^2/g) and a narrow pore radius distribution of around 2.4 nm. This surface area is about 40 % larger than the corresponding value of bulk synthesized silica using the same chemical recipe. The efficient mixing of reagents within the combined droplets and the short diffusion paths for the liquids in the droplets during the drying process help to achieve this improved quality. Without any modification in the microfluidic reaction scheme, the same protocol was further used to produce platinum doped silica catalyst particles by modifying the sol-gel recipe. The surface areas of the produced Pt doped particles are reduced to ~ 590 m^2/g (± 20 m^2/g) compared to pure silica spheres but still large compared to other synthesize schemes and are assumed to result in excellent catalytic activities.

From the obtained results we conclude that droplet-based microfluidic reaction schemes offer great possibilities to perform complex chemical reactions and to synthesize materials with superior properties which were not possible with existing microfluidic approaches. This type of droplet-based microfluidic scheme is thus expected to have a large potential not only for synthesis reactions but also for drug screening purposes or even combinatorial approaches when being combined with techniques to manipulate, redistribute and split merged droplets.

7. Acknowledgments

We thank the group of U. Hartmann for help with SEM imaging and R. Birringer and J. Schmauch for TEM measurements.

8. References

Abate A., Hung T., Mary P., Agresti J., and Weitz D. (2010), Proc. Natl. Acad. Sci., 107, 19163-19166.

Anderson K. T., Martin J. E., Odinek J. G., and Newcomer P. P., (1998) Chem. Mater. 10, 311-321.

Andersson N., Kronberg B., Corkery R., Alberius P. (2007), Langmuir, 23, 1459.

Atencia J. and Beebe D. J. (2005), Nature, 437, 648–655.

Barrett E. P., Joyner L. G., and Halenda P. P. (1951), Journal of the American Chemical Society, 73, 373-380.

Brinker C. J., Scherer G. W. (1990), Sol-Gel Science: The Physics and Chemistry of Sol-Gel Processing, Academic Press, Inc., Boston, ISBN: 0121349705.

Broekhoff J. C. P. and de Boer J. H. (1968), Journal of Catalysis, 10, 368-376.

Brunauer S., Emmett P. H. and Teller E. (2002), Journal of the American Chemical Society, 60, 309-319.

Bruus H. (2007), Theoretical Microfluidics, Oxford University Press, USA, ISBN 0199235090.

Carroll N. J., Rathod S. B., Derbins E., Mendez S., Weitz D. A. and Petsev D. N. (2008) Langmuir, 24, 658-661.

Chen D. L., Gerdts C. J., Ismagilov R. F. (2005) J. Am. Chem. Soc., 127, 9672.

Chen Y., Wang Y. J., Yang L. M. and Luo G. S. (2008), AIChE Journal, 54, 298-309.

Chokkalingam V., Herminghaus S. and Seemann R. (2008), Appl. Phys. Lett., 93, 254101-254103

Chokkalingam V., Weidenhof B., Krämer M., Herminghaus S., Seemann R., and Maier W. F. (2010), ChemPhysChem, 11, 2091 - 2095.

Eddings M., Johnson M., and Gale B. (2008), J. Micromech Microeng., 18, 067001.

Evans H., Surenjav E., Priest C., Herminghaus S., Seemann R. and Pfohl T. (2009), Lab Chip, 9, 1933-1941.

Günther A., Khan S. A., Thalmann M., Trachsel F., Jensen K. F. (2004), Lab on a Chip, 4, 278-283.

Langmuir I. (1917), J. Am. Chem. Soc., 39 (9), 1848-1906.

Lee I., Yoo Y., Cheng Z., Jeong H.-K. (2008), Adv. Funct. Mater., 18, 4014-4021.

Li L., Boedicker J. Q., Ismagilov R. F. (2007), Anal. Chem., 79, 2756.

Khan S. A., Günther A., Schmidt M. A. and Jensen K. F. (2004), Langmuir, 20, 8604–8611.

Oddy M. H., Santiago J. G., and Mikkelsen J. C. (2001), Anal. Chem. 73, 5822–5832.

Pouxviel J. C., Boilot J. P., Beloeil J. C. and Lallemand J. Y. (1987), Journal of Non-Crystalline Solids, , 89, 345-360.

Priest C., Herminghaus S., and Seemann R. (2006-1), Appl. Phys. Lett., 88, 024106.

Priest C., Herminghaus S., and Seemann R. (2006-2), Appl. Phys. Lett., 89, 134101

Solomon T. H., Mezic I. (2003) Uniform resonant chaotic mixing in fluid flows. Nature 425: 376–380.

Song H., Tice J. D. and R. F. Ismagilov (2003) Angew. Chem. Int. Edt., 42, 768-772.

Srinivasan V., Pamula V. K., and Fair R. B. (2004), Analytica Chimica Acta, 507, 145-150.

Stöber W., Fink A., Bohn E. (1968), J. Colloid Interf. Sci., 26, 62-69.

Stone H. A., Stroock A. D. and Ajdari A. (2004), Annual Review of Fluid Mechanics, 36, 381-411.

Storck S., Bretinger H., and Maier W. F. (1998), Applied Catalysis A: General, 174, 137-146.

Tabling P. (2006), Introduction to Microfluidics, Oxford University Press., ISBN: 0198568649.

Taylor G. (1954), Proceedings of the Royal Society of London, Series A. Mathematical and Physical Sciences, 225, 473-477.

Teh S.-Y., Lin R., Hung L.-H. and Lee A. P. (2008), Lab on a Chip, 8, 198-220.

Tilgner I. C., Fischer P., Bohnen F. M., Rehage H., Maier W. F. (1995), Microporous Mater., 5, 77-90.

Whitesides G. M., Ostuni E., Takayama S., Jiang X., and Ingber D. E. (2001), Annu. Rev. Biomed. Eng., 335–373.

Wong I. and Ho C.-M. (2009), Microfluidics and Nanofluidics, 7, 291.

Xia Y. and Whitesides G. M. (1998), Annu. Rev. Mater. Sci., 28,153–84.

Yazawa Y., Takagi N., Yoshida H., Komai S.I., Satsuma A., Tanaka T., Yoshida S., Hattori T., (2002) Appl. Catal. A 233, 103.

Yao Y. F. (1980), Ind. Eng. Chem . Prod. Res. Dev. 19, 293.

5

Mesoscopic Simulation Methods for Studying Flow and Transport in Electric Fields in Micro- and Nanochannels

Jens Smiatek[1] and Friederike Schmid[2]
[1]*Institut für Physikalische Chemie, Westfälische Wilhelms-Universität Münster, Münster*
[2]*Institut für Physik, Johannes-Guttenberg-Universität Mainz, Mainz*
Germany

1. Introduction

In the past decades, several mesoscale simulation techniques have emerged as tools to study hydrodynamic flow phenomena on scales in the range of nano- to micrometers. Examples are Dissipative Particle Dynamics (DPD), Multiparticle Collision Dynamics (MPCD), or Lattice Boltzmann (LB) methods. These methods allow one to access time and length scales which are not yet within reach of atomistic Molecular Dynamics (MD) simulations, often at relatively moderate computational expense. They can be coupled with particle-based (*e.g.*, molecular dynamics) simulation methods for thermally fluctuating nanoscale objects, such as colloids or large molecules. This makes them particularly attractive for the application in microfluidic or nanofluidic research.

Mesoscale coarse-grained simulation models are characterized by (i) simple potentials and (ii) simple, often stochastic dynamics. On large length scales, chemical details like the explicit structure of the water molecules can be neglected, such that the water can either be replaced by a simpler model fluid ("explicit solvent") or altogether omitted ("implicit solvent"). A suspended particle undergoes stochastic collisions with the solvent resulting in Brownian motion. This process can be described by a Markovian process on the timescale of microseconds. Mesoscale coarse-grained techniques exploit this behavior. As long as the relevant conservation laws are obeyed (mass, momentum, energy, charge etc.), the actual solvent dynamics may be replaced by a computationally cheaper artificial dynamics. For example, transport mechanisms as well as structure formation processes are often dominated by hydrodynamic interactions, and these can be reproduced by a variety of dynamical models. The essence of mesoscopic simulation methods is to combine a significant reduction of degrees of freedom with an artificial dynamical model that reproduces the main dynamical features on long time scales. This results in greatly reduced computation times compared to atomistic simulations.

In this chapter, we focus on the problem of molecular transport in microfluidic geometries, with particular emphasis on transport driven by electric fields. Due to the emerging technology of microfluidic devices like micro-arrays and microchips and their applications for polymer separation and micromanipulation, the fluid mechanics on these micro- and nano length scales has been an extensive subject of research over the last years. Mesoscopic simulation methods provide useful tools for understanding basic physical mechanisms in these confined systems, and for designing optimal microfluidic setups for given applications.

We intend to give a pedagogical introduction, focusing on the fundamentals of modelling microchannel flow phenomena using coarse-grained techniques. We briefly present the theory behind the most important mechanisms and discuss practical applications. Popular mesoscale simulation methods will be introduced and some examples will be given that illustrate the use and reliability of the presented methods. For further reading we refer to recent reviews in the field such as (Pagonabarraga et al. (2010); Slater et al. (2009); Sparreboom et al. (2010); Viovy (2000)).

2. Physics of flow in micro- and nanochannels: A brief introduction

In the following, we will give a brief overview over important factors that govern flow and transport phenomena on the nanoscale. Nanochannels are characterized by an exceedingly large surface-to-volume ratio. Therefore, *boundary effects* become important and dominate the flow properties to a large extent. A second factor that strongly influences the transport properties of molecules in flow is the presence of fluid-mediated *hydrodynamic interactions*. They in turn depend on the nature of the driving force (electric field vs. pressure driven transport or sedimentation) and are affected by the confinement in the channel. Here we can only touch on these issues. For further information we refer to the references.

2.1 Hydrodynamics

Throughout this chapter, we will restrict the discussion to Newtonian fluids, *i.e.*, isotropic fluids with a linear stress-strain relationship such as water. Flows in micro- and nanochannels are characterized by low Reynolds numbers. Even though driven flows can reach fluid velocities of meters per second in extreme cases and shear rates as high as $10^6 s^{-1}$ (Kuang & Wang (2010)), typical values of the Reynolds numbers are still much smaller than unity due to the tiny characteristic length scales. Therefore, it is almost impossible to enter the non-laminar regime, and the hydrodynamic behavior of incompressible Newtonian fluids follows the Stokes equation,

$$\eta \, \Delta \vec{u} = -\rho \vec{f}^{ext} + \vec{\nabla} P, \tag{1}$$

where $\vec{u}$ is the local velocity of fluid, $\vec{f}^{ext}$ an external body force, P the local pressure, and η the dynamic viscosity of the fluid. We will now discuss the two main factors that determine transport in microchannels in this regime, namely (i) the properties of the channel boundaries and (ii) the hydrodynamic interactions mediated by the fluid.

2.1.1 Hydrodynamic boundary conditions

Due to the large surface-to-volume ratio, flows in microfluidic channels strongly depend on the interaction between the fluid and the boundaries. Surface properties such as the local roughness or wettability can significantly influence the shape and amplitude of flow profiles. On the mesoscale, such effects can often be incorporated in an effective hydrodynamic boundary condition. The most popular is the so-called 'stick' or 'no-slip' boundary condition,

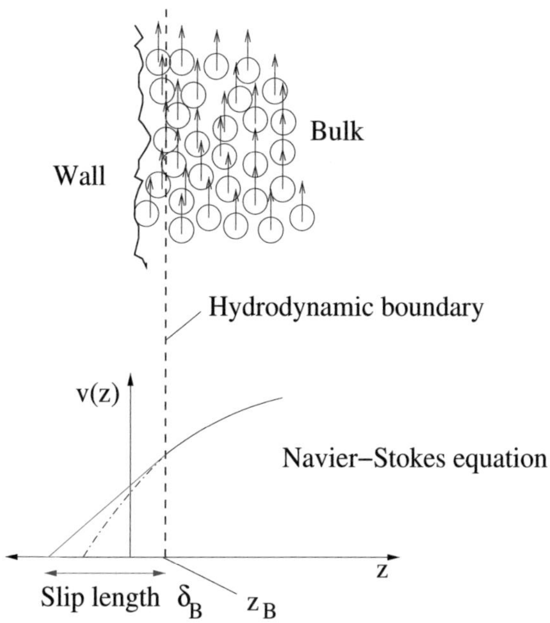

Fig. 1. **Top:** Coarse-grained schematic illustration of fluid flow in close vicinity to the boundaries. The hydrodynamic boundary is not identical to the fluid-solid interface due to atomistic roughness. **Bottom:** Flow profile in the direction normal to the hydrodynamic boundary. In close vicinity to the boundary, the Navier-Stokes equations are not valid. An extrapolation of the flow profile determines the slip length.

where the relative flow velocity is taken to vanish at the boundary (Landau & Lifshitz (1987)),

$$\vec{u}(z)|_{z=z_B} = 0 \tag{2}$$

The coordinate z is normal to the surface. Due to the atomistic roughness of the boundary, the location of the boundary is not obvious. It constitutes an effective parameter in the boundary condition (2).

The no-slip boundary condition is suitable for describing most macroscopic flows, but it is too restrictive on the nanoscale. Since the friction between the solid surface atoms and the fluid particles is finite, one expects some slippage at the physical boundaries, *i.e.*, the fluid velocity will not strictly vanish. This has been confirmed by a number of experiments on different surfaces (Neto et al. (2005); Pit et al. (2000); Tretheway & Meinhart (2002)). The amount of slippage can be estimated by considering the balance of stresses at the surface. (Kennard (1938)): Let x be the direction of flow and z the direction normal to the flow. The viscous stress σ^s_{xz} (Kennard (1938); Landau & Lifshitz (1987)) exerted by microscopic friction between the fluid particles and the solid channel walls

$$\sigma^s_{xz} = \zeta_s u_x(z) \tag{3}$$

is proportional to an *a priori* unknown boundary friction coefficient ζ_s and the fluid velocity $u_x(z)$ relative to the velocity of the wall. The complementary viscous stress generated within the fluid itself, σ^f_{xz}, is approximated by the corresponding expression in the bulk

$$\sigma^f_{xz} = \eta \frac{\partial}{\partial z} u_x(z). \tag{4}$$

The stability condition (Landau & Lifshitz (1987)) requires that these two stresses are equal, *i.e.*, $\sigma^f_{xz}|_{z=z_B} = \sigma^s_{xz}|_{z=z_B}$ at the hydrodynamic boundary z_B. After combining Eqns. (3) and (4) one obtains an effective boundary condition

$$u_x(z)|_{z=z_B} = \delta_B \left(\frac{\partial v_x(z)}{\partial z} \right)_{z=z_B}, \tag{5}$$

which depends on two effective parameters, the hydrodynamic boundary z_B and the *slip length*

$$\delta_B = \frac{\eta}{\zeta_B}. \tag{6}$$

This so-called *partial-slip boundary condition* (Eqn. (5)) represents the most general hydrodynamic boundary condition on isotropic surfaces which is consistent with all requirements of symmetry. It includes the no-slip boundary condition as the special case $\delta_B = 0$. Eq. (6) suggests that the no-slip limit can only be reached for vanishing shear viscosity or infinite wall friction coefficient. It should however be noted that the above "derivation" of (6) relied on the assumption that the Stokes equation is valid directly at the boundary, which is not obvious on rough surfaces. Thus Eq. (6) should not be taken too literally: In practice, slip lengths may become zero or even negative. As long as the actual flow velocities at the physical boundaries stay positive, this is not unphysical. A geometrical interpretation of the slip boundary condition is illustrated in Fig. 1: The slip length is the distance between the hydrodynamic boundary position z_B and the point where the tangent to the Stokes flow profile at z_B hits zero.

2.1.2 Hydrodynamic interactions

Next we discuss the transport of molecules or nanoparticles in a fluid under the influence of external forces. Consider a particle or a system of particles made of several idealized point-like "monomers". If a force acts on a monomers, it will respond by moving and drag the surrounding fluid along. This generates a flow field and convective motion at the positions of the other monomers. Thus *all* monomers will respond to a force $\vec{f}_i$ acting on one monomer i. This fluid-mediated nonlocal dynamical response is termed hydrodynamic interaction (Dhont (1996)). Specifically, the force $\vec{f}_i$ on a pointlike particle at position $\vec{r}_i$ excites a convective flow with velocity

$$\vec{u}(\vec{r}_j) = \mathbf{T}(\vec{r}_j, \vec{r}_i)\ \vec{f}_i \tag{7}$$

at the position $\vec{r}_j$ of the pointlike particle j, which is proportional to $\vec{f}_i$ with a tensorial coupling function $\mathbf{T}$, the 'Oseen tensor'. Since the Stokes equation is linear, all these flow velocities add up linearly. The Oseen tensor depends on the boundary conditions. In the bulk, it is a function of the difference $\vec{r} = \vec{r}_i - \vec{r}_j$ only and given by

$$T_{\alpha\beta}(\vec{r}) = \frac{1}{8\pi\eta\, r} \Big(\delta_{\alpha\beta} + \hat{r}_\alpha\, \hat{r}_\beta \Big). \tag{8}$$

with $\hat{r} = \vec{r}/r$. This corresponds to a long-range "interaction" which decays like $1/r$.

Due to their long-range nature, the hydrodynamic interactions strongly influence the dynamics and transport properties of single molecules and molecule assemblies. For example, polymer coils in solution move like a Stokes sphere due to hydrodynamic interactions, with an effective friction coefficient ζ which is proportional to their radius ("Zimm dynamics"), whereas ζ is simply proportional to the chain length in the absence of hydrodynamics ("Rouse dynamics") (Doi & Edwards (1986)).

In nanochannels, the presence of boundaries further complicates the situation. Since the boundaries break translational symmetry, the assumption $\mathbf{T}(\vec{r}_i,\vec{r}_j) = \mathbf{T}(\vec{r}_i - \vec{r}_j)$ is no longer justified and the Oseen tensor depends on both arguments, $\vec{r}_i$ and $\vec{r}_j$, independently. In homogeneous channels, one may still assume $\mathbf{T}(\vec{r}_i,\vec{r}_j) = \mathbf{T}(\vec{r}_{i\|} - \vec{r}_{j\|};\vec{r}_{i\perp},\vec{r}_{j\perp})$, where $\vec{r}_\|$ is the component of $\vec{r}$ along the channel and $\vec{r}_\perp = \vec{r} - \vec{r}_\|$. In structured channels, even this last symmetry assumption breaks down.

A second important effect of boundaries is that they constrain the flow. This affects the hydrodynamic interactions dramatically. For example, the hydrodynamic interactions between two particles close to a single no-slip wall decay like $1/r^3$ instead of $1/r$ (Dufresne et al. (2000)). The hydrodynamic interactions between two particles confined in a thin planar slit with no-slip boundaries decay like $1/r^2$, *i.e.*, much faster than in the bulk, albeit still long-range. (Diamant et al. (2005); Liron & Mochon (1976)). Even these remaining long-range contribution may effectively average out in many applications, since the two-dimensional Oseen tensor in slits is traceless and its orientational average vanishes. It has been argued that polymers confined in thin sheets should therefore exhibit Rouse-like dynamics (Tlusty (2006)). In quasi one dimensional thin channels with stick boundaries, hydrodynamic interactions are altogether screened on distances larger than the channel width (Cui et al. (2002)).

2.2 Electrokinetics

If external electric fields are involved, flow and transport processes in microchannels are governed by the interplay of hydrodynamic and electrostatic driving forces. Both the hydrodynamic and the electrostatic interactions are long-range, thus none of the two necessarily dominates over the other. We consider the transport of charged macromolecules or colloids ("macroions") in solution due to electric fields. On the one hand, typical buffers used in (bio)technological applications have high ionic strengths (*i.e.*, high salt concentrations), such that macrocharges are efficiently screened with Debye screening lengths in the nanometer range. On the other hand, the (micro-)ions in solution respond to external fields just like the macroions. Furthermore, external fields may induce flow in microchannels by means of an effect called electroosmosis, which also contributes to the effective transport.

2.2.1 Basic concepts

In the following we will take the correlations between microions to be negligible. Correlation effects become important at high surface charges, low temperatures, and/or if microions have a high valency (Boroudjerdi et al. (2005)). In most practical applications, they can be neglected and one may apply mean-field approaches. At thermal equilibrium, the mean-field theory of choice is the *Poisson-Boltzmann* theory (Hunter (1989); Israelachvili (1991); Lyklema (1995); Viovy (2000)), which describes the distribution of the microions by a combination of the Poisson equation and the Boltzmann distribution. The theory neglects the excluded-volume of the counterions and regards them as pure uncorrelated point-like particles. Ions of type i are distributed in space according to $\rho_i(\vec{r}) \propto \exp(-Z_i e \psi(\vec{r})/k_B T)$, where Z_i is the valency of the ion i, e the elementary charge, $k_B T$ the thermal energy, and $\psi(\vec{r})$ the mean electrostatic potential at position $\vec{r}$. Combining this with the Poisson equation, one obtains the self-consistent equation

$$\epsilon_r \, \Delta\psi(\vec{r}) = -\rho_{\text{macro}}(\vec{r}) - \sum_i Z_i e \, \rho_i(\vec{r}) = -\rho_{\text{macro}(\vec{r})} - \sum_i Z_i e \, \rho_{0,i} \, \mathrm{e}^{-\frac{Z_i e \psi(\vec{r})}{k_B T}} \tag{9}$$

with the dielectric constant ϵ_r, the macroion charge density $\rho_{\text{macro}}(\vec{r})$, and the bulk microion densities $\rho_{0,i}$ (the hypothetical densities at infinite distance from the macroions). Here the ρ_{macro} may also refer to a charged surface.

Due to its nonlinear character, the Poisson-Boltzmann equation (9) can usually not be solved in closed form. If the potential ψ is small, it can be linearized by expanding the exponent. This yields the famous *Debye-Hückel* equation

$$(\Delta - \lambda_D^{-2})\psi(\vec{r}) = -\frac{\rho_{\text{macro}}(\vec{r})}{\epsilon_r} \tag{10}$$

with the Debye-Hückel screening length

$$\lambda_D = \sqrt{\frac{\epsilon_r \, k_B T}{\sum_i \rho_{0,i}(Z_i e)^2}}. \tag{11}$$

Here we have used $\sum_i \frac{Z_i e}{\epsilon_r} \rho_{0,i} = 0$. According to the Debye-Hückel theory, the effective interactions between macroions decay exponentially with the characteristic length λ_D. The range of validity of the Debye-Hückel approximation is very limited, it breaks down already for moderate macroion charges and/or for highly concentrated ion solutions. Nevertheless, the exponential behavior often persists in systems where the Debye-Hückel approximation is not valid. For highly charged macroions, the Debye-Hückel theory can still be applied far from the macroions, but the macroion charge must be replaced by an effective "renormalized charge" (Alexander et al. (1984); Trizac et al. (2002)). For high ion concentrations, detailed theoretical studies have lead to the conclusion that the results of the Debye-Hückel can still be used in a wide parameter range if λ_D is replaced by a modified effective screening length (Kjellander & Mitchell (1994); McBride et al. (1998); Mitchell & Ninham (1978)).

So far, these are equilibrium considerations. At non-equilibrium and in the presence of flow, the time evolution of the ion distributions at the mean-field level are described by a convection-diffusion equation, the *Nernst-Planck* equation

$$\frac{\partial \rho_i}{\partial t} + \nabla \cdot (\rho_i \vec{u}) = \nabla \cdot D_i (\nabla \rho_i + \frac{Z_i e}{k_B T} \rho_i \, \nabla \psi). \tag{12}$$

Here D_i are the diffusivities of the ion species i, and $\vec{u}(\vec{r})$ is the local fluid velocity. The Nernst-Planck equation, the Stokes (or Navier-Stokes) equation and the Poisson equation constitute a complete set of equations that fully determine the dynamics of ionic fluids. In stationary systems without flows, they are solved by the Poisson-Boltzmann distribution.

2.2.2 Boundary layers and electroosmotic flow

In contact with a solvent, many materials acquire charges due to the ionization or dissciation of surface groups or the adsorption of ions from solution (Israelachvili (1991)). Take for example PDMS (polydimethylsiloxane), a material commonly used for the fabrication of microfluidic devices. At PDMS/water interfaces, $(OH)^-$ groups dissociate at the end of the PDMS chains. As a result, the walls of PDMS channels become charged and an ion layer builds up in front of the surfaces to screens their charge: An *electrical double layer* forms.

The structure of the ion layer is commonly described by an effective picture that distinguishes between two regions: In the outer region, the *diffuse layer*, the ions are mobile, they can thermally be removed and act as if they are part of the solution. Under the influence of force, they will drag the fluid along. The inner region, the *Stern layer*, includes ions which stick to the surface. The two regions are taken to be separated by a "plane of shear" for the fluid flow. In the coarse-grained mean-field picture adopted in this chapter, the motion of the ions in the diffuse layer is described by the Nernst-Planck equation (12), the plane of shear corresponds to the hydrodynamic boundary, and the complex of solid surface and Stern layer is replaced by an effective boundary condition which is characterized by an effective surface charge and an effective slip length.

At equilibrium, the distribution of ions in the diffuse layer is determined by the Poisson-Boltzmann equation. In the case of symmetric planar slit channels confined by homogeneous surfaces, and in the absence of salt (*i.e.*, the fluid contains only the counterions that neutralize the surface charges), the problem can be solved analytically (Israelachvili (1991)). The potential in the direction z perpendicular to the surface is then given by

$$\psi(z) = \frac{k_B T}{Ze} \log(\cos^2(\kappa z)) \tag{13}$$

with the screening constant

$$\kappa^2 = \frac{(Ze)^2 \rho_0}{2\epsilon_r k_B T}. \tag{14}$$

Here ρ_0 denotes the counterion density in the middle of the channel, and the z-direction is perpendicular to the channel. In the presence of salt an approximate solution can be obtained within the Debye-Hückel approximation, giving

$$\psi(z) \propto (e^{z/\lambda_D} + e^{-z/\lambda_D} - 2), \tag{15}$$

which results in an exponential decay of the ion distribution close to the boundaries with the decay length λ_D, the Debye-Hückel screening length λ_D from Eq. (11). The equilibrium results (13) and (15) remain valid in stationary situations with constant flow parallel to the surface, since the parallel and the perpendicular directions in Eq. (12) then decouple. In such cases, they describe the z-dependent part of the potential ψ.

If an external electric field is applied to the solution, it generates a net volume force in the diffuse ion layer, which induces flow, the 'electroosmotic flow' (EOF) (Hunter (1989)).

This phenomenon provides one with a convenient tool to manipulate and control flows in microchannels. Furthermore, it significantly influences the effective transport of nanoobjects in such channels.

The EOF flow profiles can be calculated by combining the Stokes equation (Eq. (1)) with the Poisson-Boltzmann theory. For simplicity, we consider again our symmetric slit channel. The net charge density in the channel, $\rho(z) = \sum_i (Z_i e)\rho_i(z)$, generates a force density $f_{\parallel}(z) = \rho(z)E_{\parallel}$ in the fluid, where $E_{\parallel}$ is a parallel external electric field. Comparing the Poisson equation for the electrostatic potential,

$$\frac{\partial^2 \psi(z)}{\partial z^2} = -\frac{\rho(z)}{\epsilon_r}, \tag{16}$$

with the Stokes equation for the parallel component of the fluid velocity,

$$\eta \frac{\partial^2 u_{\parallel}(z)}{\partial z^2} = -f_{\parallel}(z) = -\rho(z)E_{\parallel}, \tag{17}$$

one finds immediately

$$\frac{\partial^2}{\partial z^2} u_{\parallel}(z) = \frac{\epsilon_r E_{\parallel}}{\eta} \frac{\partial^2}{\partial z^2} \psi(z). \tag{18}$$

In a symmetric channel, the profiles $u_{\parallel}(z)$ and ψ satisfy the boundary condition $\partial_z u_{\parallel}|_{z=0} = \partial_z \psi|_{z=0} = 0$ at the center of the channel. This gives the relation (Smiatek & Schmid (2010))

$$u_{\parallel}(z) = \frac{\epsilon_r E_{\parallel}}{\eta} (\psi(z) - \psi(0)) + u_{\text{EOF}} \tag{19}$$

The constant $u_{\text{EOF}} = u_{\parallel}(0)$ in Eq. (19) is determined by the hydrodynamic boundary conditions at the surface of the channel. For example, the counterion-induced electroosmotic flow in the presence of slip resulting from Eqs. (19) and (13) is given by (Smiatek et al. (2009))

$$u_{\parallel}(z) = \frac{\epsilon_r k_B T}{Ze\eta} E_{\parallel} \Big(\log(\cos^2(\kappa z)) - \log(\cos^2(\kappa z_B)) - 2\kappa\delta_B \tan(\kappa z_B) \Big), \tag{20}$$

where δ_B is the slip length and z_B the position of the hydrodynamic boundary (cf. Eq. (6). In the presence of salt, the EOF profiles vary mostly at the boundaries and the velocity profile saturates in the center of the channel (plug flow). In that case, the constant u_{EOF}, can be identified with the EOF velocity. Setting $\psi(0) = 0$, Eq. (19) combined with the partial-slip boundary condition (5) results in the general simple expression (Bouzigues et al. (2008); Smiatek & Schmid (2010))

$$u_{\text{EOF}} = -\frac{\epsilon_r \, \psi_B}{\eta} (1 + \kappa_s \, \delta_B) \, E_{\parallel} =: -\frac{\epsilon_r \psi_\zeta}{\eta} \, E_{\parallel}. \tag{21}$$

where ψ_B is the electrostatic potential at the hydrodynamic boundary and we have defined a 'surface screening parameter'

$$\kappa_s := \mp \frac{\partial_z \psi}{\psi}\Big|_{z=\mp z_B}. \tag{22}$$

In the Debye-Hückel regime, κ_s is simply the inverse Debye-Hückel length (Joly et al. (2004)). More generally, the relation (21) also remains valid in the regime where nonlinear effects become important. On no-slip surfaces, Eq. (21) reduces to the well-known Smoluchowski result for electroosmotic flow (Hunter (1989)). In the presence of slippage, the amplitude of the induced flow can be enhanced significantly. This is also found experimentally (Bouzigues et al. (2008)). The combined effect of slip, surface potential, and surface screening parameter can be incorporated into one single parameter $\psi_\zeta = \psi_B(1 + \kappa_s \, \delta_B)$, the so-called "Zeta-potential", which fully characterizes the electroosmotic response of the surface to an applied electric field. (Hunter (1989); Lyklema (1995))

So far, we have considered simple slit channels. However, our results can also be used to assess EOF profiles in more complex channel geometries. We assume that the diffuse layer is thin compared to the channel dimensions. Then, we make two important observations: First, Eq. (21) still can be applied close to the boundaries and gives an effective *boundary condition* for a steady-state velocity field $\vec{u}_{\text{EOF}}(\vec{r})$ inside the channel, which depends linearly on the applied external field $\vec{E}^{(ext)}$,

$$\vec{u}_{\text{EOF}} = \mu_{\text{EOF}} \vec{E}^{(ext)} \quad \text{with} \quad \mu_{\text{EOF}} = -\frac{\epsilon_r \, \psi_\zeta}{\eta}. \tag{23}$$

(In a steady-state situation, the external field in the vicinity of a wall is necessarily parallel to the wall.) Cummings and coworkers have shown that for laminar incompressible flow, the boundary condition (23) effectively defines the flow inside the channel (Cummings et al. (2000)). Second, the EOF flow field inside the channel obeys the Laplace equation, $\Delta \vec{u}_{\text{EOF}} = 0$. The same holds for the external electric field $\vec{E}^{(ext)}$. Provided all channel walls are made of the same material (same slip length, same surface potential ψ_B) and provided Eq. (23) also holds a the inlet and outlet boundaries of the channel, Eq. (23) thus becomes valid everywhere in the channel. Due to this remarkable 'similitude' property, the EOF velocity profile in such channels is simply proportional to the applied electric field in the channel, which can be calculated by solving the Laplace equation in the channels with von-Neumann boundary conditions. The proportionality factor is μ_{EOF}, the so-called 'electroosmotic mobility'.

We note that the condition on the inlet and outlet is less restrictive than it may seem: One can always separate a steady-state velocity field $\vec{u}(\vec{r})$ into an EOF component $\vec{u}_{\text{EOF}}(\vec{r})$ which satisfies the EOF boundary conditions, and a residual component with no-slip boundary conditions at the channel walls and arbitrary boundary conditions at the inlet and outlet.

2.2.3 Electrophoresis

Last in this section we discuss the transport of charged nanoobjects (*i.e.*, colloids or polyelectrolytes) in microchannels under the influence of constant electric fields. In the linear regime, the velocity of the nanoobject is proportional to the field,

$$\vec{v}_P = \mu_t \vec{E}, \tag{24}$$

The proportionality constant μ_t denotes the total electrophoretic mobility and results from two contributions (Smiatek & Schmid (2010); Streek et al. (2005)): The bare electrophoretic mobility μ_e in a fluid at rest and the background EOF velocity $\vec{u}_{\text{EOF}}$. Due to the similitude discussed above between the steady-state velocity field of an EOF and the external field $\vec{E}$, the EOF flow field can be expressed in terms of an electroosmotic mobility μ_{EOF} (Eq. (23)), and the total electrophoretic mobility of the nanoparticle can be written as

$$\mu_t = \mu_{EOF} + \mu_e. \tag{25}$$

The bare electrophoretic mobility μ_e subsumes the intrinsic response of the nanoobject to the external field. For large colloids with a thin Debye layer, this response is dominated by the same mechanisms than the EOF at the walls: The electric field induces EOF at the surface of the colloid, which results in a velocity difference (23) between the colloid and the surrounding fluid. Thus μ_e is again given by the Smoluchowsky result in this case,

$$\mu_e = \epsilon_r \, \psi_\zeta / \eta, \tag{26}$$

where the Zeta-potential ψ_ζ characterizes the surface of the colloid, *i.e.*, it incorporates the effects of the Stern layer as well as possible slippage at the hydrodynamic boundary.

For small nanosized particles or ions, the situation is more complicated. The effective friction of a charged particle in an electric field mainly results from the combination of three factors: (i) The direct friction of the particle with the surrounding fluid, which can be described by a bare mobility μ^0. It determines the mobility of the nanoparticle in the absence of any other ions. (ii) The solvent-mediated friction of the particle with the surrounding cloud of oppositely charged ions (the "ionic atmosphere"). They are dragged in the opposite direction by the external field, which generates additional friction. This effect is called "electrophoretic retardation". (iii) The distortion of the ion cloud in the electric field. Since the cloud is a dynamic phenomenon, ions move in and out, it cannot adjust instantaneously to the motion of the nanoparticle. Moreover, the electric field drives the nanoparticle and the cloud apart. The cloud is somewhat behind the nanoparticle and pulls it back, thus slowing it down. This is called the "relaxation effect". The interplay of these three factors and their effect on the electrophoretic mobility of ions has been discussed already in 1926 by Onsager (Onsager (1926; 1927)). He predicted that the bare mobility of ions μ^0 in electrolytes should be renormalized according to

$$\mu_e = \mu^0 - (A\mu^0 + B)\sqrt{\sum_j \rho_j Z_j^2}, \tag{27}$$

where the factors A and B represent the relaxation effect and the electrophoretic effect. Explicit expressions for A and B for electrolytes having any number of ionic species were derived by Onsager and Fuoss in (Onsager & Fuoss (1932)). More details can be found in (Wright (2007)).

In systems of several nanoparticles or of extended molecules, hydrodynamic interactions become important. As discussed in Section 2.1.2, they can crucially affect the dynamics and transport properties in external fields. In the case of electrophoresis, one must not only consider the direct hydrodynamic interactions between the macroions, but also the effect of the surrounding microion cloud. In an external electric field, the polyelectrolyte and its surrounding ion cloud move in opposite directions, and the net momentum transferred to the solvent adds up to zero. As a result, the hydrodynamic interactions are largely screened (Manning (1981)). Within the Debye-Hückel approximation, the corresponding modified Greens function can be calculated analytically. The convective flow created by a constant electric field $\vec{E}$ acting on a particle at position $\vec{r}_i$ is given by (Barrat & Joanny (1996);

Long & Ajdari (2001))

$$\vec{u}(\vec{r}_j) = \mathbf{G}(\vec{r}_j, \vec{r}_i)\,\vec{E} \tag{28}$$

with a Greens function that depends on the Debye screening length λ_D *via*

$$G_{\alpha\beta}(\vec{r}) = \frac{e^{-r/\lambda_D}}{4\pi\eta\, r}\Big(\frac{2}{3}\delta_{\alpha\beta} + \frac{\lambda_D^2}{r^2}\big(1 + \frac{r}{\lambda_D} + \frac{r^2}{3\lambda_D^2} - e^{r/\lambda_D}\big)\,\big(\delta_{\alpha\beta} - 3\hat{r}_\alpha\,\hat{r}_\beta\big)\Big) \tag{29}$$

These expressions take the place of Eqs. (7) and (8) in electrophoresis. The modified Greens function $\mathbf{G}$ contains an exponentially decaying part, and an algebraic contribution which decays with $1/r^3$ (instead of $1/r$) and vanishes upon averaging over all directions. Thus, the effect of hydrodynamic interactions is greatly reduced. We note that this only applies for the response of charged particles to a constant electric field. The *diffusive* behavior of charged particles is still governed by the usual, long-range hydrodynamic interactions.

The remarkable electrohydrodynamic screening effect in electrophoresis discussed above has important practical consequences. A common task in bioanalytics is length-dependent separation and sorting of charged polymers, *e.g.*, DNA fragments. Strategies to sort molecules in micro- or nanofluidic channels are highly desirable as cheap and versatile alternative to standard gel electrophoresis. In the absence of hydrodynamic interactions, however, the electrophoretic mobility of polyelectrolytes does not depend on the chain length. Thus electric fields cannot be used to separate polymers in simple straight channel geometries, and one has to resort to more sophisticated approaches.

3. Mesoscale simulation methods

After this brief overview over physical mechanisms governing flow in microchannels and the transport of nanoobjects on mesoscopic scales, we will now introduce different methods to study these phenomena by computer simulations. We focus on length scales which are large compared to the solvent particles (water), but small enough that thermal fluctuations are important and nanobjects can be resolved and traced individually. On even smaller length scales, the method of choice is atomistic Molecular Dynamics. It is used, *e.g.*, to study the microscopic factors that determine surface slip on specific surfaces (Bocquet & Barrat (1994); Joly et al. (2004; 2006)). On much larger length scales, deterministic approaches can be used and one may resort to the numerical methods that have been developed in the context of computational fluid dynamics to solve the Navier-Stokes equations in various geometries. A software package for studying electrohydrodynamic phenomena in colloidal suspensions at this level is freely available at (Kim et al. (2006))

`http://www-tph.cheme.kyoto-u.ac.jp/kapsel/.`

Electrokinetic simulations are challenging, because one has to deal with two types of long-range interactions simultaneously: Long-range hydrodynamics and long-range electrostatics. In the following, we will first discuss mesoscale methods for simulating hydrodynamic phenomena in nanoparticle suspensions. Then we will present a selection of simulation methods for studying electrokinetic phenomena.

3.1 Simulation methods for hydrodynamic flows

In mesoscale simulations, one is typically not interested in the explicit dynamical behavior of the solvent molecules, as typical time scales are much smaller than the time scales of interest. For example, the relaxation time of water is 15 GHz, whereas the fastest time scale for reorganization of ionic layers in aqueous solutions is in the MHz regime (ionic migration in the double layer sets in at 1-10 MHz). A coarse-grained simulation model must therefore mainly capture the hydrodynamics of the solvent. To make the simulations more efficient, the molecular solvent can thus be replaced by something simpler. In the case of Newtonian fluids, such simplified model solvents must fulfill three main requirements:

- It must reproduce the Navier-Stokes Stokes equation. This means that it is a fluid (*i.e.*, it yields to stress), that it is "simple" (no long-range interactions, no memory effects, no internal order, no phase transitions), and that it preserves momentum locally.
- Since fluctuations are important, it should have a temperature. Simulations of microfluidic transport are typically nonequilibrium simulations where energy is constantly fed into the system. Therefore, algorithms should be available that remove this energy from the system, *i.e.*, coupling to a thermostat should be possible.
- Strategies to implement arbitrary hydrodynamic boundary conditions and interactions with nanoparticles should exist.

Other symmetry properties such as isotropy of space or translational invariance would also be desirable. However, they are not strictly mandatory. We shall see that some of the most successful models do not meet these symmetry requirements.

The reader may have noted that the requirements sketched above are already met by ideal gases. In fact, many mesoscopic fluid models in use today are actually ideal gas models. This implies that the fluid is not incompressible, which is acceptable as long as typical velocities V are much smaller than the speed of sound c in the model. Ideal gas models can be applied to study incompressible fluid dynamics if the *Mach number* $M = V/c$ is small.

3.1.1 Dissipative Particle Dynamics (DPD)

Among the mesoscopic simulation methods for fluids, the DPD method comes closest to classical Molecular Dynamics (MD). It was originally developed as a method for fluid simulations that combines elements of Lattice Gas Automata and MD (Hoogerbrugge & Koelman (1992); Koelman & Hoogerbrugge (1993)). In its modern version (Español & Warren (1995)), it is basically a momentum-conserving, Galilean invariant thermostat which creates a well-defined canonical ensemble. In DPD simulations, particles i obey Newton's equation of motion with the forces

$$\vec{F}_i^{DPD} = \vec{F}_i^C + \sum_{i \neq j} (\vec{F}_{ij}^D + \vec{F}_{ij}^R), \tag{30}$$

where $\vec{F}_i^C$ are the regular conservative forces acting on i, and $\vec{F}_{ij}^D$, $\vec{F}_{ij}^R$ are pairwise dissipative and stochastic forces. The dissipative forces $\vec{F}_{ij}^D$ have the form

$$\vec{F}_{ij}^D = -\gamma_{\text{DPD}}\, \omega_{\text{DPD}}(r_{ij})(\hat{r}_{ij} \cdot \vec{v}_{ij})\, \hat{r}_{ij} \tag{31}$$

with the friction coefficient γ_{DPD}, and depend on the interparticle distance r_{ij} and the unit vector $\hat{r}_{ij}$ connecting the two particles i and j. The random force $\vec{F}_{ij}^R$ is chosen such that it fulfills the fluctuation-dissipation relation

$$\vec{F}_{ij}^R = \sqrt{2\gamma_{\text{DPD}} k_B T \,\omega_{\text{DPD}}(r_{ij})}\,\theta_{ij}\hat{r}_{ij}, \tag{32}$$

with the Boltzmann constant k_B and the temperature T, where $\theta_{ij} = \theta_{ji}$ is a Gaussian distributed random number with zero mean and unit variance. The weighting function ω_{DPD} is arbitrary and often chosen as a linearly decaying function with a cutoff r_c,

$$\omega_{\text{DPD}}(r_{ij}) = \begin{cases} 1 - \frac{r_{ij}}{r_c} & : r_{ij} < r_c \\ 0 & : r_{ij} \geq r_c \end{cases} \tag{33}$$

Eq. (30) can be integrated by an ordinary MD integration scheme like the Velocity-Verlet algorithm (Frenkel & Smit (2001)).

In DPD simulations, arbitrary equation of states can be implemented through appropriate conservative interactions $\vec{F}_i^C$. In practice, the conservative potentials are often chosen pairwise and soft and with linearly decaying forces (Groot & Warren (1997)). When studying fluid flow, one may also simply turn them off. For such an "ideal gas" DPD fluid, approximate expressions for the viscosity have been derived *via* a Fokker-Planck formalism (Marsh et al. (1997)). The shear viscosity of the fluid can be tuned by varying the density, the friction parameter γ_{DPD} or the cutoff parameter of the dissipative interactions, r_c. Since the DPD method was first introduced by Hoogerbrugge and Koelman, various variants and extensions have been proposed. A recent review can be found in (Pivkin et al. (2010)).

Compared to other mesoscale simulation methods, DPD simulations are relatively expensive because they require the evaluation of pairwise forces in every time step. Nevertheless, they present many advantages due to the fact that they are so similar to standard MD simulations. They can easily be mixed with MD schemes, to the point that the viscous DPD interaction can simply be used to thermostat MD simulations (Soddemann et al. (2003)). Interactions with nanoparticles can be implemented in a straightforward manner through the conservative interactions. Arbitrary hydrodynamic boundary conditions with slip lengths ranging from full slip to no-slip or even slightly negative can be implemented using a recently developed "tunable slip" algorithm (Smiatek et al. (2008)). This algorithms introduces a DPD-like friction interaction with the wall. Slip lengths can be adjusted by varying the amplitude of the friction force or the interaction range, and an approximate expression for the resulting slip length can be derived analytically which is in excellent agreement with simulations. The method can also be used to implement an EOF boundary condition with fixed surface velocity $\vec{u}_{\text{EOF}}$ (by choosing a no-slip boundary condition on a hypothetically moving boundary). Alternative methods for implementing no-slip boundaries in DPD simulations can be found, *e.g.*, in Refs. (Revenga et al. (1998; 1999)).

3.1.2 Multiparticle collision dynamics (MPCD)

One drawback of the DPD method described in the previous section is that DPD fluids typically have low Schmidt numbers. The dimensionless Schmidt number $\text{Sc} = \eta/\rho D$

measures the ratio of momentum diffusivity (viscosity η) and mass diffusivity (diffusion constant D). In gases, particle transport dominates and Schmidt numbers are typically of order unity. In liquids, dissipative transport dominates and typical Schmidt numbers are high For example, the Schmidt number of water is Sc $\sim$ 700. In DPD fluids, typical Schmidt values are of order unity like in gases, hence the momentum transport ist mostly coupled to particle transport. High, fluid-like Schmidt numbers are not accessible in DPD simulations.

This problem can be overcome in the MPCD method, which is another increasingly popular method for simulating fluid flow. MPCD, also known as stochastic rotation dynamics (SRD) is a particle-based approach for simulating fluid flows which is motivated by the Boltzmann equation for transport in gases (Malevanets & Kapral (1999; 2000)). An extensive recent review can be found in (Gompper et al. (2009)).

The basic algorithm is an alternating sequence of streaming and a collision steps. In the streaming step the positions of the particles i are updated *via*

$$\vec{r}_i(t + \Delta t) = \vec{r}_i(t) + \Delta t \vec{v}_i(t). \tag{34}$$

In the collision step, the particles are sorted into cells (typically cubic with cell size a chosen such that one has about 3-20 particles per cell), and the center-of mass velocity $\vec{u}$ is evaluated in each cell. The collision update then reads

$$\vec{v}_i(t + \Delta t) = \vec{u} + \mathcal{R}\, \delta\vec{v}_i(t) \tag{35}$$

where $\mathcal{R}$ is a collision operator acting on the local velocity $\delta\vec{v}_i(t) = \vec{v}_i(t) - \vec{u}$. In the classical MPCD scheme, $\mathcal{R}$ is a rotation by a fixed angle α about a random axis. Malevanets and Kapral have shown that this scheme generates a Maxwellian velocity distribution at equilibrium and reproduces the hydrodynamic equations with an ideal gas equation of state (Malevanets & Kapral (1999)). The shear viscosity can be calculated analytically, *e.g.*, *via* a Green-Kubo formalism, and it can be tuned through tuning $a, \Delta t$, and/or the rotation angle α. In addition, one can also tune the Schmidt number by adjusting the time step Δt. Large fluid-like Schmidt numbers can be realized by choosing small time steps.

Several variants of MPCD have been proposed. The classical version of the MPCD algorithm is not Galilean invariant due to the underlying grid structure. While this is often acceptable, it may cause problems if the mean free path of particles is smaller than the cell size. Galilean invariance can be restored by performing a random shift of the cells before each collision step (Ihle & Kroll (2003)). Another issue is isotropy of space. The classical MPCD scheme does not strictly conserve angular momentum. This can be fixed by choosing a stochastic collision step, where new relative velocities $\delta\vec{v}_i$ are generated randomly from a Maxwell distribution subject to additional constraints (Noguchi et al. (2007)). The stochastic variant also provides one with a natural implementation of a thermostat, which is necessary for nonequilibrium simulations.

Since MPCD is a particle-based method, interactions with nanoparticles can be implemented in a straightforward manner, just like in DPD simulations. Likewise, the tunable slip method for implementing arbitrary hydrodynamic boundary conditions in DPD simulations (Smiatek et al. (2008)) should also work for MPCD models, but this has not yet been explored. Instead, hydrodynamic boundary conditions in MPCD are usually realized by implementing suitable *reflection rules* on hard surfaces. Full-slip boundaries can be obtained with specular reflections. No-slip boundaries are obtained with "bounce-back" reflections, where the velocities of particles are reversed when they hit the boundary. Even though these reflections take place during the streaming steps, the collision steps are also affected by the presence of hard walls, due to the fact that the number of MPCD particles is reduced in cells that contain boundaries. Since the local viscosity of the fluid depends on the local density, a local reduction may cause artefacts. One common strategy to overcome this problem is to randomly insert virtual MPCD particles in underfilled cells (Lamura et al. (2001)), with velocities drawn randomly from a Maxwell distribution which is centered about the mean velocity of the solid in no-slip cases, or the fluid in full-slip cases. Based on these realizations of full-slip and no-slip boundaries, one can also implement arbitrary partial-slip boundary conditions by randomly switching between specular and bounce-back reflections in the streaming step, and suitably interpolating the mean velocity of the pseudoparticles in the collision step (Whitmer & Luijten (2010)).

3.1.3 Lattice-Boltzmann method (LB)

Similar to MPCD, the LB method realizes a Boltzmann equation, but this time fully on the lattice (Dünweg & Ladd (2009); Succi (2001); Yeomans (2006)). Each lattice site $\vec{r}$ carries a set of discrete velocity distribution functions $n_i(\vec{r},t)$, which describe the partial density of fluid particles with velocity $\vec{c}_i$ at time t. The finite set of velocities $\{\vec{c}_i\}$ contains only lattice vectors. For example, the velocity set of the popular D3Q19 model contains the zero velocity and the vectors that connect a node with its nearest and next-nearest neighbors on a cubic lattice. The velocity unit is the "lattice velocity" $c = a/\Delta t$ with the lattice parameter a and the time step Δt.

As in MPCD, the LB dynamics is a sequence of alternating collision and streaming steps. The distributions are first reshuffled in the collision step,

$$n_i^*(\vec{r},t) = n_i(\vec{r},t) + L_{ij}\big(n_j(\vec{r},t) - n_j^{\rm eq}(\rho,\vec{u})\big) \tag{36}$$

and then evolve in the streaming step according to

$$n_i(\vec{r} + \vec{c}_i\Delta t, t + \Delta t) = n_i^*(\vec{r},t) \tag{37}$$

The collision step describes particle scattering events which relax the distributions towards a local equilibrium distribution $\{n_j^{\rm eq}\}$ with a linear collision operator L_{ij}. The simplest choice for L is $L_{ij} = -\delta_{ij}/\tau$, corresponding to a lattice Bhatnagar-Gross-Krook operator with a single relaxation time (Bhatnagar et al. (1954)). Multiple relaxation times schemes have been conceived as well and have the advantage that shear and bulk viscosity can be tuned separately (Dünweg & Ladd (2009)). The local pseudo equilibrium distribution $\{n_i^{\rm eq}\}$ depends on the local mass density $\rho(\vec{r},t)$ and momentum density $\vec{j}(\vec{r},t)$, which are related to the moments of the distribution function through

$$\rho(\vec{r},t) = \sum_i n_i(\vec{r},t), \qquad \vec{j}(\vec{r},t) = \rho(\vec{r},t)\,\vec{u}(\vec{r},t) = \sum_i n_i(\vec{r},t)\,\vec{c}_i. \tag{38}$$

It is typically constructed as a second order expansion in the velocity,

$$n_i^{\rm eq}(\rho,\vec{u}) = \rho(A_q + B_q(\vec{c}_i\cdot\vec{u}) + C_q u^2 + D_q(\vec{c}_i\cdot\vec{u})^2) \tag{39}$$

with coefficients A_q, B_q, C_q, D_q chosen such that $\{n_i^{\text{eq}}\}$ satisfies the constraints $\sum_i n_i^{\text{eq}} = \rho$, $\sum_i n_i^{\text{eq}} \vec{c}_i = \rho \vec{u}$, and the stress condition $\sum_i n_i^{\text{eq}} c_{i\alpha} c_{i\beta} = \rho u_\alpha u_\beta + P_{\alpha\beta}$. Here $P_{\alpha\beta}$ is the local stress tensor, which takes the form $P_{\alpha\beta} = \delta_{\alpha\beta} c_s^2 \rho$ in an ideal gas with the isothermal speed of sound c_s. Other equations of state can be implemented as well.

This summarizes the basic structure of LB algorithms. By means of a Chapman-Enskog equation, one can show that LB fluids reproduce the Navier-Stokes equations on sufficiently large length and time scales and can thus be used as Navier-Stokes solver (Benzi et al. (1992); Chen & Doolen (1998)). Such expansions also give analytical expressions for the shear viscosity as a function of the density and the relaxation time(s). The pseudo equilibrium distribution defines an intrinsic fluid temperature (which enters through the stress condition). Nevertheless, the original formulation of the LB algorithm is fully deterministic. Fluctuations can be incorporated by adding a stochastic term in the collision step, which must to be chosen carefully such that the fluctuation-dissipation relation is fulfilled (Dünweg et al. (2007); Yeomans (2006)).

LB strategies for implementing hydrodynamic boundary conditions are very similar to those used in MPCD. No-slip boundary conditions can be realized by "bounce-back" boundary conditions where the velocities are reversed at the walls. In practice, one introduces a layer of buffer nodes behind the wall, which contains distributions with reversed velocities. This generates a no-slip boundary condition at a hypothetical wall-plane which is half way between the real nodes and the buffer nodes. Generalizations for moving boundaries are available (Ladd (1994a)). Full-slip boundaries can be realized with specular reflection rules. A simple way to generate partial-slip boundary conditions is to blend bounce-back and specular boundary conditions with a weighting factor r. This generates finite slip lengths which depend on r in a complex nonlinear manner (Ahmed & Hecht (1999); Succi (2002)). Alternative approaches generate wall slip by implementing conservative wall-fluid interactions which mimick the effect of rough surfaces (Kunert & Harting (2008)) or impose fixed velocities at the boundaries (Hecht & Harting (2010); Zou & He (1997)). The latter approach is particularly interesting in the context of EOF.

Since the LB method is not a particle-based method, coupling LB fluids to nanoparticles is less straightforward than in the case of DPD or MPCD. One popular option is to treat the nanoparticles as extended objects with a well-defined hard surface which interact with the LB fluid *via collisions* at the boundaries. The nanoparticle surface is first mapped on a lattice grid. Then suitable boundary conditions are implemented, which take into account the local velocity of the fluid. From the transfer of momentum during the reflections of the LB fluid, one can evaluate the forces and torques on the nanoparticle and update its velocity and angular velocity accordingly (Ladd (1994a;b)).

An alternative approach which is more appropriate for studying macromolecules is to couple the nanoobjects to the LB fluids *via* the *forces* that they exert on each other (Ahlrichs & Dünweg (1999)). We will illustrate this method at the example of a single point particle with velocity $\vec{V}$ and position $\vec{R}$. In a fluid flow $\vec{u}$, this particle experiences a force

$$\vec{F} = \vec{F}^C - \zeta(\vec{V} - \vec{u}(\vec{R}, t)) + \vec{F}^R. \tag{40}$$

The first term $\vec{F}^C$ subsumes the conservative forces acting on the particle, the second represents the frictional drag in the LB fluid, and the last one ($\vec{F}^R$) a stochastic random force which satisfies the fluctuation dissipation relation

$$\langle F_\alpha^R(t)\, F_\beta^R(t')\rangle = 2\zeta\, k_B T\, \delta_{\alpha\beta}\, \delta(t-t'). \tag{41}$$

The dissipative and stochastic terms describe the forces on the particle that stem from the particle-fluid interactions. To ensure momentum conservation (actio=reactio), the opposite forces must be transferred to the fluid. This is done by introducing an additional forcing term in the collision step. Extended nanoobjects are taken to consist of several pointlike monomers. This is clearly a reasonable picture in the case of polymers. The approach can also be used for studying colloids, which are then constructed from point-like interaction sites (raspberry model) (Lobaskin & Dünweg (2004)). In that case, one has the somewhat unphysical situation that the LB fluid may flow through the particle. In practice, however, the dissipative coupling ζ ensures that the LB fluid locally moves at exactly the same speed than the colloid – which effectively generates no-slip boundary conditions at the surfaces.

3.1.4 Brownian Dynamics simulations (BD)

As discussed in section 2.1, fluid flow is laminar on the nanoscale. Therefore, mesoscale simulation methods must not necessarily be able to reproduce the full Navier-Stokes equations, it suffices if they reproduce the Stokes equation. Nanoscale hydrodynamic flow patterns build up almost instantly, on time scales much smaller than the motion of the nanoparticles: The fluid follows the particles quasi adiabatically. This motivates BD approaches where the solvent is omitted altogether. In BD simulations, the particles undergo overdamped Brownian motion and the effect of hydrodynamics are mimicked by the implementation of suitable hydrodynamic interactions (7). The equations of motion for a system of particles i subject to the forces $\vec{F}_i$ in an external flow field $\vec{u}(\vec{r})$ then have the general form (Ermak & McCammon (1978))

$$\dot{r}_{i\alpha} = u_\alpha(\vec{r}_i) + \sum_j \sum_\beta D_{i\alpha,j\beta}\, F_{j\beta} + \sum_j \sum_\beta \xi_{i\alpha,j\beta}\, \theta_{j\beta}, \tag{42}$$

where $\mathbf{D}$ is a symmetrical mobility matrix, θ is an uncorrelated Gaussian distributed white noise with

$$\langle \theta_{i\alpha}\rangle = 0 \quad \text{and} \quad \langle \theta_{i\alpha}(t)\theta_{j\beta}(t')\rangle = 2\, k_B T\, \delta_{ij}\, \delta_{\alpha\beta}\, \delta(t-t'), \tag{43}$$

and the matrix ξ is chosen such that $\xi\, \xi^T = \mathbf{D}$. The indices i, j denote particles and α, β cartesian coordinates.

The mobility matrix $\mathbf{D}$ incorporates the effect of hydrodynamic interactions in the given channel geometry. In free solution, it would seem natural to construct it from the Oseen tensor

$$D_{i\alpha,j\beta} = \delta_{ij}\, \delta_{\alpha\beta} \mu_i^{(0)} + (1-\delta_{ij})\, T_{\alpha\beta}(\vec{r}_i - \vec{r}_j), \tag{44}$$

where $\mathbf{T}$ is the Oseen tensor of a point particle given by Eq. (8) and $\mu_i^{(0)}$ the mobility of a single particle, which can be approximated by the inverse Stokes friction $\mu^{(0)} = (6\pi\eta a)^{-1}$ for spherical particles of radius a. Unfortunately, the naive implementation of this expression leads to numerical problems due to the fact that **D** may become singular. Therefore, the Oseen tensor $\mathbf{T}(\vec{r})$ in Eq. (44) is usually replaced by the more refined Rotne-Prager tensor

$$T^{RP}_{\alpha\beta}(\vec{r}) = \frac{1}{8\pi\eta\, r}\Big(\delta_{\alpha\beta} + \hat{r}_\alpha\hat{r}_\beta + 2(\frac{a}{r})^2(\frac{1}{3}\delta_{\alpha\beta} - \hat{r}_\alpha\hat{r}_\beta)\Big), \tag{45}$$

which describes approximately the hydrodynamic interactions between spheres of radius a with no-slip surfaces (Dhont (1996)).

In confined geometry, the hydrodynamic interactions depend on the boundary conditions at the walls. General analytical expressions for **D** are not available. In some cases, however, solutions for **D** can be constructed from the knowledge of the mobility matrix in free solution. For example, **D** can be calculated in slit geometries calculated by a summation over a series of image charges (Kekre et al. (2010)).

Even though BD simulations disregard the solvent degrees of freedom, they quickly become very expensive for large macroion numbers n. This is not only due to the long-range nature of the hydrodynamic interactions, but also because of the stochastic contribution to Eq. (42), which has to be determined from the mobility matrix. For symmetric and positively definite matrices **D**, a matrix ξ satisfying $\xi\xi^T = \mathbf{D}$ can be calculated in a straightforward manner, *e.g.*, *via* a Cholesky decomposition. Unfortunately, the costs for this operation scale with n^3. This prohibits exact large scale simulations of many-particle systems with BD, and schemes for treating the noise at an approximate level are necessary (Banchio & Brady (2003); Fixman (1986); Geyer & Winter (2009)). The evaluation of the hydrodynamic interactions themselves can be accelerated using methods developed for electrostatics such as fast multipole techniques.

3.2 Simulation of electrokinetic phenomena

Simulations of hydrodynamic phenomena are challenging. Simulations of electrohydrodynamic phenomena are even worse due to the long-range nature of the Coulomb interactions. Even though the static equilibrium interactions between macroions are exponentially screened in fluids of high ionic strength, the microions in solution still influence the hydrodynamic flows and can thus not be ignored altogether.

In electrophoresis, we have discussed in Sec. 2.2.3 that the microions also screen hydrodynamic interactions in some sense. Early simulation approaches have therefore resorted to local BD simulation models where both hydrodynamic and electrostatic interactions are neglected (Viovy (2000)) (see Sec. 3.2.2). Such models can be quite successful. Nevertheless, they clearly disregard important physics. For example, hydrodynamic interactions are "screened" only with respect to the response of particles to slowly varying external electric fields, but not with respect to other external forces, internal forces, or diffusion. Moreover, time-dependent phenomena where ion relaxation becomes important cannot be studied within such simplified BD schemes.

To go beyond local BD models, one must account for both hydrodynamics and electrostatics simultaneously. Three types of approaches have been pursued in the past: Fully explicit models with explicit solvent and explicit microions, fully implicit models with implicit solvent and implicit microions, and mixed schemes where either the solvent or the microions are treated at an implicit level. In the following, we shall give a brief overview over these approaches with a few examples. More detailed reviews can be found in (Pagonabarraga et al. (2010); Slater et al. (2009)).

3.2.1 Fully explicit models

In fully explicit simulation approaches, the fluid is treated by explicit mesoscale methods such as DPD, MPCD, or LB, and the charges are represented by explicit charged point particles which are coupled to the fluid following the prescriptions described in the Sections 3.1.1–3.1.3. This is computationally expensive, but feasible. To give an example, the electrophoresis of short polyelectrolytes in free solutions has been studied recently both by fully explicit LB and MPCD simulations (Grass et al. (2008); Frank & Winkler (2008)). The computationally most challenging part of the simulations is the evaluation of the electrostatic interactions, especially in systems with high salt concentrations. Large efforts are currently dedicated to developing efficient methods for treating Coulomb interactions in many-particle systems (Sutmann (2009)), which will undoubtedly also benefit electrokinetic simulations in the future.

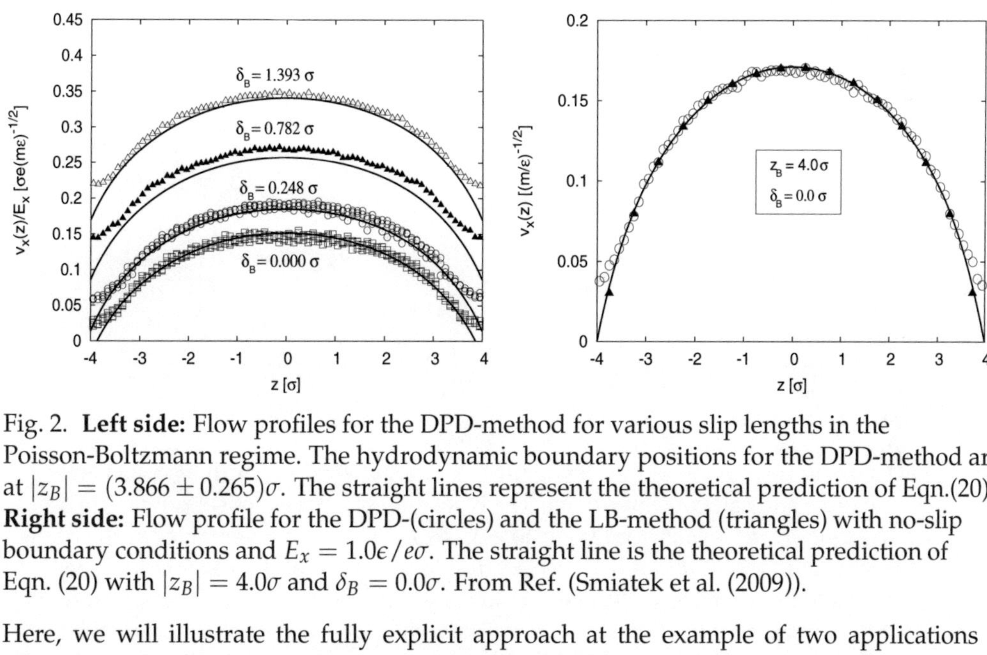

Fig. 2. **Left side:** Flow profiles for the DPD-method for various slip lengths in the Poisson-Boltzmann regime. The hydrodynamic boundary positions for the DPD-method are at $|z_B| = (3.866 \pm 0.265)\sigma$. The straight lines represent the theoretical prediction of Eqn.(20). **Right side:** Flow profile for the DPD-(circles) and the LB-method (triangles) with no-slip boundary conditions and $E_x = 1.0\epsilon/e\sigma$. The straight line is the theoretical prediction of Eqn. (20) with $|z_B| = 4.0\sigma$ and $\delta_B = 0.0\sigma$. From Ref. (Smiatek et al. (2009)).

Here, we will illustrate the fully explicit approach at the example of two applications in microchannels. The first one is a comparative study of counterion-induced EOF in symmetric planar slit channels using fully explicit DPD and LB simulations (Smiatek et al. (2009)). The theoretical prediction for the EOF profiles in this case is given by Eq. (20). We have studied such flows by two complementary approaches, DPD and LB, and devised schemes how the model parameters could exactly be mapped onto each other. Fig. 2 demonstrates that the results of DPD and LB simulations are in excellent agreement with each other and with theory. The left graph illustrates remarkable flow enhancement due to slippage already in the presence of small surface slip. The examples shown in Fig. (2) refer to situations where the Poisson-Boltzmann theory is applicable, *i.e.*, the microions are monovalent and the charges on the channel surfaces are moderate. With fully explicit simulations, one can also study regimes where the mean-field assumption breaks down. In that case, explicit simulations show that the relation between flow and electrostatic potential, Eq. (19), remains valid even though the ion distribution can no longer be calculated analytically (Smiatek et al. (2009)).

The electrophoretic response of polyelectrolytes in symmetric slit channels is considered in our second example. As discussed in Section 2.2.3, it results from the combination of bare electrophoresis and background EOF. From Eqs. (21), (23) and (25), one can derive the relation

$$\frac{\mu_t}{\mu_{\mathrm{EOF}}} = 1 + \frac{\mu_e}{\mu^0_{\mathrm{EOF}}(1+\delta_B \kappa_s)}. \tag{46}$$

This relation is verified in Fig. 3 (left). We have studied the electrophoresis of a charged polyelectrolyte in several electrolytes with different salt concentrations, confined by charged planar walls with different slip lengths. The behavior of the electrophoretic mobility is in very good agreement with the theoretical prediction. Most notably, the EOF flow may reverse the sign of the electrophoretic mobility, such that the polyelectrolyte migrates *against* the applied field. Following the discussion in Section 2.2.3, we expect the bare mobility of the polyelectrolytes to be independent of chain length in solvents of high ionic strength. Fig. 3 (right) demonstrates that this is indeed the case.

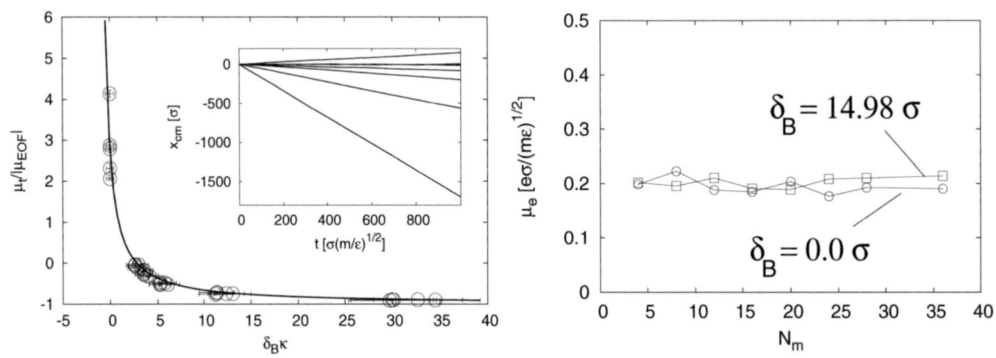

Fig. 3. **Left side:** Ratio $\mu_t/|\mu_{EOF}|$ plotted against $\delta_B\,\kappa_s$ for chains of length $N_m = 20$ and several slip lengths and salt concentrations. The black line is the theoretical prediction of Eqn. (46). The ratio μ_e/μ^0_{EOF} has been fitted to -3.778 ± 0.128. Negative values of $\mu_t/|\mu_{EOF}|$ indicate negative total mobilities of the polyelectrolyte. **Inset:** Total displacement of the polyelectrolyte's center of mass for different boundary conditions and a salt concentration of $\rho_s = 0.05625\sigma^{-3}$. in units of the ion diameter σ. The total mobility becomes negative for $|\mu_e| \ll |\mu_{EOF}|$. The lines correspond from top to bottom to the slip lengths $\delta_B \approx (0.00, 1.292, 1.765, 2.626, 5.664, 14.98)\sigma$. Thus larger slip lengths indirectly enhance the total mobility of the polyelectrolyte. **Right side:** Bare electrophoretic mobility for different monomer lengths N_m in confined geometry in the presence of salt for two slip lengths.

Finally in this section, we mention a special boundary effect which has not been discussed so far and is often neglected in electrokinetic simulations: The dielectric contrast between water and typical microchannel materials is very high. Therefore, the surfaces are polarized and charges are induced at the boundaries. Tyagi et al. have recently developed a general method that allows one to account for this effect (Tyagi et al. (2007; 2010)). First preliminary studies seem to indicatethat it does not influence the transport properties very much on length scales larger than the Debye length.

3.2.2 Fully implicit models

In fully implicit approaches, both solvent and microion degrees of freedom are disregarded and their effect on the particles is incorporated in effective interactions and an effective mobility matrix (see Eq. (42)). This implies an adiabatic approximation: Both the hydrodynamic flows and the local microion distributions are assumed to adjust instantaneously to the macroion configuration. Whereas this approximation can be justified in the case of flows at low Reynolds numbers (as discussed in Section 3.1.4), it is somewhat questionable in the case of the ions. We have discussed important phenomena in Section 2.2 which cannot be captured within such an approach, *e.g.*, the relaxation effect in electrophoresis. Relaxation processes in the ionic double layer of charged colloids have characteristic time scales ranging from microseconds (Maxwell-Wagner relaxation, ionic migration sets in) to 10 milliseconds (the thickness of the Debye layer adjusts), which is close to the characteristic time scales of many mesoscale simulations. Thus fully implicit methods are hardly suited for studying dynamical processes with a time evolution. They can however be used to study quasi-stationary situtaions such as electrophoresis or pressure-driven transport at a semi-quantitative level.

A second drawback of implicit schemes is that they usually rely on the Debye-Hückel approximation, which has a limited range of validity as discussed in Section 2.2.1. Nevertheless, it can often be applied successfully even in the nonlinear regime as long as one accounts for nonlinear effects by appropriate charge renormalization or renormalization of the local Debye Hückel screening length (see 2.2.1).

Fully implicit simulations are stochastic simulations with the structure of BD simulations (Eq. (42)). Setting up BD schemes for nanoparticles in external electric fields is slightly subtle due to the fact that one has two mobility matrices: One ($\mathbf{D}^C$) which describes the hydrodynamic interactions in response to conservative forces and is related to the unscreened Greens function Eq. (8) or (45) in free solution, and one ($\mathbf{D}^E$) which describes the hydrodynamic interactions in response to external electric fields, and can be related to the screened Greens function (29) in free solution if the external fields vary slowly compared to the Debye Hückel screening length. A clean way of dealing with this situation is to treat both types of responses on separate footings: In a constant or slowly varying electric field $\vec{E}$ and external flow field $\vec{u}(\vec{r})$, the equation of motion of a system of particles i with charges q_i (the equivalent of Eq. (42)) is then given by (Kekre et al. (2010))

$$\dot{r}_{i\alpha} = u_\alpha(\vec{r}_i) + \sum_j \sum_\beta D^C_{i\alpha,j\beta} F^C_{j\beta} + \sum_j \sum_\beta \xi_{i\alpha,j\beta}\, \theta_{j\beta} + \sum_j \sum_\beta D^E_{i\alpha,j\beta}\, q_j E_\beta, \tag{47}$$

where $\mathbf{D}^C$ and $\mathbf{D}^E$ are given by

$$D^C_{i\alpha,j\beta} = \delta_{ij}\,\delta_{\alpha\beta}\,\mu_i^{(0)} + (1-\delta_{ij})\,T_{\alpha\beta}(\vec{r}_i - \vec{r}_j)$$
$$D^C_{i\alpha,j\beta} = \delta_{ij}\,\delta_{\alpha\beta}\,\mu_i^{\text{el}} + (1-\delta_{ij})\,G_{\alpha\beta}(\vec{r}_i - \vec{r}_j)$$

and the matrix ξ meets the requirement $\xi\xi^T = \mathbf{D}^C$. Here $\mu_i^{(0)}$ is the bare frictional mobility of the particle i, μ_i^{el} its electrophoretic mobility, and $\mathbf{T}$, $\mathbf{G}$ are the screened and unscreened hydrodynamic Greens functions, which can be approximated by the Rotne-Prager tensor (45) and the screened Greens function (29) (or the corresponding Rotne-Prager version (Fischer et al. (2008))) in free solution. It is important to note that $\vec{F}_j^C$ subsumes all effective conservative interactions between ions, including those of electrostatic origin. One can show rigorously that a procedure where (microion) degrees of freedom are integrated out adiabatically only renormalizes the conservative potentials of the remaining macroions and not their mobility coefficients. Thus, the Coulomb potentials are replaced by screened Yukawa potentials, but the hydrodynamic interactions remain long-range. The situation is different in electrophoresis since this is an inherently nonequilibrium process.

Solving Eq. (47) for many-particle systems is expensive due to the long-range hydrodynamic interactions and the noise issue discussed in Section 3.1.4. If the dominating forces in a system are due to external electric fields, it may become acceptable to neglect the long-range hydrodynamic contributions. In nanochannels, this approximation can further be justified by the fact that hydrodynamic interactions are screened in confinement (see Section 2.1.2). For these reasons, simulation approaches have been popular already early on where both hydrodynamics and electrostatics were neglected entirely. The motion of macroions in an external electric field $\vec{E}(\vec{r})$ and flow field $\vec{u}(\vec{r})$ is then described by simple local Langevin dynamics, where the force $\vec{F}_i$ on a particle i at position $\vec{r}_i$ is given by the sum

$$\vec{F}_i = \vec{F}_i^C + \vec{F}_i^D + \vec{F}_i^R \tag{48}$$

of conservative contributions $\vec{F}_i^C = q_i\vec{E}(\vec{r}_i) + \vec{F}_{i,\text{int}}^C$, the dissipative drag force $\vec{F}_i^D = -\zeta(\vec{v}_i - \vec{u}(\vec{r}_i))$ and a Gaussian distributed random force $\vec{F}_I^R$ which satisfies the fluctuation dissipation relation $\langle F_{i\alpha}^R(t)\, F_{j\beta}^R(t')\rangle = 2\,\zeta\, k_B T\, \delta_{ij}\, \delta_{\alpha\beta}\, \delta(t-t')$. The internal contributions $\vec{F}_{i,\text{int}}^C$ subsume the screened electrostatic interactions between particles as well as other forces with non-electrostatic origin.

The situation is particularly simple if the external flow $\vec{u}(\vec{r})$ has been generated by EOF and the channel has the similitude property (Section 2.2.2), *i.e.*, $\vec{u}(\vec{r}) = \mu_{\text{EOF}}\vec{E}(\vec{r})$. In that case, the total force on the particle i is given by

$$\vec{F}_i = \vec{F}_{i,\text{int}}^C + (q_i + \mu_{\text{EOF}}\zeta)\vec{E}(\vec{r}_i) - \zeta\vec{v}_i + \vec{F}_i^R. \tag{49}$$

Thus the EOF flow $\vec{u}(\vec{r})$ simply has the effect of renormalizing the charge q_i (Streek et al. (2005)).

Simple local BD approaches such as Eq. (49) have been used quite successfully to study electrophoresis in complex geometries, and they can yield good qualitative and even semiquantitative agreement with experiments (Streek et al. (2004); Viovy (2000)). As an example, Fig. 4 shows results from such a BD simulation of DNA electrophoresis in a structured microchannel. Both in experiments and simulations, one observes a two-state behavior where the DNA switches between a "slow" and "fast" migration mode. This is reflected by an inhomogeneous average monomer distribution in the channel (Streek et al. (2005)).

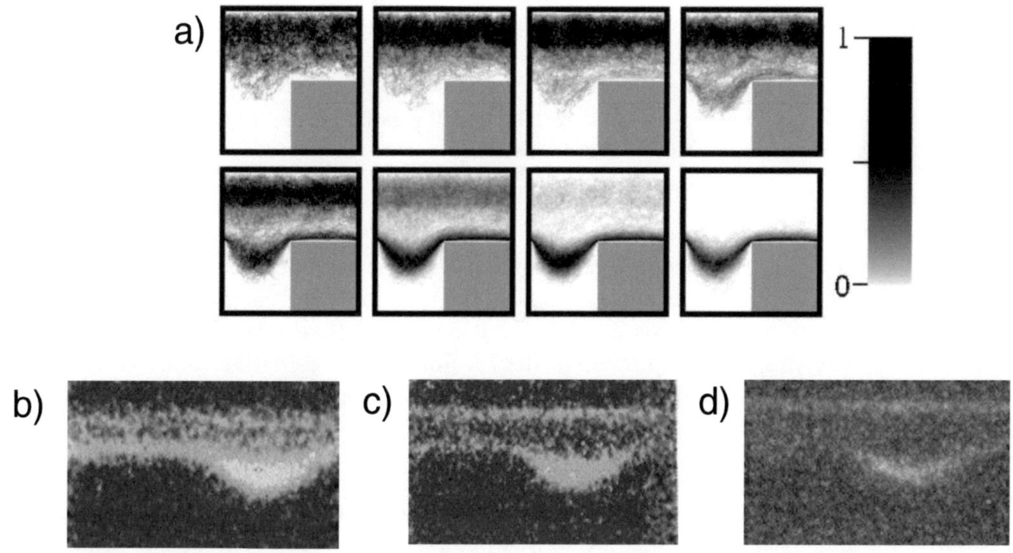

Fig. 4. **Top panel** (a): Monomer density histograms obtained from local BD simulation of electrophoresis of charged chains with different lengths $N = 50, 100, 120, 140$ (top, left to right) and $N = 170, 200, 250$ and 320 (bottom) at. **Bottom panel**: Monomer density histograms obtained from experiment at applied voltages (b) $E = 43V/cm$, (c) $E = 86V/cm$, and (d) $E = 130V/cm$ for λ-DNA in 5μm constrictions. The fluorescence intensity decreases because the illumination time was kept constant. Note the depletion zone in the center of the channel. From (Streek et al. (2005))

Despite these successes, the complete neglect of hydrodynamic interactions sometimes simplifies the situation too much even in simulations of electrophoretic transport. In these cases, one may still choose to work with the screened hydrodynamic Greens function **G** for reasons of efficiency and neglect the unscreened contributions. Löwen et al have demonstrated that such a scheme can be used successfully to describe the effect of hydrodynamic interactions in systems of oppositely charged colloids subject to an external field. Even the screened hydrodynamic interactions have a noticeable effect on structure formation and "laning" in such systems (Rex & Löwen (2008)).

3.2.3 Mixed schemes

Mixed schemes for electrohydrodynamic simulations treat either the solvent or the microions at an implicit level.

In the first approach, both microions and macroions are represented by explicit particles which interact with each other by means of the full Rotne-Prager tensor, Eq. (45) (Fischer et al. (2008); Kim & Netz (2006)). These methods are motivated by the fact mentioned earlier that hydrodynamic flows adjust very rapidly, whereas ion rearrangement processes can be slow on mesoscopic time scales. Unfortunately, they suffer from the bad scaling properties of BD simulations and have therefore not been used to study very large systems.

In the second type of approach, the solvent is modeled as an explicit fluid, but microions are no longer represented by individual particles. Many kinetic processes in electrolyte solutions do not depend on ion correlations and can thus be described satisfactorily at the mean-field level. To access larger length scales in simulations, one may therefore resort to methods that treat the ions at the level of the Nernst-Planck equation (12). So far, efforts to develop such methods have mainly focused on LB approaches (Capuani et al. (2004); Horbach & Frenkel (2001); Pagonabarraga et al. (2005); Warren (1997)).

A number of authors have proposed simplified schemes that capture basic electrokinetic phenomena at a heuristic level in explicit fluid simulations. For example, one simple way to deal with EOF is to impose a velocity boundary condition (Eq. (23)) at the walls of a microfluidic channel (Duong Hong et al. (2008)). Similarly, the electrohydrodynamic screening effect in electrophoresis can be mimicked in a LB simulation by adding a slip term to the Stokes drag on the particle in the LB flow. The dissipative contribution $F^D = -\zeta(\vec{V} - \vec{u}(\vec{R},t))$ to Eq. (40) is then replaced by a modified drag force (Hickey et al. (2010))

$$F^D = -\zeta(\vec{V} - \vec{u}(\vec{R},t) - \mu_0 \vec{E}) \tag{50}$$

Unfortunately, this elegant idea cannot easily be transferred to DPD and MPCD simulations. Duong-Hong et al. have proposed an alternative approach where charges are assigned to DPD or MPCD particles that are within a certain distance (the Debye length) of the macroions or charged walls. These particles are then dragged along by the electric field, which can be used to mimick electrophoretic retardation, electrohydrodynamic screening effects, and EOF generation on channel surfaces in a unified manner (Duong Hong et al. (2008b)).

4. Summary and outlook

With the advent of faster computers and improved simulation methods for electrostatics and hydrodynamics, electrohydrodynamic simulations become an increasingly attractive tool for studying and predicting electrokinetic phenomena in microgeometries. In this chapter we have introduced the main concepts behind coarse-grained mesoscale simulations of micro- and nanofluidic flow and transport. We have focused on boundary effects and mean-field phenomena, *e.g.*, phenomena that can be rationalized within the Poisson-Boltzmann theory. Several important phenomena such as counterion condensation, overcharging, shape transformations of the polyelectrolytes, confinement effects, to name a few, have not been discussed in this work. Also, important problems of current interest such as electrophoresis with obstacle collisions or polymer nanopore translocations, have not been mentioned. For further reading, we refer the interested reader to the reviews (Boroudjerdi et al. (2005); Graham (2011); Pagonabarraga et al. (2010); Slater et al. (2009); Viovy (2000)).

We have shown that mesoscopic simulation methods are powerful tools to explore fluid behavior on the micro- and nanoscale. Such simulations not only help to understand and optimize transport processes in micro- and nanofluidic channels, they can also be used to study basic mechanisms for many physical processes of life, since such processes often take place in confinement (*e.g.*, cells) and in aqueous solutions with high ion concentrations. We expect that the mesoscale techniques introduced in this chapter, combined with microscopic and macroscopic approaches, will open new routes towards a better understanding of transport phenomena on the micro- and nanoscale.

5. Acknowledgments

We have benefitted from discussions with Mike P. Allen, Burkhard Dünweg, Ralf Eichhorn, Kai Grass, Christian Holm, Sebastian Meinhardt, Ulf D. Schiller, Marcello Sega, Martin Streek, and Tatiana Theis. The fully explicit simulations presented here have been carried out using the software package ESPResSo

(`http://www.icp.uni-stuttgart.de/~icp/ESPResSo`)

(Arnold et al. (2006)). We thank the Volkswagen Stiftung and the Deutsche Forschungsgemeinschaft for funding and the HLR Stuttgart, NIC Jülich and Arminius PC^2-cluster in Paderborn for computing times.

6. References

Ahlrichs, P. & Dünweg, B. (1999). Simulation of a single polymer chain in solution by combining Lattice Boltzmann and molecular dynamics. *Journal of Chemical Physics* Vol. 111, 8225.

Ahmed, N.K. & Hecht, M. (2009). A boundary condition with adjustable slip length for Lattice Boltzmann simulations. *Journal of Statistical Physics* Vol. 2009, P09017.

Alexander, S.; Chaikin, P.M.; Grant, P.; Morales, G.H. ; Pincus, P. & Hone, D. (1984). Charge renormalization, osmotic pressure, and bulk modulus of colloidal crystals: Theory. *Journal of Chemical Physics*, Vol. 80, 5776.

Arnold, A.; Mann, B.A.; Limbach, H.-J. & Holm, C. (2006). ESPResSo - An Extensible Simulation Package for Research on Soft Matter Systems. *Computer Physics Communications*, Vol. 174, 704.

Banchio, A.J. & Brady, J.F. (2003). Accelerated Stokesian Dynamics: Brownian motion. *Journal of Chemical Physics* Vol. 118, 10323.

Barrat, J.-L. & Joanny, J. F. (1996). Theory of polyelectrolyte solutions. *Advances in Chemical Physics* Vol. 94, 1.

Bocquet, L. & Barrat, J.-L. (1994). Hydrodynamic boundary conditions, correlation functions, and Kubo relations for confined fluids. *Physical Review E* Vol. 49, 3079.

Benzi, R.; Succi, S. & Vergassola, M. (1992). The Lattice Boltzmann equation: Theory and applications. *Physics Reports* Vol. 22, 145.

Bhatnagar, P. L., Gross, E. P. & Krook, M. (1954). A model for collision processes in gases. 1. Small amplitude proceesses in charged and neutral one-component systems. *Physical Review* Vol. 54, 511.

Boroudjerdi, H.; Kim, Y. -W.; Najim A.; Netz, R. R.; Schlagberger, X. & Serr, A. (2005). Statics and dynamics of strongly charged soft matter. *Physics Reports* Vol. 416, 129.

Bouzigues, C.I.; Tabeling, P. & Bocquet, L. (2008). Nanofluidics in the Debye Layer at Hydrophilic and Hydrophobic Surfaces. *Physical Review Letters*, Vol. 101, Article-No. 114503.

Capuani, F.; Pagonabarraga, I. & Frenkel, D. (2004). Discrete solution of the electrokinetic equations. *Journal of Chemical Physics* Vol. 121, 973.

Chen, S. & Doolen, G. D. (1998). Lattice-Boltzmann method for fluid flows. *Annual Review of Fluid Mechanics* Vol. 30, 329.

Cui, B.; Diamant, H. & Lin, B. (2002). Screened Hydrodynamic Interactions in a Narrow Channel. *Physical Review Letters* Vol. 89, Article-No. 188302.

Cummings, E.B.; Griffiths, S. K.; Nilson, R. H.; & Paul, P. H. (2000). Conditions for similitude between the fluid flow and electric field in electroosmotic flow. *Analytical Chemistry* Vol. 72, 2526.

Diamant, H.; Cui, B.; Lin, B. & Rice, S.A. (2005). Hydrodynamic interactions in quasi-two-dimensional suspensions. *Journal of Physics: Condensed Matter*, Vol. 17, S2787.

Dhont, J.K.G. (1996). *An Introduction to Dynamics of Colloids*. Elsevier, Amsterdam.

Doi, M. & Edwards, S. F. (1986). *The Theory of Polymer Dynamics*. Oxford Science Publications, Oxford.

Dufresne, E.R.; Squires, T.M.; Brenner, M.P. & Grier, D.G. (2000). Hydrodynamic coupling of two Brownian spheres to a planar surface. *Physical Review Letters*, Vol. 85, 3317.

Duong-Hong D.; Wang, J.-S.; Liu, G.R.; Chen, Y.Z.; Han, J. & Hadjiconstantinou, N.G. (2008). Dissipative simulations of electroosmotic flow in nanofluidic devices. *Microfluidics and Nanofluidics* Vol. 4, 219.

Duong-Hong, D.; Han, J.; Wang, J.-S.; Hadjiconstantinou, N.G.; Chen, Y.Z. & Liu, G.-R. (2008). Realistic simulations of combined DNA electrophoretic flow and EOF in nanofluidic devices. *Electrophoresis* Vol. 29, 4880.

Dünweg, B., Schiller, U. & Ladd, A.J.C (2007). Statistical mechanics of the fluctuating Lattice Boltzmann equation. *Physical Review E* Vol. 76, 036704.

Dünweg, B. & Ladd, A.J.C (2009). Lattice Boltzmann simulations of soft matter systems. *Advances in Polymer Science* Vol. 221, 89.

Ermak, D. L. & McCammon, J. A. (1978). Brownian dynamics with hydrodynamic interactions. *Journal of Chemical Physics* Vol. 69, 1352.

Español, P. & Warren, P. B. (1995). Statistical mechanics of dissipative particle dynamics. *Europhysics Letters* Vol. 30, 191.

Fischer, S.; Naji, A. & Netz, R.R. (2008). Salt-induced counterion-mobility anomaly in polyelectrolye electrophoresis. *Physical Review Letters*, Vol. 101, 176103.

Fixman, M. (1986). Construction of Langevin forces in the simulation of hydrodynamic interactions. *Macromolecules* Vol. 19, 1204.

Frenkel, D. & Smit, B. (2001). *Understanding molecular simulation*. Academic Press, Orlando.

Frisch, U.; Hasslacher, B. & Pomeau, Y. (1986). Lattice Automata for the Navier-Stokes Equation. *Physical Review Letters* Vol. 56, 1505.

Geyer, T. & Winter, U. (2009). An $\mathcal{O}(N^2)$ approximation for hydrodynamic interactions in Brownian dynamics simulations. *Journal of Chemical Physics* Vol. 130, 114905.

Gompper, G.; Ihle, T.; Kroll, K. & Winkler. R. G. (2009). Multi-particle collision dynamics: A particle-based mesoscale simulation approach to the hydrodynamics of complex fluids. *Advances in Polymer Science* Vol. 221, 1.

Graham, M. D. (2011). Fluid dynamics of dissolved polymer molecules in confined geometry. *Annual Review of Fluid Mechanics* Vol. 43, 273.

Grass, K.; Böhme, U.; Scheler, U.; Cottet, H. & Holm, C. (2008). Importance of hydrodynamic shielding for the dynamic behavior of short polyelectrolyte chains. *Physical Review Letters* Vol. 100, Article-No. 096104.

Frank, S. & Winkler, R.G. (2008). Polyelectrolyte electrophoresis: Field effects and hydrodynamic interactions. *Europhysics Letters* Vol. 83, 38004.

Groot, R. D. & Warren, P. B. (1997). Dissipative particle dynamics: Bridging the gap between atomistic and mesoscopic simulations. *Journal of Chemical Physics* Vol. 107, 4423.

Happel, J.; Brenner, H. (1983). *Low Reynolds Number Hydrodynamics.* Martinus Nijhoff. Dordrecht.

Hecht, M.; Harting, J. (2010). Implementation of on-site velocity boundary condition for D3Q19. *Journal of Statistical Mechanics* Vol. 2010, P01018.

Hickey, O.A.; Holm, C.; Harden, J.L. & Slater, G.W. (2010). Implicit method for simulating electrohydrodynamics of polyelectrolytes. *Physical Review Letters* Vol. 105, 148301.

Hoogerbrugge, P. J. & Koelman, J. M. V. A. (1992). Simulating microscopic hydrodynamic phenomena with dissipative particle dynamics. *Europhysics Letters* Vol. 19, 155.

Horbach, J. & Frenkel, D. (2001). Lattice-Boltzmann method for the simulation of transport phenomena in charged colloids. *Physical Review E* Vol. 64, Article-No. 051607.

Hunter, R. J. (1991). *Foundations of Colloid Science.* Clarendon Press, Oxford.

Ihle, T. & Kroll, D.M. (2003). Stochastic rotation dynamics. I. Formalism, Galilean invariance, and Green-Kubo relations. *Physical Review E* Vol. 67, 066705.

Israelachvili, J. (1991). *Intermolecular and surface forces.* Academic Press. London.

Joly, L.; Ybert, C.; Trizac, E. & Bocquet, L. (2004). Hydrodynamics within the electric double layer on slipping surfaces. *Physical Review Letters,* Vol. 93, Article-No. 257805.

Joly, L.; Ybert, C.; & Bocquet, L. (2006). Probing the nanohydrodynamics at liquid-solid interfaces using thermal motion. *Physical Review Letters,* Vol. 96, Article-No. 04601.

Kekre, R.; Butler, J.E.; Ladd, A.J.C. (2010). A comparison of lattice-Boltzmann and Brownian dynamics simulations of a polymer migration in confined flows. *Physical Review E,* Vol. 82, Article-No. 011802.

Kekre, R.; Butler, J.E.; Ladd, A.J.C. (2010). Role of hydrodynamic interactions in the migration of polyelectrolytes driven by a pressure gradient and an electric field. *Physical Review E,* Vol. 82, Article-No. 050803(R).

Kennard, E. H. (1938). *Kinetic theory of gases.* McGraw-Hill, New York.

Kim, K.; Nakayam, Y. & Yamamoto R. (2006). Direct Numerical Simulations of electrophoresis of Charged Colloids. *Physical Review Letters,* Vol. 79, 031401.

Kim, Y.W. & Netz, R.R. (2006). Electroosmosis at inhomogeneous charged surfaces: Hydrodynamic versus electric friction. *Journal of Chemical Physics,* Vol. 124, 114709.

Kjellander, R. & Mitchell D.J. (1994). Dressed-ion theory for electrolyte solutions: A Debye-Hückel like reformulation of the exact theory for the primitive model. *Journal of Chemical Physics* Vol. 101, 603.

Koelman, J. M. V. A. & Hoogerbrugge P. J. (1993). Dynamic simulation of hard sphere suspensions under steady shear. *Europhys. Lett.* Vol. 21, 363.

Kuang, C. & Wang, G. (2010). A novel far-field nanoscopic velocimetry for nanofluidics. *Lab Chip* Vol. 10, 240.

Kunert, C. & Harting, J.D.R. (2008). Simulation of fluid flow in hydrophobic rough microchannels. *International Journal of Computational Fluid Dynamics* Vol. 22, 475.

Ladd, A.J.C. (1994). Numerical simulations of particulate suspensions via a discretized Boltzmann equation Part I: Theoretical foundations. *Journal of Fluid Mechanics* Vol. 271, 285.

Ladd, A.J.C. (1994). Numerical simulations of particulate suspensions via a discretized Boltzmann equation Part II: Numerical Results. *Journal of Fluid Mechanics* Vol. 271, 311.

Landau, L. D.; Lifshitz, E. M. (1987). *Fluid Mechanics. Volume 6 (Course of Theoretical Physics).* Butterworth-Heinemann. Oxford.

Lamura, A.; Gompper, G.; Ihle, T.; Kroll, D. (2001). Multi-particle collision dynamics: Flow around a circular and a square cylinder. *Europhysics Letters* Vol. 56, 319.

Liron, N. & Mochon, S. (1976). Stokes flow for a stokeslet between two parallel flat plates. *Journal of Engineering Mathematics*, Vol. 10, 287.

Lobaskin, V. & Dünweg, B. (2004). A new model for simulating colloidal dynamics. *New Journal of Physics* Vol. 6, 54.

Long, D. & Ajdari, A. (2001). A note on the screening of hydrodynamic interactions, in electrophoresis, and in porous media. *European Physical Journal E* Vol. 4, 29.

Lyklema, J. (1995). *Fundamentals of Interface and Colloid Science*. Academic Press. New York.

Malevanets, A. & Kapral, R. (1999). Mesoscopic model for solvent dynamics. *Journal of Chemical Physics* Vol. 110, 8605.

Malevanets, A. & Kapral, R. (2000). Solute dynamics in mesoscale solvent. *Journal of Chemical Physics* Vol. 112, 7260.

Manning, G. S. (1981). Limiting laws and counterion condensation in polyelectrolyte solutions. 7. Electrophoretic mobility and conductance. *Journal of Physical Chemistry* Vol. 85, 1506.

Marsh, C.; Backx, G. & Ernst, M.H. (1997). Static and dynamic properties of dissipative particle dynamics. *Physical Review E* Vol. 56, 1976.

McBride, A.; Kohonen, M. & Attard, P. (1998). The screening length of charge-asymmetric electrolytes: A hypernetted chain calculation. *Journal of Chemical Physics* Vol. 109, 2423.

McNamara, G. R. & Zanetti, G. (1988). Use of the Boltzmann equation to simulate Lattice-Gas Automata. *Physical Review Letters* Vol. 61, 2332.

Mitchell, D.J. & Ninham, B.W. (1978). Range of the screened Coulomb interaction in electrolytes and double layer problems. *Chemical Physics Letters* Vol. 53, 397.

Neto, C.; Evans, D. R.; Bonaccurso, E.; Butt, H.-J. & Craig V. S. J. (2005). Boundary slip in Newtonian liquids: A review of experimental studies. *Report on Progress in Physics* Vol. 68, 2859.

Noguchi, H, ; Kikuchi, R. & Gompper, G. (2007). Particle-based mesoscale hydrodynamic techniques. *Europhysics Letters* Vol. 78, 10005.

Onsager, L. (1926). Zur Theorie der Elektrolyte 1. *Physikalische Zeitschrift* Vol. 27, 388.

Onsager, L. (1927). Zur Theorie der Elektrolyte 2. *Physikalische Zeitschrift* Vol. 28, 277.

Onsager, L. & Fuoss, R. M. (1932). Irreversible processes in electrolytes. Diffusion, conductance, and viscous flow in arbitrary mixtures of strong electrolytes. *Journal of Physical Chemistry* Vol. 36, 2689.

Pagonabarraga, I.; Capuani, F. & Frenkel, D. (2005). Mesoscopic lattice modeling of electrokinetic phenomena. *Computer Physics Communications* Vol. 169, 192.

Pagonabarraga, I.; Rotenberg, B. & Frenkel, D. (2010). Recent advances in the modelling and simulation of electrokinetic effects: Bridging the gap between atomistic and macroscopic descriptions. *Physical Chemistry Chemical Physics* Vol. 12, 9566-9580.

Pit, R.; Hervet, H. & Leger, L. (2000). Direct experimental evidence of slip in hexadecane:solid interfaces. *Physical Review Letters* Vol. 85, 980.

Pivkin, I.V., Caswell, B. & Karniadakis, G.E. (2010). Dissipative particle dynamics. *Reviews in Computational Chemistry*, Vol. 27 (ed. K.B. Lipkowitz), John Wiley & Sons, Ind., Hoboken, NJ, USA.

Revenga, M.; Zuniga, I.; Español, P. & Pagonabarraga, I. (1998). Boundary models in dissipative particle dynamics. *International Journal of Modern Physics C* Vol. 9, 1319.

Revenga, M.; Zuniga, I. & Español P. (1999). Boundary Conditions in Dissipative Particle Dynamics. *Computer Physics Communications* Vol. 121-122, 309.

Rex, M. & Löwen, H. (2008). Influence of hydrodynamic interactions on lane formation in oppositely charged driven colloids. *European Physical Journal E* Vol. 26, 143

Rotne, J. & Prager, S. (1969). Variational treatment of hydrodynamic interactions in polymers. *Journal of Chemical Physics* Vol. 50, 4831.

Slater, G. W.; Holm, C.; Chubynsky, M. V.; de Haan, H. W.; Dube, A.; Grass, K.; Hickey, O. A.; Kingsburry, C.; Sean, D.; Shendruk, T. N. & Nhan, L. X. (2009). Modeling the separation of macromolecules: A review of current computer simulation methods. *Electrophoresis* Vol. 30, 792.

Smiatek, J.; Allen, M. P. & Schmid, F. (2008). Tunable slip boundaries for coarse-grained simulations of fluid flow. *European Physical Journal E* Vol. 26, 115.

Smiatek, J. (2009). *Mesoscopic simulations of electrohydrodynamic phenomena*. PhD thesis. Bielefeld Unversity, Bielefeld.

Smiatek, J.; Sega, M.; Schiller, U. D.; Holm, C. & Schmid, F. (2009). Mesoscopic simulations of the counterion-induced electro-osmotic flow: A comparative study. *Journal of Chemical Physics* Vol. 130, 244702.

Smiatek, J.; Schmid, F. (2010). Polyelectrolyte electrophoresis in nanochannels: A dissipative particle dynamics simulation. *Journal of Physical Chemistry B* Vol. 114, 6266.

Smiatek, J. & Schmid, F. (2011). Mesoscopic simulations of electroosmotic flow and electrophoresis in nanochannels. *Computer Physics Communications* Vol. 182, 1941.

Soddemann, T.; Dünweg, B. & Kremer, K. (2003). Dissipative particle dynamics: A useful thermostat for equilibrium and nonequilibrium molecular dynamics simulations. *Physical Review E* Vol. 68, 046702.

Sparreboom, W.; van den Berg, A. & Eijkel, J.T.C. (2010). Transport in nanofluidic systems: A review of theory and applications. *New Journal of Physics* Vol. 12, 015004.

Streek, M.; Schmid, F.;, Duong, T. T. & Ros, A. (2004). Mechanisms of DNA separation in entropic trap arrays: A Brownian dynamics simulation. *Journal of Biotechnology* Vol. 112, 79.

Streek, M.; Schmid, F.;, Duong, T. T.; Anselmetti, D. & Ros, A. (2005a). Two-state migration of DNA in a structured microchannel. *Physical Review E* Vol. 71, 011905.

Succi, S. (2001). *The Lattice Boltzmann equation for fluid dynamics and beyond*, Oxford University Press, USA.

Succi, S. (2002). Mesoscopic modeling of slip motion at fluid-solid interfaces with heterogeneous catalysis. *Physical Review Letters* Vol. 89, 064502.

Sutmann, G. (2009). ScaFaCoS – when long range goes parallel. *Innovative Supercomputing in Deutschland* Vol. 7, No. 1.

Tlusty, T. (2006). Screening by symmetry of long-range hydrodynamic interactions of polymers confined in sheets. *Macromolecules*, Vol. 39, 3927.

Tretheway, D.& Meinhart, C. (2002). Apparent fluid slip at hydrophobic microchannel walls. *Physics of Fluids* Vol. 14, 9.

Trizac, E.; Bocquet, L. & Aubouy, M. (2002). Simple approach for charge renormalization in highly charged macroions. *Physical Review Letters*, Vol. 89, Article-No. 248301.

Tyagi, S.; Arnold, A. & Holm, C. (2007). ICMMM2D: An accurate method to include planar dielectric interfaces via image charge summation. *Journal of Chemical Physics* Vol. 127, 154723.

Tyagi, S.; Süzen, M.; Sega, M.; Barbosa, M.; Kantorovich, S. & Holm, C. (2010). An iterative, fast, linear-scaling method for computing induced charges on arbitrary dielectric boundaries. *Journal of Chemical Physics* Vol. 132, 154112.

Viovy, J.-L. (2000). Electrophoresis of DNA and other polyelectrolytes: Physical mechanisms. *Review of Modern Physics*, Vol. 72, 813.

Warren, P.B. (1997). Electroviscous transport problems via Lattice Boltzmann. *International Journal of Modern Physics C* Vol. 8, 889.

Whitmer, J, & Luijten, E. (2010). Fluid-solid boundary conditions for multiparticle collision dynamics. *J. Phys.: Cond. Matter* Vol. 22, 104106.

Wright, M. R. (2007). *An introduction to aqueous electrolyte solutions*, John Wiley & Sons, Chichester.

Yamakawa, H. (1970). Transport properties of polymer chains in dilute solution: Hydrodynamic interaction. *Journal of Chemical Physics* Vol. 53, 436.

Yeomans, J. M. (2006). Mesoscopic simulations: Lattice Boltzmann and particle algorithms. *Physica A* Vol. 369, 159.

Zou, Q. & He, X. (1997). On pressure and velocity boundary conditions for the Lattice Boltzmann BGK model. *Physics of Fluids* Vol. 9, 1591.

Smart Microfluidics: The Role of Stimuli-Responsive Polymers in Microfluidic Devices

Simona Argentiere[1], Giuseppe Gigli[2], Mariangela Mortato[3], Irini Gerges[1] and Laura Blasi[4]
[1]*Fondazione Filarete Srl, Viale Ortles, Milano,*
[2]*Dipartimento di Ingegneria dell'Innovazione, Universita del Salento, Lecce,*
[3]*Superior School ISUFI, Università del Salento, Lecce,*
[4]*Nanoscience Institute of CNR, Lecce, Italy*

1. Introduction

1.1 The emerging market of lab-on-chip microfluidics

Microfluidic technologies actually represent one of the most expanding fields in the market of commercial instruments. With an estimated world market of about 3 billion dollars in the year 2014, (www.yole.fr/) it is clear that microfluidic devices will continue to find new applications, and to generate interest from a wide range of industries and research fields.

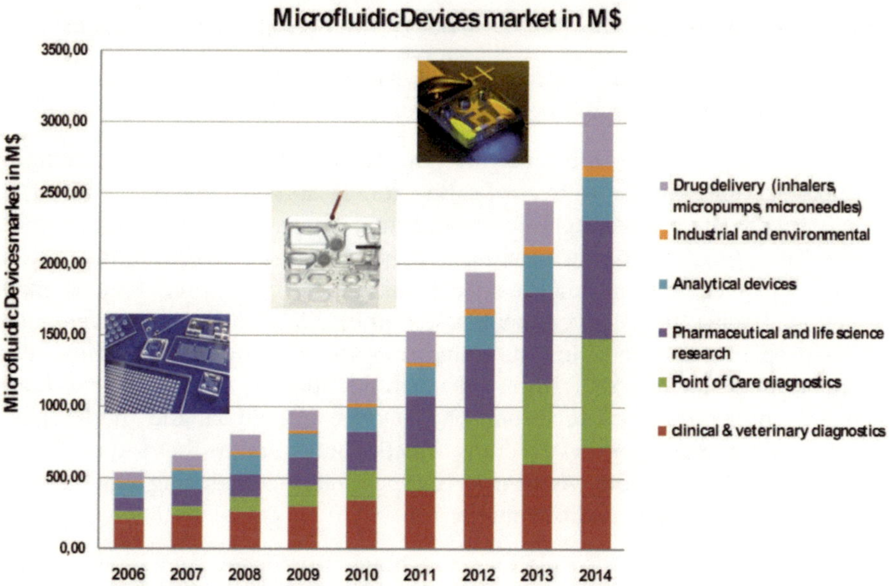

Fig. 1. Trend in microfluidics market. Reprinted from http://www.yole.fr/.

The past few years have seen rapid growth within the microfluidics field, whose applications include pharmaceuticals, life sciences, biotechnology, industry, environment and agriculture. The applications of microfluidics in food, agriculture and biosystems have been recently reviewed.(Neethirajan et al., 2011) However, the market of microfluidics in the life sciences is considered to possess by far the greatest potential in terms of products and commercial outlets, and is markedly increased in the last years. Accordingly, most efforts by both academic groups and private companies have been focused on the development of microfluidic devices for point-of-care (POC) diagnostics, biosensors and cell biology.

Microfluidics covers a set of multidisciplinary technologies such as physics, chemistry, engineering and biotechnology that deal with the development of miniaturized devices to manipulate and process minute volumes of fluids. Essentially, a 'lab on a chip' is a microfluidic device comprising tiny channels and components (*i.e.,* few to hundreds of micrometres) connected to liquid reservoirs that control the flow of nano-/pico-liters of liquids. The intrinsic features of microfluidic devices ensure low consumption of reagents and sample for each process, and rapid and repeatable analysis.(Sia & Kricka, 2008) Indeed, the increased surface-to-volume ratio in microfluidic channels shortens reaction times, and make the reactions more accurate and effective, whereas microfluidic chambers and channels measure volumes more consistently than human hands can, thus reducing error rates.(Beebe et al, 2002; Dittrich & Manz, 2006) Microfluidics technologies can be also easily automated to do routine assay and sample preparation, with the advantages of high-throughput assays, multistage automation, and parallel processing of multiple analytes.(Beebe et al, 2002) Finally, unlike conventional equipment, these miniaturized systems possess the dimensions and volume handling capacities to manipulate and sample single cells, which make them particularly attractive for biomedical applications.(Kim et al, 2008)

1.2 The need for highly integrated LOC devices: The new challenge in microfluidics

To date, extensive biological and chemical tasks are currently required outside the microfluidic devices to prepare and pre-process samples prior to their detection.

These tasks include sampling, pre-concentration, fluorescence labelling, filtration, mixing, reaction modules (for example, different heating zones), and analytes purification. Further, they usually involve high consumption of sample and expensive reagents, and require both instruments and time-consuming manual manipulations in which human error and contamination can be introduced.(Toner & Irimia, 2005) Therefore, there is increasing interest in downscaling these conventional tasks within LOCs devices and developing novel systems for preparing and detecting the tested samples "on chip". To this aim, integrated microfluidic systems have recently received much attention. However, their technological development requires the miniaturization of analytical techniques and the integration of multiple microcomponents with different functionalities onto the same chip. These integrated microfluidic systems would offer the possibility of reducing time and costs, as well as increasing throughput and automation.

To realize integrated, multifunctional LOCs, two main approaches can be pursued. The first one involves the fabrication of miniaturized functional units and their assembly level by level to achieve desired functionalities. Indeed, microfluidic devices have planar structures that can be easily integrated with mechanical, electronic, fluid functions and optical elements (Balslev et al., 2006; Verpoorte, 2003a) According to this approach, a great variety of microfabrication techniques have been developed for manufacturing microfluidic networks integrating advanced functions, such as pumps,(Laser & Santiago, 2004) valves,(Oh & Ahn, 2006) mixers,(Nguyen & Wu 2005) membranes,(de Jong et al., 2006) reactors(McCalla & Tripathi, 2011) and heating and cooling elements.(Chen et al., 2007) The discussion on the lab-on-chip technologies is beyond the scope of this review and the readers are addressed on recent reviews.(Abgrall & Gué, 2007; Fiorini & Chiu, 2005; Ng et al., 2002; Craighead, 2006) Several papers are also available on the integration of some laboratory tasks, including sample pre-processing(C.-Y. Lee et al., 2005; Kraly et al., 2009; Abdelgawad et al., 2009) and pre-concentration,(S. Song & Singh, 2006; Anson et al., 2006) filtration and separation,(Broyles et al., 2003; Freira & Wheeler, 2006) as well as fluorescence staining,(S.D.H. Chan et al., 2003; Buhlmann et al., 2003) chemical processing and analysis.(Sims & Allbritton, 2007)

A great variety of microchip fabrication techniques and materials are available for producing highly sophisticated two- and three-dimensional microstructures with integrated modules.(Erikson & D. Li, 2004) However, techniques developed to date according to this approach generally require complicated control systems and are by far from being trivial.

The second approach to achieve multifuctional LOCs is recently emerging and consists in incorporating stimuli-responsive hydrogels into microfluidic devices, to obtain smart functional components and simple, highly controllable systems.

Stimuli-responsive hydrogels are polymers demonstrating considerable changes in properties in response to small variations of environmental conditions. To date a variety of polymers have been developed that are sensitive to environmental and biochemical stimuli, including light, pH, temperature, electric and magnetic fields, chemical analytes and biological components.(S.-k. Ahn et al., 2008; Kumar et al., 2007; Chaterji et al., 2007)

The use of an external stimulus in LOC systems is appealing because it can be applied precisely in different sections within a microdevice, thus allowing the microfluidic functions to be integrated in a geometrical confinement. Finally, stimuli-responsive hydrogels have many of the advantages of polymeric materials, such as high response to weak stimuli, versatility, low cost, and compatibility with biological media.

In general, the formation of covalent or non-covalent bonds upon applying a certain stimulus induces changes in polymer chain's conformation. This could lead to some consequences such as the increase/reduction of both pore size and permeability to analytes, as well as significant changes in volume and hydrophilic/hydrophobic properties. These switching properties might be exploited to introduce into the chip "on-demand" changes in volume, permeability and surface chemistry, which in turn activate different functions such as the capability of molecular recognition (including capture, release and detection of biomolecules), autonomous flow rate and wettability switching. It is noteworthy that light and electric field are particularly attractive as stimuli for lab-on-chip applications, since they allow tight control with both high spatial and temporal resolution.

Multiresponsive surfaces are also appealing in microfluidics because they mimic the cooperativity of responsive systems found in nature.(Pasparakisa & Vamvakaki, 2011)

1.3 How to integrate smart polymers into microfluidics - The need of downscaling to micro/nano size

In the development of hydrogel-based microfluidic devices, an important issue is to downscale the hydrogel size up to micro- or nano-meter scale. This enables not only to obtain components with size comparable to that of microchannels, but above all to overcome the diffusion limitations suffered by larger hydrogels. Indeed, the hydrogel response is typically driven by diffusion of ions and molecules through the polymer network, according to the Donnan equilibrium. The time scales for the volumetric change is highly dependent on the distance of diffusion or the size of the hydrogel, therefore a decrease in hydrogel size at the micro/nanoscale can enhance the hydrogel response time up to few seconds or subseconds,(Eichenbaum et al., 1998; S.K. De et al., 2002) with interesting applications in microfluidic device.(Peppas et al., 2006)

Smart polymers have been essentially deposited as thin films(Tokarev & Minko 2009) and colloidal particles(Malmstadt et al., 2003, 2004; Peterson, 2005) on solid surfaces for microfluidics. Hydrogel structures have been mainly patterned using photolithographic approaches similar to those employed in semiconductor industry.(Ebara et al., 2006; Khademhosseini & Langer, 2007; Seong et al., 2002) On the other hand, colloidal particles are excellent candidates for the fabrication of stimuli-responsive surfaces and coatings.

A great variety of applications would ideally take advantage of the unique properties of such smart surfaces. Some examples include microfluidic flow control, (bio)sensors, microscale chromatographic separation, solid-phase extraction, biological assays, and cell culture.

The idea of this review is to present and discuss recent trends in the development of stimuli-responsive polymers for the design and fabrication of multi-integrated microfluidic devices. This review is structured in four major parts, which correspond to the main functions of smart polymers in integrated microfluidic devices: "Smart polymers as actuators", "Smart polymers as sensors", "Controlled absorption and release of (bio)molecules" and "Wetting properties of the microchannels". Finally, a paragraph is dedicated to summarize the immobilization methods of smart polymers and the structural forms by which they are used in microfluidic systems.

2. Smart polymers as actuators

Due to their switching properties, stimuli-responsive polymers act as elementary machines (actuators), able to convert environmental signals into a mechanical response. Their ability to undergo abrupt volumetric changes in response to the surrounding environment without the requirement of external power sources provides the polymeric integrated microfluidic components to be autonomous. To date, the autonomous functionality of these components has been achieved by exploiting mainly volume changes exhibited by pH- and temperature-responsive hydrogels.

Microfluidic actuators can be divided in those having micromechanical properties, such as microvalves, micromixers, and micropumps, and micro-optical properties, such as microlenses.

2.1 Micropumps and micromixers

In microfluidic systems, micromixers and micropumps are essential components for fluidic handling. Even though numerous designs and materials have been developed,(S.-H. Chiu et al., 2009; S.-M. Ha et al., 2009; Graf & Bowser, 2008; Tovar & A.P. Lee, 2009) many of these systems require on-chip power and costly and time-consuming fabrication processing.

In general, micropumps can be classified as either mechanical or non-mechanical. Mechanical micropumps need physical actuators or mechanisms to achieve pumping; they include electrostatic, piezoelectric, thermopneumatic and electromagnetic type.

Non-mechanical micropumps have the ability to transform non-mechanical energy into movement, so that the fluid can be driven. They include magnetohydrodynamic, electrohydrodynamic, electroosmotic, electrowetting, bubble type, electrochemical and evaporation-based micropump.(Tay, 2002; Nisar et al., 2008)

A diffusion micropump intended for low performance applications and a displacement pump which provides higher performance, both based on the temperature-sensitive hydrogel poly(N-isopropylacrylamide) (PNIPAAm) have been reported by Ritcher et al. In a diffusion pump, if the swollen PNIPAAm actuator is heated, it shrinks causing the solution to be released into the outlet and generate pumping pressure; when the hydrogel actuator is cooled down, it swells by absorbing the liquid, leading to the deformation of an elastic membrane which acts as pressure accumulator. In displacement micropumps, unlike in diffusion types, the polymeric actuator generates pumping pressure when it is swelled, and the pressure is reduced when the pumping fluid fills the chamber, i.e. when the hydrogel is in the shrunk phase. Generally, these polymeric micropumps have a simple electrothermic control by means of resistive heating elements, which locally heat the thermo-responsive hydrogel. Therefore, they can be easily integrated on lab-on-a-chip devices.(Richter et al., 2009)

Nestler and co-workers reported a micropump chip that employed a hydrogel based on poly(acrylic acid) sodium salt (PAAS) for generating gas molecules by electrolysis. As almost no heat is generated by electrolysis, this pumping principle is well suited for protein-sensing applications. The PAAS hydrogel was used as electrolyte instead of a liquid, and placed on electrodes. Then a voltage was applied to the electrodes, leading to gas generation by electrolysis. The generated gas was directly used to drive liquids through the channel system in a well-controlled manner.(Nestler et al., 2010)

Micropumps with simple structures and actuated by hydrogel drying have been demonstrated by Choi and co-workers. After exposure to ambient air, the gel block was dehydrated and became absorbent, leading to the suction of the liquid from the connected microchannel and its diffusion into the gel. As a result of the negative pressure, the liquid sample flowed through the microchannel. This kind of pump finds application in microchips that need slow but stable and long-lasting actuation, as well as in sampling proteins from viscous liquids such as blood without filtration or centrifugation.(Y.H. Choi et al., 2009)

Good et al. described a fluid-responsive micropump, which was activated by the addition of water to fluid-responsive polymeric particles. The addition of water induces a significant particle volume expansion and pushes a stored fluid from an adjacent reservoir at a predicted flow rate. The experiments were run on two particle systems to investigate how polymer properties can affect the rate and amount of fluid delivered. (Good et al., 2004)

Temperature-sensitive micropumps or micromixers, based on the concept of recirculation and able to autonomously decide when to pump fluid, were developed by Agarwal et al. These systems exploited a poly(HEMA-co-DMAEMA) hydrogel to control a Nickel rotor which could stop pumping as the temperature decreased, whereas for higher temperatures the pumping started. Because of the rotating actuation of the impeller, a change in pressure was induced within the microchannels, thus creating the driving force necessary to pump the liquid.(Agarwal et al., 2005)

An approach aimed to achieve cyclical changes without applying repeated stimuli is based on the use of Belousov–Zhabotinsky (BZ) reaction to control swelling and collapsing of hydrogels in a spatiotemporal manner. The BZ reaction is well known as a non-equilibrium dissipative reaction and generates autonomous oscillations in the redox potential. For example, Murase et al. fabricated gels by copolymerizing temperature-responsive N-isopropylacrylamide with ruthenium tris(2,20-bipyridine) ($Ru(bpy)_3$) as the catalyst for the Belousov–Zhabotinsky (BZ) reaction, and obtained autonomous peristaltic motion without external stimuli.(Murase et al., 2008)

Hara and co-workers reported another intriguing approach based on BZ reaction to achieve smart micropumps using a self-oscillating gel actuator, which generates a pendulum motion by fixing one edge of the gel.(Hara & Yoshida, 2008; Maeda et al., 2008)

Kwon and co-workers reported a valveless micropump system based on the electroactive hydrogel 4-hydroxybutyl acrylate (4-HBA), which shows extremely low energy consumption and high durability. Such pumping system integrated into a compact and portable microfluidic system should represent a promising candidate for an implantable drug delivery system.(G.H. Kwon et al., 2011)

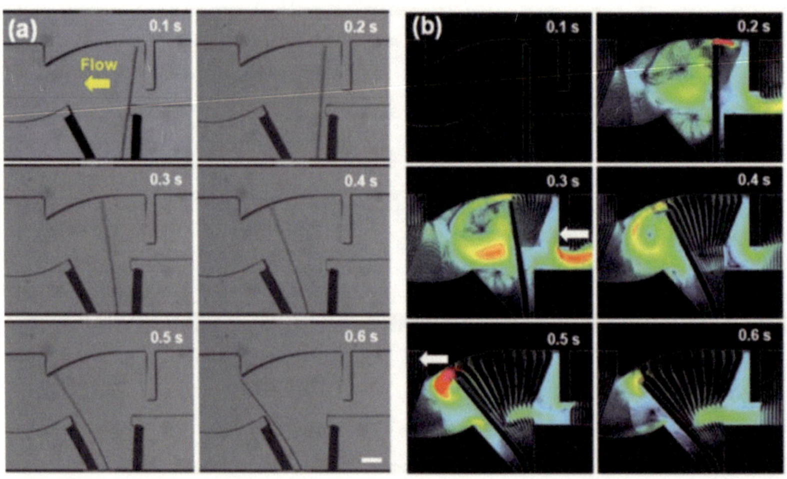

Fig. 2. Sequential photographs showing the gradual development of pumping at 1V. The scale bar was 500 µm. (b) microchannel shapes and control signals were simulated using computational fluid dynamics (CFD) analysis to improve the hydrodynamic performance. At 0.3 s and 0.5 s, arrowheads indicate strong flow. Reprinted from (G.H. Kwon et al., 2011).

2.2 Microvalves

Valves are one of the most crucial components for flow control in microfluidic systems. A conventional active valve consists of a deformable diaphragm coupled to an actuator that controls the on/off state. Micrometer-scale valves, more commonly known as microvalves, can offer important advantages over macroscale valves, as they can operate using small sample volumes, providing rapid response time and low power consumption.

The mechanical or electrical control of these systems relies on their intrinsic responsiveness to thermal, chemical, or electroptical stimuli. Therefore, smart polymers show promising ways of flow control for microfluidic devices based on stimuli-responsive properties.(Dong & Jiang, 2007; Eddington & Beebe, 2004) A number of hydrogel microvalves have been developed to date. For example, stimuli-responsive membranes with valve function have been reviewed by Yang et al.(Yang et al., 2011) Membrane valves were realized as a combination of a rigid porous membrane with a soft stimuli-responsive hydrogel, either by depositing a thin layer on the top surface of membrane or by suitable functionalization of the pores. The specific topology - hydrogel on top of or within the porous barrier - was important for the type of on/off response, as shown in Fig. 3.

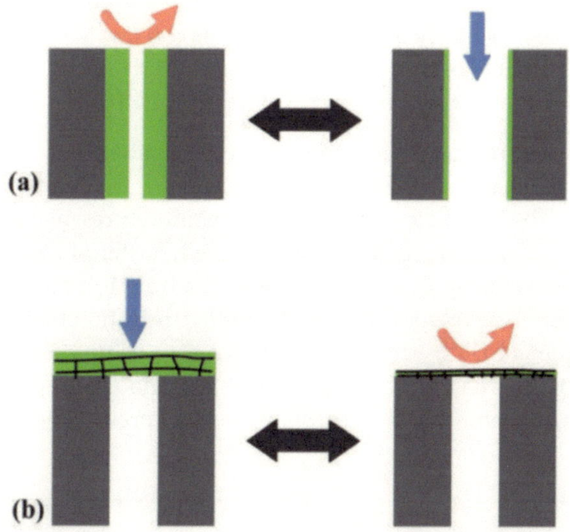

Fig. 3. Schematic overview on two types of membrane valves with stimuli-responsive hydrogel layer at different locations: (a) layer on the pore wall (hydrogel deswelling leads to increasing permeability); (b) crosslinked layer on the top of membrane covering the pore (hydrogel deswelling leads to decreasing permeability). Reprinted from (Yang et al., 2011)

Thus, the selection of the appropriate hydrogel can lead to the design of membranes responsive to different stimuli such as temperature, pH, light, and magnetic forces.

Geiger et al. presented a highly functional microfluidic device exhibiting an integrated thermally sensitive hydrogel valve. The valve was normally closed at room temperature. Upon heating above the lower critical solution temperature (LCST) of 32°C, the polymer valve became hydrophobic, shrank and formed large pores, thus allowing the solution to flow.(Geiger et al., 2010)

Chunder et al. developed a superhydrophobic/hydrophilic switchable surface through the combination of layer-by-layer self-assembly and microfabrication techniques. This smart surface realized in a microfluidic channel could therefore act as thermosensitive valve able to control fluid flow by changing the temperature.(Chunder et al., 2009)

Chen et al. developed a light-actuated microvalve by photoinitiated patterned polymerization of NIPAAm within microfluidic channels. In contrast to conventional microvalves activated by heater elements, the novelty of this system relies on heating from the absorption of light provided by a quartz halogen illuminator. The mechanical strength and the resistance to pressure of these microvalves can be tuned by properly choosing the amount of monomer and crosslinker.(G. Chen et al., 2008)

Electronically controllable microvalves based on temperature sensitive hydrogel actuators have been described by Richter et al. Because of the direct placement into the channel, the elastic properties of the hydrogel actuator were exploited to improve the pressure insensitivity, achieve high particle tolerance and avoid a leakage flow.(Richter et al., 2003) However, the utility of electrically driven hydrogel actuators is severely limited in some applications by biocompatibility issues and hydrolytic bubbles generation when the electric field strength exceeds 1.2 V. In particular, bubbles distort the electrical ion flux and degrade the fidelity of the system. To prevent this problem, Kwon and co-workers implemented an electro-responsive hydrogel sorter based on 4-Hydroxybutyl acrylate (4-HBA), which was able to operate at low driving voltages.(G.H. Kwon et al., 2010)

Beebe et al. designed and realized microvalves by either patterning a pH-sensitive hydrogel along the walls of a channel or creating an array of the same hydrogel inside the microfluidic devices. The swelling of these hydrogel structures blocked a channel when a high pH solution flowed into the channel, whereas when the pH value was appropriately decreased, the contracted state of hydrogel allowed the fluid to pass.(Beebe et al., 2000; R.H. Liu et al., 2002) Kim et al. developed a hydrodynamic fabrication method for pH-responsive microspheres housed in a PDMS-based microfluidic valve, as reported in Fig. 4. The analysis of volume-changes by alternating application of acidic and basic solutions showed a large and fast volume transition, which was stable and reproducible even under repeated motions.(D. Kim et al., 2007)

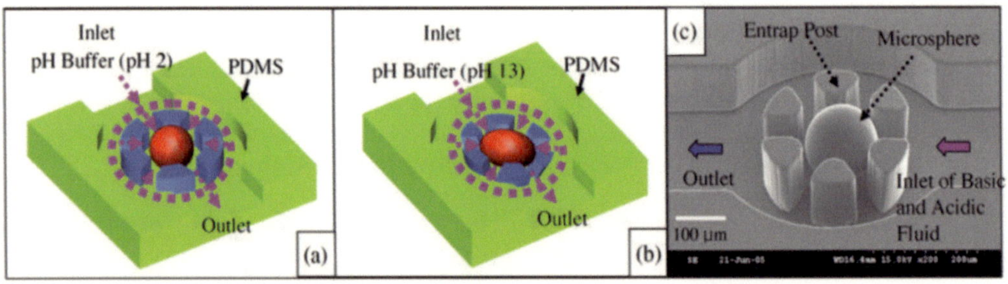

Fig. 4. (a) Conceptual schematic (three-dimensional) of the shrinking motion of the microsphere inside the entrap posts (valve is in the "On" state). (b) Conceptual schematic (three dimensional) of the swelling motion inside the entrap posts (valve is in the "Off" state). (c) SEM image of the microsphere positioned inside the entrap posts. Reprinted from (D. Kim et al., 2007)

2.3 Microlenses

Micro-optical components such as microlenses have been recently benefited from the use of stimuli-responsive hydrogels. In contrast to traditional optical systems, microlenses based on smart polymers have the potential to allow for autonomous focusing without the need for mechanical parts, and above all to achieve a higher degree of integration with other optical components. To date, microlenses have been developed that are actuated by hydrogels responsive to temperature, pH and light. Dong and his colleagues constructed smart liquid microlenses able to adjust their shape and focal length by taking advantage of temperature and pH-responsive hydrogels. Microlenses were integrated into the chip by filling the microchannel with water and oil, so that the water-oil interface formed the microlens, and placing a stimulus-sensitive hydrogel ring between a glass plate and an aperture ring. The hydrogel ring was able to control the curvature and the focal length of the water-oil meniscus upon environmental changes.(Dong & Jiang, 2007a, 2007b; Dong et al., 2006)

Gold-PNIPAAm nanocomposites have been also demonstrated to be tunable light-sensitive microlenses. In an attractive demonstration, multiple micropost structures of gold-PNIPAAm hydrogels were patterned around a lens aperture. The hydrogel swelling state was controlled by IR light irradiation, which in turn changed the curvature of liquid-liquid interface and, thus, the focal length of the lens. Light controlled microlenses have the potential to replace current technology that uses mechanical or electrical signals.(Zeng & Jiang, 2008)

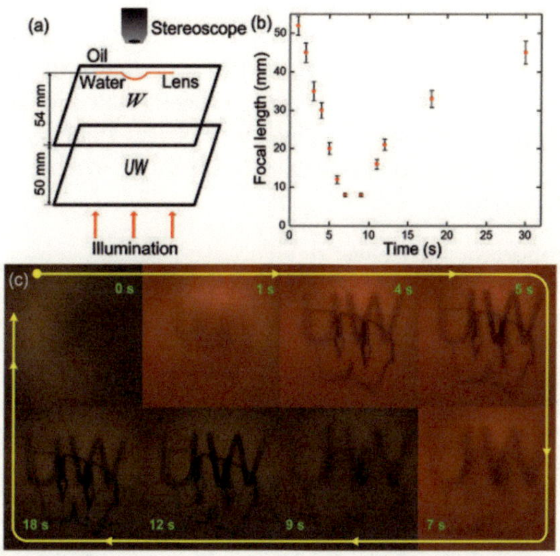

Fig. 5. (a) Schematic of scanning image planes using a liquid tunable microlens. Two logos, W and UW, are printed on transparency films and are 54 and 104 mm, respectively, below the glass substrate with the microlens. A CCD-coupled stereoscope is placed above the microlens to monitor and record the images. (b) Dynamic change in the positive focal length of the microlens (convergent) in one scanning cycle as a function of time. (c) Frame sequence of the focused images in one scanning cycle obtained by tuning the microlens. Reprinted from (Zeng & Jiang, 2008).

3. Smart polymers as sensors

A biosensor can be viewed as a combination of a selective detection/recognition unit, a transducing unit and a readout part. The detection unit is designed to react in the presence of the desired analyte. The function of the transducing part of the biosensor is to convert the presence of the relevant analyte into a measured output.(Sugiura et al., 2009) Transduction systems can be divided in three categories: electrochemical, mechanical and optical.

Within the field of biosensors, hydrogels have been applied for two main purposes: to increase the loading capacity of an analyte, and/or to take advantage of hydrogel specific properties including swelling volume, weight, mechanical properties, and refractive index.

Electrochemical sensors represent an important subclass of chemical sensors in which an electrode is used as the transduction element. In this case, hydrogels are mainly exploited as supports for biomolecules, to increase the loading capacity or to stabilize the embedded bioactive material on electrode surface.(Heller, 2006; B. Yu et al., 2008) Some examples of electrochemical sensors are reported in literature. Fernandez-Barbero et al. exploited cross-linked PAA microgels to immobilize Glucose Oxidase (GOx), and studied the modifications induced by the enzyme in the microstructure of the microgels. The use of microgels with entrapped GOx as biological component of an amperometric biosensor allowed evaluating the enzymatic activity of the GOx entrapped in the polymer network. The electrode was prepared by depositing microgel particles on the surface of a platinum electrode, which was then covered with a dialysis membrane to fix the microgels to the electrode.(Fernandez-Barbero et al., 2009) Huang et al. realized an affinity biosensor based on changes of dielectric properties of a polymer in response to its specific, reversible binding with an analyte. The approach is demonstrated using a synthetic polymer with specific affinity to glucose. The presence of glucose in the tested solution induces changes in the permittivity of the polymer, which can be correlated with specific glucose detection.(X. Huang et al., 2010) Hydrogels of ferrocene-modified acrylamide(Calvo et al., 1993) and poly(vinyl pyridine) containing an osmium complex to electrically wire GOx to carbon electrodes(Gregg & Heller, 1991) were also developed.

Sridhar & Takahata reported a wireless device based on *L*–*C* circuitry that consists of a variable inductor and a fixed capacitor, coupled with a poly(vinyl alcohol)–poly(acrylic acid) hydrogel for biomedical and chemical sensing.(Sridhar & Takahata, 1993)

A recent and comprehensive review reported on electroconductive hydrogels, polymeric blends or co-networks (that combine inherently conductive electroactive polymers with highly hydrated hydrogels) and their applications as biorecognition membranes for implantable biosensors.(Guiseppi-Elie, 2010)

Mechanical transducing systems which involve stimuli-responsive polymers exploited changes in volume, weight, mechanical properties and distribution of mechanical stresses. These changes occurring in polymer materials in response to either chemical/physical stimuli or binding events are converted into readable electrical signals. Most studies have been conducted on polymers in an equilibrium state at constant pressure. Under such conditions, the hydrogel polymeric network can expand freely to accommodate an imbalance in osmotic pressure between the inside and the outside of the gel and reach a new equilibrium state. However, if the gel is encapsulated inside a solid structure, the polymer network will exert a pressure over its casing.(I.S. Han et al., 2002)

Guenther and co-workers realized chemical sensors containing PNIPAAm-based hydrogels as chemo-mechanical transducers. To this aim, thermo-shrinking photo cross-linkable polymers were deposited onto the backside of commercially available pressure sensor chips with a flexible thin silicon membrane. The hydrogel swelling leads to a bending of the silicon membrane and a therefore to an electrical output signal.(Guenther et al., 2007, 2008)

Trinh et al. reported the time-dependent response of a chemo-mechanical sensor by placing a pH-responsive hydrogel to the backside of a piezoresistive membrane, and estimated the degree of hydrogel constraint within the sensor.(Trinh et al., 2006)

A new type of sensor which combines piezoresistive-responsive elements as mechano-electrical transducers and the phase transition behaviour of hydrogels as a chemo-mechanical transducer has been recently developed. The sensor consists of a pH-responsive poly(acrylic acid)/poly(vinyl alcohol) (PAA/PVA) hydrogel and a standard pressure sensor chip.(Sorber et al., 2008; Trinh et al., 2006)

Herber et al. presented a sensor based on a pH-sensitive hydrogel for the detection of carbon dioxide gas inside the stomach, in order to diagnose gastrointestinal ischemia.(Herber et al., 2003, 2004, 2005a, 2005b)

Further, a novel hydrogel piezoresistive sensor array has been developed for *in vitro* monitoring of pH, ionic strength, and glucose concentration. The sensor consists of three components: hydroxypropyl methacrylate (HPMA), N,N-dimethylaminoethyl methacrylate (DMA) and the crosslinker tetra-ethyleneglycol dimethacrylate (TEGDMA) hydrogel. (G. Linet al., 2009; Orthner et al., 2010a, 2010b)

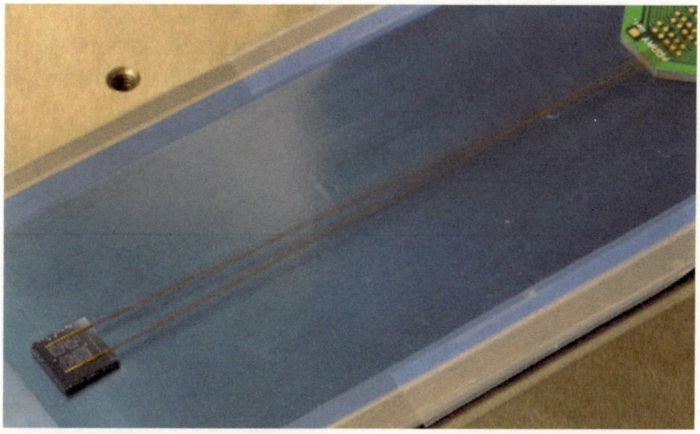

Fig. 6. Photograph of topside of a solid diaphragm sensor assembly with hydrogel inserted and mounted to a perforated backing plate (out of view). Wire bonding was performed using gold insulated wire (d=50 μm). Reprinted from (Orthner et al., 2010b)

Tierney et al. bound a hydrogel to the tip of an optical fiber constituting the environmental sensing element for high-resolution detection. When light was sent through the fiber, it was reflected at the fiber–gel and gel–solution interfaces. The interference wave generated by reflected light enabled the detection of the optical path length within the gel and therefore the degree of gel swelling. These sensors were used in solutions of varying ionic strength and pH, showing a good reproducibility and resolution in the millimolar range.(Tierney et al., 2008)

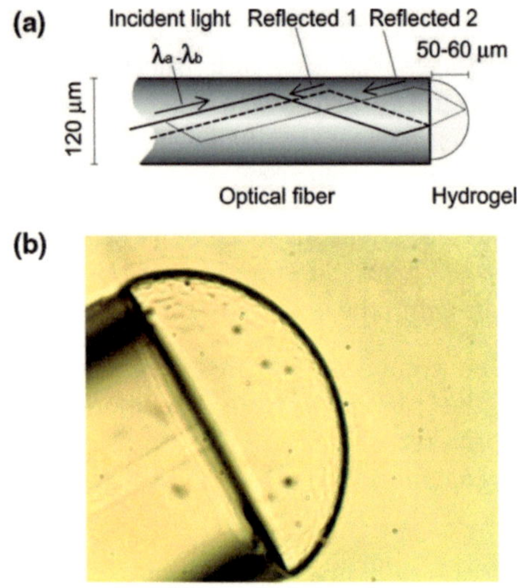

Fig. 7. (a) Schematic illustration of the instrumental technique. The incident light (wavelength range λ_a–λ_b from 1530 to 1560 nm) sent through the optical fiber (diameter 120 μm) is reflected at both the fiber–gel (reflected 1) and the gel–solution (reflected 2) interfaces. The interference wave is the basis for the measuring technique. The hydrogel covalently linked to the end of the optical fiber adopts a near half-spherical geometry with radius of the order 50–60 μm. (b) Micrograph of a polymerized gel attached to a fiber. Reprinted from (Tierney et al., 2008)

Recently, great interest has been shown in surface plasmon resonance (SPR) techniques to detect small changes in refractive indexes close to metallic surfaces. In these systems, changes in the volume density of the hydrogel due to interaction between the analyte and the polymeric network affect the refractive index at the interface.

Huang et al. implemented evanescent wave affinity biosensors based on hydrogel binding matrix for ultra-sensitive detection of molecular analytes. Highly swollen carboxylated poly(N-isopropylacryamide) (NIPAAm) hydrogel was grafted to a sensor surface, functionalized with antibody recognition elements and employed for immunoassay-based detection of target molecules in the tested sample. Molecular binding events were detected by long range surface plasmon (LRSP) and hydrogel optical waveguide (HOW) field-enhanced fluorescence spectroscopy. The results demonstrated that the analyte detection was improved by properly design the hydrogel structure.(C.J. Huang et al., 2010)

Hydrogels have been also integrated in microlenses for biosensor application. In the system described by Kim and co-workers, microlenses are formed by direct absorption of microgel on a substrate surface; upon absorption, the microgel adopts a lens shape. Changes in the hydrogel swelling affect the shape of the lens and, hence, its focal length. Microlenses made of poly(N-isopropylacrylamide-co-acrylic acid) (PNIPAAm-co-AAc) were properly functionalized to detect antibody-antigen binding, which resulted in an increased local refractive index of the hydrogel matrix. The change in optical properties of the gels can be measured qualitatively: first by the appearance of 'dark rings' in the lenses and second by using the hydrogel lenses to focus a square image; the higher the concentration of analytes, the larger the increase in refractive index, the more focused the image.(J. Kim, 2006, 2007)

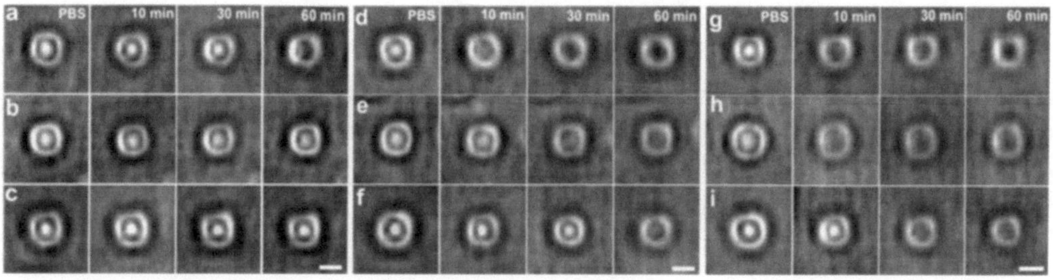

Fig. 8. Microlens response as a function of sensitivity (*i.e.*, bound antibody concentration). Microlens were incubated and photo-cross-linked in anti-biotin solutions with concentrations of 0.7 μM (top row a, d, g), 1.0 μM (middle row b, e, h), and 2.0 μM (bottom row c, f, i). Contact times are indicated at the top of each column, with the initial image in PBS shown in the first column. The scale bar is 2 μm. Reprinted from (J. Kim, 2007)

Asher and co-workers have developed attractive approaches to make hydrogel based sensors. In one example, they realized a 3-D polymerized colloidal crystal array (PCCA) containing hydrogel sensing units functionalized with molecular recognition agents to detect analytes such as creatinine,(Sharma et al., 2004) glucose,(Alexev et al., 2004) pH,(K. Lee & Asher, 2000) and organo-phosphorus compounds.(Walker & Asher, 2005; Tokarev & Minko, 2010) Afterwards, they reported a new sensing approach that exploited 2-D monolayer arrays of particles attached to molecular recognition polymer hydrogel networks. The 2-D array of photonic crystals, because the hydrogel network swells or shrinks in response to analyte concentration changes, enables to visually identify and quantitate chemical species inside the sample.(K. Lee & Asher, 2011)

4. Controlled absorption/release of (bio)molecules within microfluidic devices

In the development of integrated microfluidics, it is very appealing the miniaturization of multiple components able to achieve separation, purification, and delivery of (bio)molecules on-demand. For instance, many emerging LOC systems – including protein and cell analyses - usually require the reversible, controlled absorption and release of both chemical analytes and biomolecules. Accordingly, the development of active coatings for the uptake and release of (bio)molecules on-chip in response to external stimuli would provide novel functionalities within microdevices towards fully automated LOC systems for sophisticated biomedical analyses.

To this aim, smart polymers have found many applications in microfluidic systems due to hydrophilicity and biocompatibility properties, a wide range of available polymeric materials and the relative ease of chemical modification. Indeed, hydrophilic smart polymers, once immobilized on a solid support, form a soft three-dimensional (3D) matrix able to encapsulate bioactive molecules, provide them with a protective environment and preserve their native status. Above all, they are appealing for microfluidics because of their ability to change their volume and permeability in response to environmental stimuli, which makes them able to encapsulate and release biomolecules on demand, as well as to separate or detect the target analytes.(Peppas et al., 2006)

However, in the development of microfluidic diagnostics, only few works reported the use of stimuli-responsive polymers for the encapsulation and release of (bio)molecules.

For example, the functionalization of microchannel walls with smart polymers could lead to reversible hydrophilic-hydrophobic surfaces, which allows the physical or covalent immobilization of biomolecules such as proteins. This property has been exploited to integrate protein separation/pre-concentration on chip. Huber et al. developed a microfluidic device in which proteins were adsorbed from solution with negligible denaturation, and released on demand. The key element in this device was an end-tethered monolayer of PNIPAM, which was thermally activated from a hydrophilic state to a hydrophobic, high protein adsorbing state. The PNIPAM film was realized by *in situ* polymerization onto surfaces previously functionalized with silane coupling agents. Further, heating/cooling transitions were controlled by realizing an array of gold or platinum heater lines deposited onto a silicon nitride membrane.(Huber et al., 2003)

The immobilization of biomolecules by means of stimuli-responsive coating would also allow the implementation of smart protein nanoarrays with well-defined size, shape and interfeature spacing. In particular, Hyun et al. fabricated elastin-like polypeptide (ELP) nanostructures grafted onto ω-substituted thiolates, which in turn were patterned onto gold surfaces by dip-pen nanolithography. These structures underwent a reversible hydrophilic-hydrophobic phase transition in response to environmental changes such as temperature or ionic strength. According to this phase transition behaviour, two different proteins, namely a thioredoxin-ELP fusion protein and its related monoclonal antibody, were surface-captured, thus forming a nanoarray.(Hyun et al., 2004)

It is noteworthy that switchable hydrophilic-hydrophobic surfaces have been employed to capture and release not only biomolecules, but also smart nanobeads carrying stimuli-responsive polymers on the surface. Malmstadt et al. described an interesting approach to graft temperature responsive PNIPAM onto the nanobeads, which underwent a temperature-induced phase transition from hydrophilic to hydrophobic. The nanobeads were blocked into a microchannel using temperatures above the lower critical solution temperature (LCST), whereas they were released and detected upon applying lower temperatures.(Malmstadt et al., 2003, 2004) The flexibility of this system allowed the reversible immobilization of biomolecules for separation and detection in microfluidic devices. Several applications of smart nanobeads include microfluidic affinity chromatography and diagnostic immunoassays. Similarly, Ebara et al. functionalized nanobeads with both stimuli-responsive polymers and affinity moieties, such as antibodies. Further, microchannel walls were modified with suitable stimuli-responsive polymers, to ensure a more selective and efficient capture of functionalized nanobeads. (Ebara et al., 2006)

Micro- and nano-particles represent one of the most promising systems for the uptake and release of biomolecules in microfluidics. Thanks to their high surface area per unit volume, they ensure enhanced transport and binding properties.(Verpoorte, 2003a) Nowadays, microparticles have been mainly deposited as close-packed monolayers using layer-by-layer (LbL) assembly for uptake and release of biomolecules and therapeutics. Several instances of LbL multilayer films are reported in literature.(Lynn, 2006; L. Wang et al., 2008; Serpe et al., 2004) Films containing microgels can be swollen in response to external stimuli, which can facilitate the incorporation of guest materials within their polymer network. For example, a new type of microgels of poly(allylamine hydrochloride) (PAH) and dextran (D) was alternatively deposited with polyanion poly(styrene sulfonate) (PSS) by Wang et al., to produce multilayer films of PAH-D/PSS able to reversibly load and release negatively charged guest materials. The PAH-D microgel could be deposited on both hydrophilic and hydrophobic surfaces, such as quartz, polytetrafluoroethylene, polystyrene, etc. without any substrate modification. (L. Wang et al., 2008)

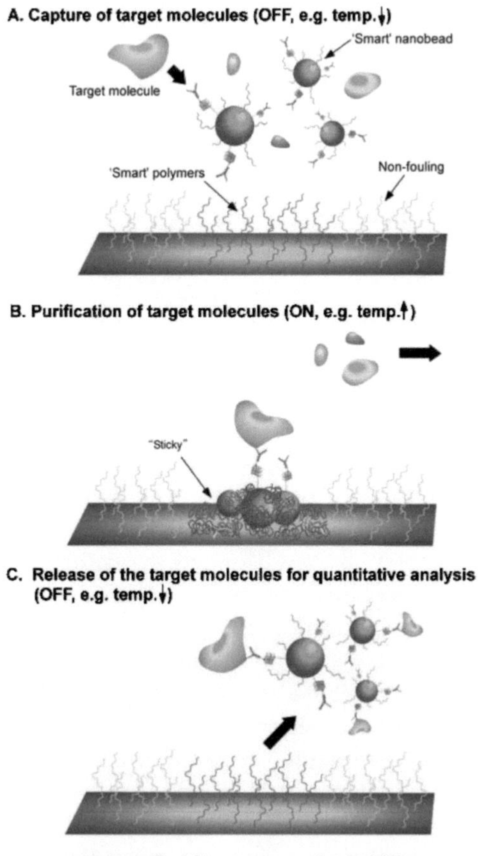

Fig. 9. Design concept for smart microfluidic system that utilizes smart nanobeads. Reprinted from (Ebara et al., 2006)

A functional use of microgels in LOCs requires that they have to be fixed within the active regions of the device channels and not flow out during the analysis. Therefore, we recently reported a novel method for *in situ* pH-controlled cell staining by means of covalently immobilized poly(methacrylic acid) (PMAA) microgels. The PMAA microgels were synthesized in solution and covalently immobilized on a glass substrate previously modified with an organosilane agent and an amine-terminated PEG. The immobilized PMAA microgels preserved their pH-sensitivity, thus making them suitable for the pH-controlled uptake of an oligothiophene-conjugated anti-human CD4 monoclonal antibody (MAb). This smart support was incubated with a Jurkat T-cell suspension, the physiological pH of the extracellular environment induced the release of the labeled MAbs, and the highly selective staining of the CD4-positive subpopulations within the Jurkat cell suspension was demonstrated by confocal microscopy.(Argentiere et al., 2009)

5. Wetting properties of the microchannels

Smart polymers are key elements to achieve switching wettability in microfluidics. When the proper stimulus is applied, they undergo a a reversible change from hydrophilic to hydrophobic, resulting in the wettability switching on a properly engineered surface. This reversible transition might be exploited in many applications, such as cell detachment and cell harvesting, and to produce self-cleaning membranes/filters within microdevices. An intriguing approach to realize switchable hydrophilic-hydrophobic surfaces has been developed in the past years and consists in grafting smart polymers onto chemically modified surfaces. Actually, the most exploited stimuli in the development of such smart coatings are temperature and light.

5.1 Thermoresponsive wettability

Poly(N-substituted) acrylamides are still being the most investigated family of thermoresponsive polymers. Their temperature-sensitive behaviour has been widely investigated and is dependent on the formation of intramolecular hydrogen bonds between C=O and N-H groups in polymer chains.(G. Chen & Hoffman, 1995) Poly(N,N'-diethyl acrylamide) (PDEAAm), poly(N-(DL)-(1-hydroxymethyl) propylmethacrylamide) (P(DL)-HMPMA), poly(dimethylaminoethyl methacrylate) (PDMAEMA) and poly(N-vinyl caprolactone) (PVCL) exhibit lower critical solution temperature (LCST) in a range of 25–32°, 35°, 50° and 35°C respectively.(Idziak et al., 1999; Qiu & Park, 2001; F. Liu & Urban, 2008; Aoki et al., 2001) However, poly(N-isopropylacrylamide) (PNIPAAm) is still being the most studied in the family of poly(N-substituted acrylamide), especially in biomedical applications, due to its sharp transition range as well as the close proximity of its LCST (around 32°C) to physiologically relevant temperatures. To date, the most common method for grafting PNIPAAm on inert microchannel surfaces such as silicon, glass, gold and polystyrene is the benzophenone-initiated photopolymerization.(D. Ma et al., 2009) Cheng et al. developed PNIPAAm based microheaters, able to carry out a selective protein binding as a function of temperature. They demonstrated that PNIPAAm films switched between adhesive and non-adhesive under the control of individual heaters. This localized change in the surface adhesive behaviour has been used to direct site-specific cell attachment.(X. Cheng et al., 2004) Further, the thermo-switchable hydrophobic/hydrophilic properties of PNIPAAm-grafted-PDMS microchannels has been exploited to develop novel techniques for cell detachment at room temperature and without trypsin digestion.(D. Ma et al., 2010)

Wetting properties have significant effects on liquid behaviour in microfluidic systems.(Handique et al., 1997) In general, a surface effect is the basis of both valve and pump functions.(Delamarche et al., 1997) Thus, the modification of channel walls with smart, reversible polymers leads to useful properties, such as reversible hydrophilic-hydrophobic surfaces, and to the development of novel microcomponents with actuating functions. (Idota et al., 2005, 2006). These applications are described in details in Section 1.

It is important to highlight that the switchable wettability of PNIPAAm grafts, in particular for multiscale micro- and nano-structures, is closely related to the substrate roughness. The introduction of rough structures into micro-areas is difficult, for this reason the enhancement of PNIPAAm performance on flat substrates, especially for applications at the microscale, represents an important challenge. For example, Fu et al. grafted PNIPAAm onto porous anodic aluminum oxide. The obtained nanostructured surfaces were demonstrated to undergo simultaneous changes in roughness and wettability by varying the temperature.(Fu et al., 2004) Finally, Sun and Qing reported that the behaviour of PNIPAAm brushes are significantly affected by the grafting density on the surface. To this aim, they introduced inert molecules, such as heptadeca-fluorodecyltrimethoxysilane, onto the surfaces, to dilute the initiator concentration in the atom-transfer radical polymerization (ATRP) process.(Sun et al., 2005)

5.2 Photoresponsive wettability

The approach of photoresponsive wettability relies on the use of photochromic molecules able to undergo reversible changes in their polarity when irradiated at specific light wavelength. For instance, Caprioli et al. demonstrated that the wettability of microchannels functionalized by cyclic olefin copolymers can be switched by irradiation. Such characteristic can be exploited to enhance the flow rate of fluids in microfluidic channels by on-off valve behaviour, allowing or blocking the liquid filling process on the base of optical control.(Caprioli et al., 2007) The reversible photoswitchable wettability of organic nanofibers based on 1',3'-dihydro-1',3',3'-trimethyl-6-nitrospiro[2H-1-benzopyran-2,2'-(2H)-indole] has been studied by Di Benedetto and co-workers. The switching properties of the nanofibers were investigated by studying the evolution of absorbance with different UV exposure times, photoluminescence lifetime, cyclic photoisomerization, and reversible wettability.(Di Benedetto et al., 2008) Athanassiou and co-workers prepared a nanopatterned spiropyran-doped polymer surface and observed the impact of surface roughness on surface wettability by alternation of UV and green laser irradiation.(Athanassiou et al., 2006) The contact angle (CA) changes were amplified on the nanopatterned grating surface, and either surface hydrophobicity or hydrophilicity was enhanced on an optimized pattern size. Spiropyran-based responsive surfaces have also attracted much attention in the exploration of stimuli-responsive surface wettability. Rosario et al. used a photoresponsive spiropyran monolayer to coat a silicon nanowire surface. Upon irradiation from UV (366 nm) to visible wavelengths (450–550 nm), the CA increased to 23° from 12° on the flat surface.(Rosario et al., 2004)

6. Immobilization methods

In microfluidic applications, smart polymers must be kept within the channel and not flushed out during analysis. Accordingly, a multitude of approaches have been developed to introduce and fix them into microfluidic devices. Ideally, smart polymers should be fixed using cheap, robust and reliable methods.

Photopolymerization has been successfully used for smart polymers immobilization. For example, multiple photopolymerized gel patches have been prepared within a microchannel, and each patch can be used as a pH sensor or to analyze enzymatic activity.(Zhan et al., 2002; Q.M. Zhang et al., 2002) PNIPAM hydrogels have been also photopolymerized in microfluidic channels, resulting in the fabrication of thermally responsive devices.(Kuckling et al., 2002; Harmon et al., 2003)

To date, smart polymers have been employed in microfluidic systems in different forms such as particles, films, membranes, fibers and monoliths. Micro- and nano-particles have been used extensively in microfluidics to produce or enhance surface coatings.(Verpoorte, 2003b) Colloidal particles have been also suspended in the medium to obtain logic control.(L. Wang et al., 2010) Actually, there are essentially two methods in order to keep fixed the beads within a microfluidic device. According to the first one, smart microparticles are trapped within a chamber on the microfluidic device. The second method involves immobilizing beads onto a surface, typically using self-assembly methods, such as layer-by-layer (LbL), and covalent immobilization.(Peterson, 2005) For instance, functionalized microparticles may be fixed onto surfaces using hydrophobic/hydrophilic interactions or assemblies of oligonucleotides.(T.T. Huang et al., 2003; McNally et al., 2003) A number of different methods have been reported for trapping particles into a microdevice, some using physical barriers to retain beads, others using magnetic fields to manipulate and trap beads.(Linder et al., 2002; J.-W. Choi et al., 2001) An interesting entrapment method is reported in the previous section by Malmstadt et al.(Malmstadt et al., 2003, 2004) As reported by Andersson et al., a series of pillars could constrain beads placed in a microfluidic chip and this approach was useful in fabricating devices for single-nucleotide polymorphism analysis and capillary electrochromatography.(Andersson et al., 2000)

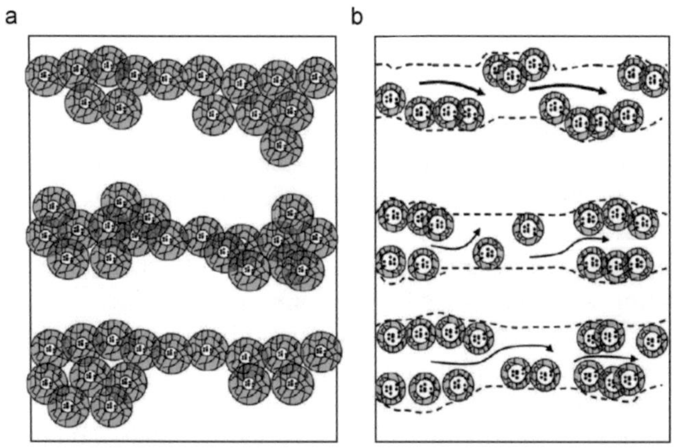

Fig. 10. Schematic representation of channels made of magnetic polystyrene latex covered with PNIPAAm and built in the PVA gel matrix: (a) "off" state below the collapse transition temperature; (b) "on" state above the collapse transition temperature. (Kumar et al., 2007)

As alternative to particles, smart polymers may be placed within microfluidic systems as films. Multilayer films of smart polymers with precise control of thickness and composition are fabricated generally by LbL deposition, a very versatile and convenient method.

The electrodeposition is another technique for the immobilization of films onto the conducting part of the surface, with the additional benefit of spatially selective cathode patterning (electropatterning).(Shacham et al., 1999)

Alternative smart polymeric supports for separating and detecting biomolecules in microfluidic systems are stimuli-responsive membranes.(Xue et al., 2008; Wandera et al., 2010) They are characterized by pores that can open and close by electronic induced stimuli such as electric or magnetic fields.(Csetneki et al., 2006) Membranes incorporating specific recognition elements after detecting the presence of a specific substance or a specific stimulus were able to translate this signal into a change of physico-chemical properties (*i.e.*, changes in the permeability).(Kumar et al., 2007) Smart hydrogel membranes functionalized with enzymes can recognized glucose, which in turn actuates other stimuli-mediated changes (*e.g.*, pH in the case of GOx-mediated glucose degradation).(Chaterji et al., 2007) Biocatalytic systems containing stimuli-responsive hydrogel membranes represent another promising approach for the uptake and release of biomolecules and drugs by receiving and processing the biochemical information.(Tokarev, 2009)

Stimuli-responsive electrospun fibers represent appealing and highly versatile tools for microfluidic devices.(Dalton et al., 2002; Jeong et al., 2004) Cylindrical polymeric structures composed of smart polymers can be employed for the encapsulation and detection of biomolecules.(C. Huang et al., 2011) Different kinds of smart fibers can be obtained, depending on the responsive polymers used. Jeong et al. fabricated fibers by employing 3-D multiple stream laminar flow and "on the fly" photopolymerization.(Tao, 2011)

Smart polymers have been also placed within porous monoliths.(Svec et al., 2003) Monoliths have been fabricated in specific areas of a microchip via photolithographic techniques based on UV-initiated polymerization methods.(Peterson et al., 2003) Smart gels and polymer monoliths have been shown to be very versatile in producing microfluidic systems. (Peterson, 2005) For instance, NIPAM was incorporated into the monolith and polymerized to realize temperature-responsive systems.(C. Yu et al., 2003)

7. Conclusion

Today, there is a growing need to integrate biological and chemical tasks within microfluidic devices, in order to achieve multifunctional systems. An overview of the applications of stimuli-responsive polymers in microfluidics literature is here presented. Stimuli responsive polymers, also known as "smart" polymers, have recently found many applications in integrated microfluidic systems due to their properties, a wide range of available polymeric materials and the relative ease of chemical synthesis. Modification of the channel walls with smart polymers leads to multilayers structure capable to encapsulate and release (bio)molecules on demand, undergo micromechanical changes for actuating and sensing functions, modify the wetting properties of the microchannels to realize for example valving elements, and generally integrate many functions in one chip, thus producing "smart" multifunctional devices.

8. References

Abdelgawad, M., Watson, M.W.L., Wheeler, A.R. (2009). Hybrid microfluidics: A digital-to-channel interface for in-line sample processing and chemical separations. *Lab Chip*, Vol. 9, No. 8, pp. 1046-1051.

Abgrall P., Gue´, A.-M. (2007). Lab-on-chip technologies: making a microfluidic network and coupling it into a complete microsystem—a review. *J. Micromech. Microeng.*, Vol. 17, pp. R15–R49.

Agarwal, A.K., Sridharamurthy, S.S., Beebe, D.J. (2005). Programmable Autonomous Micromixers and Micropumps. *Journal of Microelectromechanical Systems*, Vol. 14, pp. 1409-1421.

Ahn, S.-k., Kasi, R.M., Kim, S.-C., Sharma, N., Zhou, Y. (2008). Stimuli-responsive polymer gels. *Soft Matter*, Vol. 4, pp. 1151–1157.

Alexeev, V.L., Das, S., Finegold, D.N. & Asher, S.A. (2004). Photonic Crystal Glucose-Sensing Material for Noninvasive Monitoring of Glucose in Tear Fluid. *Clin. Chem.*, Vol. 50, pp. 2353 - 2360.

Andersson, H., van der Wijngaart, W., Enoksson, P. & Stemme, G. (2000). Micromachined flow-through filter-chamber for chemical reactions on beads. *Sensors and Actuators B: Chemical*, Vol. 67, No. 1-2, pp. 203-208.

Anson V. Hatch, Herr, A.E., Throckmorton, D.J., Brennan, J.S., Singh, A.K. (2006). Integrated Preconcentration SDS-PAGE of Proteins in Microchips Using Photopatterned Cross-Linked Polyacrylamide Gels. *Anal. Chem.*, Vol. 78, pp. 4976-4984.

Aoki, T., Muramatsu, M., Torii T., Sanui K. and Ogata N. (2001).Thermosensitive phase transition of an optically active polymer in aqueous milieu. *Macromolecules*, Vol. 34, pp. 3118-3119.

Argentiere, S., Blasi, L., Ciccarella, G., Cazzato, A., Barbarella, G., Cingolani, R., Gigli, G. (2009). Smart surfaces for pH controlled cell staining. *Soft Matter*, Vol. 5, No. 21, pp. 4101-4103.

Athanassiou, A., Lygeraki, M.I., Pisignano, D., Lakiotaki, K., Varda, M., Mele, E., Fotakis, C., Cingolani, R. & Anastasiadis, S.H. (2006). Photocontrolled Variations in the Wetting Capability of Photochromic Polymers Enhanced by Surface Nanostructuring. *Langmuir*, Vol. 22, pp. 2329- 2333.

Balslev, S., Jorgensen, A.M., Bilenberg, B., Mogensen, K. B., Snakenborg, D., Geschke, O., Kutter, J.P., Kristensen, A. (2006) Lab-on-a-chip with integrated optical transducers. *Lab Chip*, Vol. 6, pp. 213–217.

Beebe, D.J., Moore, J.S., Bauer, J.M., Yu, Q., Liu, R.H., Devadoss, C., Jo, B.-H. (2000). Functional hydrogel structures for autonomous flow control inside microfluidic channels. *Nature*, Vol. 404, No. 6778, pp. 588-590.

Beebe, D.J., Mensing, G.A., Walker, Glenn M. (2002) Physics and applications of microfluidics in biology. *Annu. Rev. Biomed. Eng.*, Vol. 4, pp. 261-286.

Broyles, B.S., Jacobson, S.C., Ramsey, J.M. (2003). Sample Filtration, Concentration, and Separation Integrated on Microfluidic Devices. *Anal. Chem.*, Vol. 75, pp. 2761-2767.

Buhlmann, C., Preckel, T., Chan, S., Luedke, G., Valer, M. (2003). A New Tool for Routine Testing of Cellular Protein Expression: Integration of Cell Staining and Analysis of Protein Expression on a Microfluidic Chip-Based System. *Journal of Biomolecular Techniques*, Vol. 14, No. 2, pp. 119–127.

Calvo, E.J., Danilowicz, C. & Diaz. L.J. (1993). Enzyme catalysis at hydrogel-modified electrodes with redox polymer mediator. *Chem. Soc. Faraday Trans.*, Vol. 89, pp. 377-384.

Caprioli, L., Mele, E., Angile, F.E., Girardo, S., Athanassiou, A., Camposeo, A., Cingolani, R. & Pisignano, D. (2007). Photocontrolled wettability changes in polymer microchannels doped with photochromic molecules. *Applied Physics Letters*. Vol. 91, pp. 113113-3.

Chan, S.D.H., Luedke, G., Valer, M., Buhlmann, C., Preckel, T. (2003). Cytometric Analysis of Protein Expression and Apoptosis in Human Primary Cells With a Novel Microfluidic Chip-Based System. *Cytom. Part A*, Vol. 55A, pp. 119-125.

Chaterji, S., Kwon, I.K., Park, K. (2007). Smart Polymeric Gels: Redefining the Limits of Biomedical Devices. *Prog. Polym. Sci.*, Vol. 32, No. 8-9, pp. 1083-1122.

Chen , G., Hoffman, A.S. (1995). Graft copolymers that exhibit temperature-induced phase transitions over a wide range of pH. *Nature*, Vol. 373, pp. 49 - 52.

Chen, G., Svec, F., Knapp, D.R. (2008). Light-actuated high pressure-resisting microvalve for on-chip flow control based on thermo-responsive nanostructured polymer. *Lab Chip*, Vol. 8, No. 7, pp. 1198-1204.

Chen, L., Manz, A., Day, P.J.R. (2007) Total nucleic acid analysis integrated on microfluidic devices. *Lab Chip*, Vol. 7, No. 11, pp. 1413-1423.

Cheng, X., Wang, Y., Hanein, Y., Böhringer, K.F. & Ratner, B.D. (2004). Novel cell patterning using microheater-controlled thermoresponsive plasma films. *J. Biomed. Mater. Res. A.*, Vol. 70, No. 2, pp. 159-168.

Chiu, S.-H., Liu, C.-H. (2009). An air-bubble-actuated micropump for on-chip blood transportation. *Lab Chip*, Vol. 9, No. 11, pp. 1524-1533.

Choi, J.-W., Oh, K. W., Han, A., Wijayawardhana, C. A., Lannes, C., Bhansali, S., Schlueter, K. T., Heineman, W. R., Halsall, H. B., Nevin, J. H., Helmicki, A. J., Henderson, H. T. & Ahn, C. H. (2001). Development and characterization of microfluidic devices and systems for magnetic bead-based biochemical detection. *Biomedical Microdevices*, Vol. 3, No. 3, pp. 191–200.

Choi, Y.H., Chung, K.H.; Lee, S.S. (2009). Microfluidic actuation by dehydration of hydrogel. *Sensors, 2009 IEEE*, pp.1370 – 1373.

Chunder, A., Etcheverry, K., Londe, G., Cho, H.J., Zhai, L. (2009). Conformal switchable superhydrophobic/ hydrophilic surfaces for microscale flow control. Colloids and Surfaces A: *Physicochemical and Engineering Aspects*, Vol. 333, No. 1-3, pp. 187-193.

Craighead, H. (2006). Future lab-on-a-chip technologies for interrogating individual molecules. *Nature*, Vol. 442, pp. 387-393.

Csetneki, I., G. Filipcsei, & M. Zrínyi (2006). Smart Nanocomposite Polymer Membranes with On/Off Switching Control. *Macromolecules*, Vol. 39, No. 5, pp. 1939-1942.

Dalton, P. D., Flynn, L. & Shoichet, M. S. (2002). Biomaterials, Vol. 23, pp. 3843.

De, S.K., Aluru, N. R., Johnson, B., Crone, W.C., Beebe, D.J., Moore, J. (2002). Equilibrium Swelling and Kinetics of pH-Responsive Hydrogels: Models, Experiments, and Simulations. *Journal of Microelectromechanical systems*, Vol. 11, No. 5, pp. 544 - 555.

de Jong, J., Lammertink, R.G., Wessling, M. (2006). Membranes and microfluidics: a review. *Lab Chip*, Vol. 6, pp. 1125–1139.

Delamarche, E., Bernard, A., Schmid, H., Michel, B. & Biebuyck H. (1997). Patterned delivery of immunoglobulins to surfaces using microfluidic networks. Science Vol. 276, pp. 779-781.

Di Benedetto, F., Mele, E., Camposeo, A., Athanassiou, A., Cingolani, R. & Pisignano, D. (2008). Photoswitchable organic nanofibers. *Adv. Mater.*, Vol. 20, pp. 314.

Dittrich, P.S., Manz, A. (2006). Lab-on-a-chip: microfluidics in drug discovery. *Nature*, Vol. 5, pp. 210-218.

Dong, L., Agarwal, A.K., Beebe, D.J., Jiang, H. (2006). Adaptive liquid microlenses activated by stimuli-responsive hydrogels. *Nature*, Vol. 442, No. 7102, pp. 551-554.

Dong, L., Jiang, H. (2007a). Autonomous microfluidics with stimuli-responsive hydrogels. *Soft Matter*, Vol. 3, No. 10, pp. 1223-1230.

Dong, L., Jiang, H. (2007b). Variable-Focus Liquid Microlenses and Microlens Arrays Actuated by Thermo-responsive Hydrogels. *Adv. Mater.*, Vol. 19, No. 3, pp. 401-405.

Ebara, M., Hoffman, J.M., Hoffman, A.S., Stayton, P.S. (2006). Switchable surface traps for injectable bead-based chromatography in PDMS microfluidic channels. *Lab Chip*, Vol. 6, No. 7, pp. 843-848.

Eddington, D.T., Beebe, D.J. (2004). Flow control with hydrogels. *Adv. Drug Deliv. Rev.*, Vol. 56, No. 2, pp. 199-210.

Eichenbaum, G.M.. Kiser, P.F., Simon S.A., Needham, D. (1998). pH and Ion-Triggered Volume Response of Anionic Hydrogel Microspheres. Macromolecules, Vol. 31, pp. 5084-5093.

Erickson, D. & Li, D. (2004). Integrated microfluidic devices. *Analytica Chimica Acta*, Vol. 507, pp. 11-26.

Fernández-Barbero, A., Suárez, I.J., Sierra-Martín, B., Fernández-Nieves, A., de las Nieves F.J., Marquez, M., Rubio-Retama, J. & López-Cabarcos, E. (2009) Gels and microgels for nanotechnological applications. *Adv. Colloid Interface Sci.*, Vol. 147-148, pp.88-108.

Fiorini, G.S., Chiu, D.T. (2005). Disposable microfluidic devices: fabrication, function, and application. *BioTechniques*, Vol. 38, pp. 429-446.

Freirea, S.L.S., Wheeler, A.R. (2006). Proteome-on-a-chip: Mirage, or on the horizon? *Lab Chip*, Vol. 6, pp. 1415-1423.

Fu, Q., Rao, G.V.R., Basame, S.B., Keller, D.J., Artyushkova, K., Fulghum, J.E., López, G.P. (2004). Reversible Control of Free Energy and Topography of Nanostructured Surfaces *J. Am. Chem. Soc.*, Vol. 126, pp. 8904 – 8905.

Geiger, E.J., Pisano, A.P., Svec, F. (2010). A Polymer-Based Microfluidic Platform Featuring On-Chip Actuated Hydrogel Valves for Disposable Applications. *J. Microelectromech. Systems*, Vol. 19, No. 4, pp. 944-950.

Good, B.T., Bowman, C.N., Davis, R.H. (2004). Modeling and verification of fluid-responsive polymer pumps for microfluidic systems. *Chemical Engineering Science*, Vol. 59, No. 24, pp. 5967-5974.

Graf, N.J., Bowser, M.T. (2008). A soft-polymer piezoelectric bimorph cantilever-actuated peristaltic micropump. *Lab Chip*, Vol. 8, No. 10, pp. 1664-1670.

Gregg, B.A. & Heller, A. (1991) Redox polymer films containing enzymes. 2. Glucose oxidase containing enzyme electrodes. *J. Phys. Chem.*, Vol. 95, pp. 5976-5980.

Guenther, M., Kuckling, D., Corten, C., Gerlach, G., Sorber, J., Suchaneck, G. & Arndt, K.F. (2007). Chemical sensors based on multiresponsive block copolymer hydrogels. *Sensors & Actuators: B. Chemical*, Vol. 126, pp. 97-106.

Guenther, G.G., Corten, C., Kuckling, D., Sorber, J. & Arndt, K.F. (2008). Hydrogel-based sensor for a rheochemical characterization of solutions. *Sensors and Actuators B: Chemical*, Vol. 132, pp. 471-476.

Guiseppi-Elie, A. (2010) Electroconductive hydrogels: synthesis, characterization and biomedical applications *Biomaterials*, Vol. 31, pp. 2701-2716.

Ha, S.-M., Cho, W., Ahn, Y. (2009). Disposable thermo-pneumatic micropump for bio lab-on-a-chip application. *Microelectron. Eng.*, Vol. 86, No. 4-6, pp. 1337-1339.

Han, I.S., Han, M.-H., Kim, J., Lew, S., Lee, Y.J., Horkay, F. & Magda, J.J. (2002) Constant-volume hydrogel osmometer: a new device concept for miniature biosensors. *Biomacromolecules*, Vol. 3, pp. 1271–1275.

Handique K., Gogoi, .B.P., Burke, D.T., Mastrangelo, C.H. & Burns, M.A. (1997). Microfluidic flow control using selective hydrophobic patterning. Proc. SPIE, Vol. 3224, pp. 185-95.

Hara, Y., Yoshida, R. (2008). Self-Oscillating Polymer Fueled by Organic Acid. *J. Phys. Chem. B*, Vol. 112, No. 29, pp. 8427-8429.

Harmon, M.E., Tang, M. & Frank, C.W. (2003). A microfluidic actuator based on thermoresponsive hydrogels. *Polymer*, Vol. 44, No. 16, pp. 4547-4556.

Heller, A., (2006). Electron-conducting redox hydrogels: design, characteristics and synthesis *Curr. Opin. Chem. Biol.*, Vol.10, pp. 664-672.

Herber, S., Olthuis, W. & Bergveld, P. (2003). A swelling hydrogel-based PCO2 sensor. *Sensors & Actuators: B. Chemical*, Vol. 91, pp. 378-382.

Herber, S., Olthuis, W., Bergveld, P. & Van Den Berg, A. (2004). Exploitation of a pH-sensitive hydrogel disk for CO2 detection. *Sensors & Actuators: B. Chemical*, Vol. 103, pp. 284-289.

Herber, S., Borner, J., Olthuis, W., Bergveld, P. & Van Den Berg, A. (2005a). A micro CO2 gas sensor based on sensing of pH-sensitive hydrogel swelling by means of a pressure sensor. *Transducers*, Vol. 2, pp. 1146-1149.

Herber, S., Bomer, J., Olthuis, W., Bergveld, P. & Van Den Berg, A. (2005b). A miniaturized carbon dioxide gas sensor based on sensing of pH-sensitive hydrogel swelling with a pressure sensor. *Biomedical Microdevices*, Vol. 7, pp. 197-204.
http://www.yole.fr/

Huang, C., Soenen, S.J., Rejman, J., Lucas, B., Braeckmans, K., Demeester J. & De Smedt S.C. (2011). Stimuli-responsive electrospun fibers and their applications. *Chem. Soc. Rev.*, Vol. 40, pp. 2417-2434.

Huang C.J., Dostalek, J. & Knoll, W. (2010). Long range surface plasmon and hydrogel optical waveguide field-enhanced fluorescence biosensor with 3D hydrogel binding matrix: On the role of diffusion mass transfer *Biosens. Bioelectron.*, Vol. 26, pp. 1425-1431.

Huang, X., Li, S., Schultz, J.S., Wang, Q. & Lin, Q. (2010). A dielectric affinity microbiosensor. *Appl. Phys. Lett.*, Vol. 96, pp. 033701.

Huang, T.T., Geng, T., Akin, D., Chang, W.-J., Sturgis, J., Bashir, R., Bhunia, A.K., Robinson, J.P. & Ladisch, M.R. (2003). Micro-assembly of functionalized particulate monolayer on C18-derivatized SiO2 surfaces. *Biotechnology and Bioengineering*, Vol. 83, No. 4, pp. 416-427.

Huber, D.L., Manginell, R.P., Samara, M.A., Kim, B.-I. & Bunker, B.C. (2003). Programmed Adsorption and Release of Proteins in a Microfluidic Device. *Science*, Vol. 301, pp. 352-354.

Hyun, J., Lee, W.-K., Nath, N., Chilkoti, A. & Zauscher, S. (2004). Capture and Release of Proteins on the Nanoscale by Stimuli-Responsive Elastin-Like Polypeptide "Switches". *J. Am. Chem. Soc.*, Vol. 126, pp. 7330-7335.

Idziak, I., Avoce, D., Lessard, D., Gravel, D. & Zhu, X. X. (1999). Thermosensitivity of Aqueous Solutions of Poly(N,N-diethylacrylamide). *Macromolecules*, Vol. 32, pp. 1260-1263

Jeong, W., Kim, J., Kim, S., Lee, S., Mensing, G., Beebe, D. J. (2004). Lab Chip, Vol 4, pp. 576.

Khademhosseini, A. & Langer, R. (2007). Microengineered hydrogels for tissue engineering. *Biomaterials*, Vol. 28, No. 34, pp. 5087-5092.

Kim, D., Kim, S., Park, J., Baek, J., Kim, S., Sun, K., Lee, T., Lee, S. (2007). Hydrodynamic fabrication and characterization of a pH-responsive microscale spherical actuating element. *Sensors and Actuators A: Physical*, Vol. 134, No. 2, pp. 321-328.

Kim, J., Singh, N. & Lyon, L.A. (2006). Label-Free Biosensing with Hydrogel Microlenses. *Angew. Chem. Int. Ed.*, Vol. 45, pp. 1446 -1449.

Kim, J., Singh, N. & Lyon, L.A. (2007). Displacement-Induced Switching Rates of Bioresponsive Hydrogel Microlenses. *Chem. Mater.*, Vol. 19, pp. 2527-2532.

Kim, S.M., Lee, S.H., Suh, K.Y. (2008) Cell research with physically modified microfluidic channels: A review. *Lab Chip*, Vol. 8, pp. 1015–1023.

Kraly, J.R., Holcomb, R.E., Qian, G., Henry, C.S. (2009). Review: Microfluidic applications in metabolomics and metabolic profiling. *Analytica Chimica Acta*, Vol. 653, No. 1, pp. 23-35.

Kumar, A., Srivastava, A., Galaev, I.Y., Mattiasson, B. (2007). Smart polymers: Physical forms and bioengineering applications. *Prog. Polym. Sci.*, Vol. 32, pp. 1205 - 1237.

Kuckling, D., Harmon, M.E. & Frank, C.W. (2002). Photo-Cross-Linkable PNIPAAm Copolymers. 1. Synthesis and Characterization of Constrained Temperature-Responsive Hydrogel Layers. *Macromolecules*, Vol. 35, No. 16, pp. 6377-6383.

Kwon, G.H., Choi, Y.Y., Park, J.Y., Woo, D.H., Lee, K.B., Kim, J.H., Lee, S.-H. (2010). Electrically-driven hydrogel actuators in microfluidic channels: fabrication, characterization, and biological application. *Lab Chip*, Vol. 10, No. 12, pp. 1604-1610.

Kwon, G.H., Jeong, G.S., Park, J.Y., Moon, J.H., Lee, S.-H. (2011). A low-energy-consumption electroactive valveless hydrogel micropump for long-term biomedical applications. *Lab Chip*, Vol. 11, No. 17, pp. 2910-2915.

Laser, D., Santiago, J.G. (2004) A review of micropumps. *J. Micromech. Microeng.*, Vol. 14, pp. R35–R64.

Lee, C.-Y., Lee, G.-B., Lin, J.-L., Huang, F.-C., Liao, C.-S. (2005). Integrated microfluidic systems for cell lysis, mixing/pumping and DNA amplification. Journal of Micromechanics and Microengineering, Vol. 15, No. 6, pp. 1215-1223.

Lee, K. & Asher, S.A. (2000). Photonic Crystal Chemical Sensors: pH and Ionic Strength. *J. Am. Chem. Soc.*, Vol. 122, pp. 9534-9537.

Lee, K. & Asher, S.A. (2011). Photonic Crystal Chemical Sensors: pH and Ionic Strength. *J. Am. Chem. Soc.*, Vol.133, pp. 9152-9155.

Lin, G., Chang, S., Kuo, C.-H., Magda, J. & Solzbacher, F. (2009). Free swelling and confined smart hydrogels for applications in chemomechanical sensors for physiological monitoring *Sens. Actuators, B Chem.*, Vol. 136, pp. 186-195.

Linder, V., Verpoorte, E., de Rooij, N.F., Sigrist, H. & Thormann, W.(2002). Application of surface biopassivated disposable poly(dimethylsiloxane)/glass chips to a heterogeneous competitive human serum immunoglobulin G immunoassay with incorporated internal standard. Electrophoresis, Vol. 23, No. 5, pp. 740-749.

Liu, F. & Urban, M.W. (2008). 3D Directional Temperature Responsive (N-(dl)-(1-Hydroxymethyl) Propylmethacrylamide-co-n-butyl Acrylate) Colloids and Their Coalescence. *Macromolecules*, Vol. 41, pp. 352 - 360.

Liu, R.H., Yu, Q., Beebe, D.J. (2002). Fabrication and Characterization of Hydrogel-Based Microvalves. *J. Microelectromech. Systems*, Vol. 11, No. 1, pp. 45-53.

Lynn, D.M. (2006). Layers of opportunity: nanostructured polymer assemblies for the delivery of macromolecular therapeutics. *Soft Matter*, Vol. 2, No. 4, pp. 269-273.

Ma, D., Chen, H.W., Shi, D.Y., Li, Z.M. & Wang, J.F. (2009). Preparation and characterization of thermo-responsive PDMS surfaces grafted with poly(N-isopropylacrylamide) by benzophenone-initiated photopolymerization. *J. Colloid Interface Sci.*, Vol. 332, pp. 85-90.

Ma, D., Chen, H., Li, Z. & He, Q. (2010). Thermomodulated cell culture/harvest in polydimethyl-siloxane microchannels with poly(N-isopropylacrylamide)-grafted surface. Biomicrofluidics, Vol. 4, No. 4, pp. 044107.

Maeda, S., Hara, Y., Yoshida, R., Hashimoto, S. (2008). Control of the Dynamic Motion of a Gel Actuator Driven by the Belousov-Zhabotinsky Reaction. *Macromolecular Rapid Communications*, Vol. 29, No. 5, pp. 401-405.

Malmstadt, N., Yager, P., Hoffman, A.S., Stayton, P.S. (2003). A Smart Microfluidic Affinity Chromatography Matrix Composed of Poly(N-isopropylacrylamide)-Coated Beads. *Anal. Chem.*, Vol. 75, No. 13, pp. 2943-2949.

Malmstadt, N., Hoffman, A.S., Stayton, P.S. (2004). "Smart" mobile affinity matrix for microfluidic immunoassays. *Lab Chip*, Vol. 4, pp. 412-415.

McCalla, S.E., Tripathi, A. (2011). Microfluidic Reactors for Diagnostics Applications. *Annual Review of Biomedical Engineering*, Vol. 13, No.1, pp. 321-343.

McNally, H., Pingle, M., Lee, S. W., Guo, D., Bergstrom, D. E. & Bashir, R. (2003). Self-assembly of micro- and nano-scale particles using bio-inspired events. *Applied Surface Science*, Vol. 214, No. 1-4, pp. 109-119.

Murase, Y., Maeda, S., Hashimoto, S., Yoshida, R. (2008). Design of a Mass Transport Surface Utilizing Peristaltic Motion of a Self-Oscillating Gel. *Langmuir*, Vol. 25, No. 1, pp. 483-489.

Neethirajan, S., Kobayashi, I., Nakajima, M., Wu, D., Nandagopal, S., Lin, F. (2011). Microfluidics for food, agriculture and biosystems industries. *Lab Chip* Vol. 11, pp. 1574-1586.

Nestler, J., Morschhauser, A., Hiller, K., Otto, T., Bigot, S., Auerswald, J., Knapp, H. F., Gavillet, J., Gessner, T. (2010). Polymer lab-on-chip systems with integrated electrochemical pumps suitable for large-scale fabrication. *The International Journal of Advanced Manufacturing Technology*, Vol. 47, No. 1, pp. 137-145.

Ng, J.M.K., Gitlin, I., Stroock, A.D., Whitesides, G.M. (2002). Components for integrated poly(dimethylsiloxane) microfluidic systems. *Electrophoresis*, Vol. 23, pp. 3461–3473.

Nguyen, N.-T., Wu, Z. (2005). Micromixers – a review. *J. Micromech. Microeng.*, Vol. 14, pp. R1–R16.

Nisar, A., Afzulpurkar, N., Mahaisavariya, B., Tuantranont, A. (2008). MEMS-based micropumps in drug delivery and biomedical applications. *Sensors and Actuators B: Chemical*, Vol. 130, No. 2, pp. 917-942.

Oh, K.W., Ahn, C.H. (2006). A review of microvalves. *J. Micromech. Microeng.*, Vol. 16, pp. R13–R39.

Orthner, M.P., Lin, G., Avula, M., Buetefisch, S., Magda, J., Rieth, L.W. & Solzbacher, F. (2010a). Hydrogel based sensor arrays (2×2) with perforated piezoresistive diaphragms for metabolic monitoring (in vitro). *Sens. Actuators, B Chem.*, Vol. 145, pp. 807-816.

Orthner, M.P., Buetefisch, S., Magda, J., Rieth, L.W. & Solzbacher, F. (2010b). Development, fabrication, and characterization of hydrogel based piezoresistive pressure sensors with perforated diaphragms. *Sensor Actuat. A-Phys.* Vol. 161, pp. 29 -38.

Pasparakisa, G., Vamvakaki, M. (2011). Multiresponsive polymers: nano-sized assemblies, stimuli-sensitive gels and smart surfaces. *Polym. Chem.*, Vol. 2, pp. 1234-1248.

Peppas, N.A., Hilt, J.Z., Khademhosseini, A., Langer, R. (2006). Hydrogels in Biology and Medicine: From Molecular Principles to Bionanotechnology. *Adv. Mater.*, Vol. 18, pp. 1345-1360.

Peterson, D.S., Rohr, T., Svec, F. & Fréchet, J.M.J. (2003). Dual-function microanalytical device by in-situ photolithographic grafting of porous polymer monolith: integrating solid phase extraction and enzymatic digestion for peptide mass mapping. *Analytical Chemistry*, Vol. 75, pp. 5328-5335.

Peterson, D.S. (2005). Solid supports for micro analytical systems. *Lab Chip*, Vol. 5, pp. 132-139.

Qiu, Y. & Park, K. (2001). Environment-sensitive hydrogels for drug delivery. *Adv. Drug Delivery Rev.*, Vol. 53, pp. 321-339.

Richter, A., Kuckling, D., Howitz, S., Gehring, T., Arndt, K. F. (2003). Electronically controllable microvalves based on smart hydrogels: magnitudes and potential applications. *J. Microelectromech. Systems*, Vol. 12, No. 5, pp. 748-753.

Richter, A., Klatt, S., Paschew, G., Klenke, C. (2009). Micropumps operated by swelling and shrinking of temperature-sensitive hydrogels. *Lab Chip*, Vol. 9, No. 4, pp. 613-618.

Rosario, R., Gust, D., Garcia, A.A., Hayes, M., Taraci, J.L., Dailey, J.W. & Picraux, S.T. (2004). Lotus Effect Amplifies Light-Induced Contact Angle Switching. *J. Phys. Chem. B*, Vol. 108, pp. 12640-12642.

Seong, G.H., Zhan, W., Crooks, R.M. (2002). Fabrication of Microchambers Defined by Photopolymerized Hydrogels and Weirs within Microfluidic Systems: Application to DNA Hybridization. *Anal Chem*, Vol. 74, No. 14, pp. 3372-3377.

Serpe, M.J., Yarmey, K.A., Nolan, C.M. & Lyon, L.A. (2004). Doxorubicin Uptake and Release from Microgel Thin Films. Biomacromolecules, Vol. 6, No. 1, pp. 408-413.

Shacham, R., Avnir, D. & Mandler, D. (1999). Electrodeposition of Methylated Sol-Gel Films on Conducting Surfaces. Advanced Materials, Vol. 11, No. 5, pp. 384-388.

Sharma, A.C., Jana, T., Kesavamoorthy, R., Shi, L., Virji, M.A., Finegold, D. N. & Asher, S. A. (2004). General photonic crystal sensing motif: creatinine in bodily fluids. *J. Am. Chem. Soc.*, Vol. 126, pp. 2971-2977.

Sia, S.K., Kricka, L.J. (2008). Microfluidics and point-of-care testing. *Lab Chip* Vol. 8, pp. 1982–1983.

Sims C.E., Allbritton N.L. (2007). Analysis of single mammalian cells on-chip. *Lab Chip*, Vol. 7, pp. 423–440.

Song, S., Singh,.A.K. (2006). On-chip sample preconcentration for integrated microfluidic analysis. *Anal. Bioanal. Chem.*, Vol. 384, pp. 41–43.

Sorber, J., Steiner, G., Schulz, V., Guenther, M., Gerlach, G., Salzer, R. & Arndt, K-F. (2008) Hydrogel-based piezoresistive ph sensors: investigations using FT-IR attenuated total reflection spectroscopic imaging. *Anal. Chem.*, Vol. 80, pp. 2957-2962.

Sridhar, V. & Takahata, K. (2009) A hydrogel-based passive wireless sensor using a flex-circuit inductive transducer. *Sensor Actuat A-Phys*,. Vol. 155, pp. 58-65.

Sugiura, S., Szilagyi, A., Sumaru, K., Hattori, K., Takagi, T., Filipcsei, G., Zrinyi, M., Kanamori, T. (2009). On-demand microfluidic control by micropatterned light irradiation of a photoresponsive hydrogel sheet. *Lab Chip*, Vol. 9, No. 2, pp. 196-198.

Sun, T., Song, W. & Jiang, L. (2005). Control over the responsive wettability of poly(N-isopropylacrylamide) film in a large extent by introducing an irresponsive molecule. *Chem. Commun.*, pp. 1723 - 725.

Svec, F., Tennikova, T.B. & Deyl, Z., (May 2003). *Monolithic Materials: Preparation, Properties and Applications (Journal of Chromatography Library)*, Elsevier Science, ISBN-10: 0444508791 ISBN-13: 978-0444508799, Amsterdam.

Tao, X. (2001). *Smart fibres, fabrics and clothing: Fundamentals and applications,* Woodhead Publishing Ltd, ISBN-10: 1855735466 ISBN-13: 978-1855735460, Hong Kong.

Tay, F. (2002). *Microfluidics and BioMEMS Applications* (1st edition), Springer, Kluwer Academic Publishers, ISBN 978-1-4020-7237-6, Boston, pp. 3-24.

Tierney, S., Hjelme, D.R. & Stokke, B.T. (2008). Determination of Swelling of Responsive Gels with Nanometer Resolution. Fiber-Optic Based Platform for Hydrogels as Signal Transducers. *Anal. Chem.*, Vol. 80, pp. 5086-5093.

Tokarev, I. (2009). *ACS Appl. Mater. Interfaces*, Vol. 1, No. 3, pp. 532-536.

Tokarev, I., Minko, S. (2009). Stimuli-responsive hydrogel thin films. *Soft Matter*, Vol. 5, No. 3, pp. 511-524.

Tokarev, I. & Minko, S. (2010). Stimuli-Responsive Porous Hydrogels at Interfaces for Molecular Filtration, Separation, Controlled Release, and Gating in Capsules and Membranes. *Adv. Mater.*, Vol. 22, No. 31, pp. 3446-3462.

Toner, M., Irimia, D. (2005). Blood-on-a-chip. *Annu. Rev. Biomed. Eng.* Vol. 7, pp. 77–103.

Tovar, A.R., Lee, A.P. (2009). Lateral cavity acoustic transducer. *Lab Chip*, Vol. 9, No. 1, pp. 41-43.

Verpoorte, E. (2003a) Chip vision – optics for microchips. Lab. Chip, Vol. 3, pp. 42N–52N.

Verpoorte, E. (2003b). Focus Beads and chips: new recipes for analysis. *Lab Chip*, Vol. 3, No. 4, pp. 60N-68N.

Trinh, Q.T., Gerlach, G., Sorber, J. & Arndt, K.-F.(2006). Hydrogel-based piezoresistive pH sensors: design, simulation and output characteristics. *Sens. Actuators, B Chem.*, Vol. 117 pp. 17-26.

Xue, J., Chen, L., Wang, H. L., Zhang, Z. B., Zhu, X. L., Kang, E. T. & Neoh, K. G. (2008). Stimuli-Responsive Multifunctional Membranes of Controllable Morphology from Poly(vinylidene fluoride)-graft-Poly[2-(N,N-dimethylamino)ethyl methacrylate] Prepared via Atom Transfer Radical Polymerization. *Langmuir*, Vol. 24, No. 24, pp. 14151-14158.

Yang, Q., Adrus, N., Tomicki, F., Ulbricht, M. (2011). Composites of functional polymeric hydrogels and porous membranes. *J. Mater. Chem.*, Vol. 21, No. 9, pp. 2783-2811.

Yu, B., Wang, Y., Ju, Y.M., West, J.H., Moussy, Y. & Moussy, F. (2008). Use of hydrogel coating to improve the performance of implanted glucose sensors. *Biosens. Bioelectron.*, Vol. 23, pp. 1278-1284.

Yu, C., Mutlu, S., Selvaganapathy, P., Mastrangelo, C.H., Svec, F. & Fréchet, J.M. (2003). Flow control valves for analytical microfluidic chips without mechanical parts based on thermally responsive monolithic polymers. *Analytical Chemistry*, Vol. 75, No. 8, pp. 1958-1961.

Walker, J.P. & Asher, S.A. (2005). Acetylcholinesterase-Based Organophosphate Nerve Agent Sensing Photonic Crystal. *Anal. Chem.*, Vol. 77, pp. 1596 -1600.

Wandera, D., Wickramasinghe, S. R. & Husson, S.M. (2010). Stimuli-responsive membranes. Journal of Membrane *Science*, Vol. 357, No. 1-2, pp. 6-35.

Wang, L., Wang, X., Xu, M., Chen, D. & Sun, J. (2008). Layer-by-Layer Assembled Microgel Films with High Loading Capacity: Reversible Loading and Release of Dyes and Nanoparticles. *Langmuir*, Vol. 24, No. 5, pp. 1902-1909.

Wang, L., Zhang, M., Li, J., Gong, X., Wen, W. (2010). Logic control of microfluidics with smart colloid. *Lab Chip*, Vol. 10, No. 21, pp. 2869-2874.

Zeng, X., Jiang, H. (2008). Tunable liquid microlens actuated by infrared light-responsive hydrogel. *Appl. Phys. Lett.*, Vol. 93, pp. 151101-3.

Zhan, W., G.H. Seong, & R.M. Crooks (2002). Hydrogel-Based Microreactors as a Functional Component of Microfluidic Systems. *Anal. Chem.*, Vol. 74, No. 18, pp. 4647-4652.

Zhang, Q.M., Li, H. F., Poh, M., Xia, F., Cheng, Z. Y., Xu, H. S. & Huang, C. (2002). *Nature* Vol. 419, pp. 284-287.

7

Robust Extraction Interface for Coupling Droplet-Based and Continuous Flow Microfluidics

Xuefei Sun[1], Keqi Tang[1], Richard D. Smith[1,2] and Ryan T. Kelly[2,*]
[1]*Biological Sciences Division*
[2]*Environmental Molecular Sciences LaboratoryPacific Northwest National Laboratory*
USA

1. Introduction

Microdroplets, in which aqueous samples are encapsulated in an immiscible phase (e.g., oil), can serve as microreactors and as vehicles for high-throughput chemical and biological studies. The oil phase separating the small (femto – nanoliter) aqueous volumes serves to eliminate dispersion, sample loss, dilution and cross-contamination. Droplet-based microfluidics also offers great promise for quantitative analysis because monodisperse microdroplets can be generated with controllable sizes and detected at well defined time points. Such platforms have been successfully applied in a number of biological research areas, for example, encapsulating and sorting single cells within droplets, studying enzyme kinetics, and performing polymerase chain reaction (PCR) with high throughput (Chiu & Lorenz, 2009; Chiu et al., 2009; Huebner et al., 2008; Song et al., 2006; Teh et al., 2008; Theberge et al., 2010; Yang et al., 2010).

A variety of methods have been developed to generate, manipulate and monitor droplets in microfluidic devices (Theberge et al., 2010). Detection of droplet contents has historically been limited to optical methods such as laser-induced fluorescence, while coupling with chemical separations and nonoptical detection has proven difficult. Combining the advantages of droplet-based platforms with more information-rich analytical techniques including liquid chromatography, capillary electrophoresis and mass spectrometry (MS) will greatly extend their reach. For example, electrospray ionization (ESI)-MS has become an essential technique for biological analysis because of its high sensitivity and ability to identify and provide structural information for hundreds or more molecules from complex samples in a given analysis (Aebersold & Mann, 2003; Liu et al., 2007). A number of techniques have been developed to couple continuous-flow microfluidic devices with MS (Kelly et al., 2008; Koster & Verpoorte, 2007; Mellors et al., 2008; Sun et al., 2011b), but the coupling of droplet-based platforms with MS remains challenging (Fidalgo et al., 2009).

ESI-MS analysis of aqueous droplets has been achieved by directly coupling a segmented flow system with MS (Pei et al., 2009), but this method has some potential risks including

* Corresponding Author

instability of the electrospray due to oil accumulation at the tip, introduction of chemical background, and contamination of the MS instrument. Hence, it is generally necessary to extract the aqueous droplet from the oil phase for further separation or MS analysis. Huck and coworkers employed electrocoalescence to extract droplets (Fidalgo et al., 2008) in which a pulsed electric field was applied over the extraction chamber to force droplets to coalesce with an aqueous stream. The droplet contents were then delivered to a capillary emitter for ESI-MS detection (Fidalgo et al., 2009). This method required careful adjustments to the flow of two immiscible phases to maintain an interface in the extraction chamber and avoid cross-contamination of the aqueous and oil streams. In addition, the severe dilution of the droplet contents resulted in high (~500 μM) detection limits. Lin and coworkers utilized an electrical-based method to control the droplet disruption and extraction at a stable oil/water interface (Zeng et al., 2011), but one reported issue in this case was the difficulty of achieving complete extraction with high efficiency, which limited its compatibility with quantitative analysis.

Another droplet extraction technique is based on surface modification at the junction between two immiscible phases in the microchannel. Kennedy and coworkers exploited selectively patterned glass microchannel surfaces to stabilize the oil-water interface and facilitate droplet extraction (Pei et al., 2010; Roman et al., 2008; Wang et al., 2009). When the segmented flow matched the aqueous flow, the entire plug was able to be extracted. In some cases only part of each droplet was transferred due to the presence of a "virtual wall" (Pei et al., 2010; Roman et al., 2008). A portion of the extracted samples were then injected into an electrophoresis channel for CE separation. Recently, Fang and coworkers employed a similar surface modification technique to obtain a hydrophilic tongue-based droplet extraction interface, which could control the droplet extraction by regulating the waste reservoir height (Zhu & Fang, 2010). The droplet contents were then detected by MS through an integrated ESI emitter. However, with a height of zero, the droplets were not extracted and aqueous buffer was occasionally transferred to the droplets. In our laboratory, a novel droplet extraction interface constructed with an array of cylindrical posts was developed (Kelly et al., 2009). When the aqueous stream and oil carrier phase flow rates were adjusted to balance the pressure at the junction, a stable oil-aqueous interface based on interfacial tension alone was formed to prevent bulk crossover of two immiscible streams, while the droplets could be transferred through the apertures between posts to the aqueous stream and finally detected by ESI-MS with virtually no dilution.

Most of the reported methods and techniques for droplet extraction, as mentioned above, need to adjust two immiscible liquid flow rates to stabilize the interface and extract entire droplets. It is desirable to perform effective and complete droplet extraction independent of the flow rates, which would provide added flexibility for device operation. In this work, we present a robust interface for reliable and efficient droplet extraction, which was integrated in a droplet-based PDMS microfluidic assembly. The droplet extraction interface consisted of an array of cylindrical posts, the same as was previously reported (Kelly et al., 2009), but the aqueous stream microchannel surface was selectively treated by corona discharge to be hydrophilic. The combination of different surface energies and small apertures (~3 μm × 25 μm) enabled a very stable liquid interface between two immiscible steams to be established over a broad range of aqueous and oil flow rates. All aqueous droplets were entirely transferred to the aqueous stream and detected by MS following ionization at a monolithically integrated electrospray emitter.

2. Experimental section

2.1 Materials

Fluorinert FC-40, leucine enkephalin, apomyoglobin, glacial acetic acid, sodium carbonate, sodium bicarbonate, and hexamethyldisilazane (HMDS) were purchased from Sigma-Aldrich (St. Louis, MO). Fluorescein was obtained from Fluka (Buchs, Switzerland). HPLC-grade methanol was purchased from Fisher Scientific (Fair Lawn, NJ). Water was purified using a Barnstead Nanopure Infinity system (Dubuque, IA). ESI buffer solution was prepared by mixing water and methanol at a 9:1 (v/v) ratio and adding 0.1% (v/v) acetic acid. Carbonate buffer (10 mM, pH 9.3) was filtered using 0.2 µm syringe filters (Pall Life Sciences, Ann Arbor, MI) before use. PDMS elastomer base and curing agent were purchased as Dow Corning Sylgard 184 from Ellsworth Adhesives (Germantown, WI).

2.2 Microfluidic design and fabrication

Figure 1A shows the pattern of an integrated droplet-based microfluidic device that was designed using IntelliCAD software (IntelliCAD Technology Consortium, Portland, OR). It is composed of two-layers, in which the control layer contains two microchannels used for valving, and the flow layer includes all other microchannels and microstructures. Droplet generation is controlled by pneumatic valves (Galas et al., 2009; Sun et al., 2011a; Zeng et al., 2009). The valving intersectional area is 100 µm × 100 µm. The length of the two side channels and the distance between them are 0.5 cm and 100 µm, respectively. The design of the droplet extraction interface is expanded in Figure 1B, which is constructed with an array of 15-µm-diameter cylindrical posts having ~3 µm apertures for fluid transfer in between. The width of both microchannels separated by the interface is 50 µm. The oil flow channel width tapers from 100 µm to 50 µm, ensuring that the droplets can contact the interface sufficiently to enable extraction. Pairs of patterned symmetrical lines with an angle of 50° are used to guide the cutting of a tapered emitter at the aqueous channel terminus (Kelly et al., 2009).

The integrated droplet-based PDMS microfluidic device was fabricated using well established multilayer soft lithography techniques (Sun et al., 2011a; Unger et al., 2000). First, two separate templates were produced on silicon substrates with a contact photomask aligner (NXQ4000-6, Neutronix-Quintel, Morgan Hill, CA). Three photomasks were printed with 50,800 dpi resolution at Fineline Imaging (Colorado Springs, CO) based on the design shown in Figure 1A. The template defining the valving control channels was formed by patterning SU-8 photoresist (Microchem, Newton, MA) with rectangular features (~ 25 µm high). The other template containing the flow layer was fabricated through two steps. Two side channels for injection of dispersed aqueous samples were first patterned with SPR 220-7 photoresist (Rohm & Haas Electronic Materials, Marlborough, MA), which was reflowed to provide rounded shape features (~ 10 µm high) and stabilized at high temperature (180 °C) for 30 min to prevent removal in subsequent processing steps. The remaining microstructures on the flow layer containing the extraction interface (Figure 1C) were then aligned with the two pregenerated side channels and patterned with SU-8 photoresist to form rectangular shaped features (~25 µm high). The silicon template defining the control layer was modified with HMDS using vapor deposition to assist in releasing the PDMS membrane from the patterned template. A 10:1 ratio (w/w) of PDMS base to curing agent

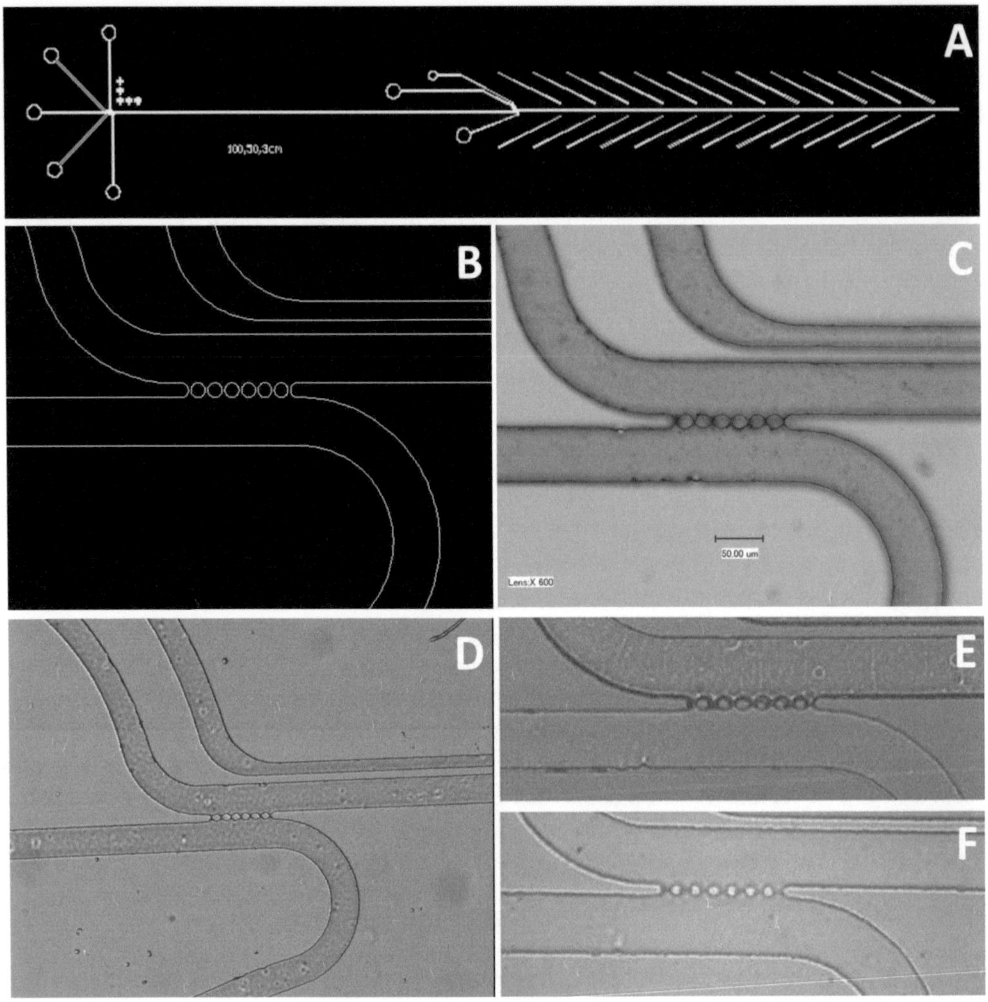

Fig. 1. Overview of device design and fabrication. (A) Design of a droplet-based microfluidic device including a valve-based droplet generator, a droplet extraction interface, and guide marks for cutting the ESI emitter. (B) Amplified view of the droplet extraction interface design. (C) Photograph of the extraction interface structure on a silicon template. (D) Photograph of the extraction interface in a completed PDMS device. (E) Photograph of the water-air interface. Top channel was empty and bottom channel filled with DI water. (F) Photograph of the oil-water interface. Top channel filled with DI water and bottom channel filled with oil (FC-40).

was then mixed, degassed under vacuum, and poured onto the patterned flow layer template to a thickness of 1–2 mm. This PDMS prepolymer mixture was also spin-coated on the HMDS-treated template at 2000 rpm for 30 s to create a control layer membrane with a thickness of ~ 50 μm. Both substrates were cured in an oven at 75 °C for at least 2 h. After removing the patterned PDMS from the flow layer template, several small through-holes were created at the ends of all flow channels by punching the substrate with a manually sharpened syringe needle (NE-301PL-C; Small Parts, Miramar, FL). The flow layer PDMS piece was then cleaned and treated with oxygen plasma (PX-250; March Plasma Systems, Concord, CA) for 30 s. It was immediately aligned on the top of the control layer PDMS membrane (still on the silicon wafer) based on alignment marks under a stereomicroscope (SMZ-U, Nikon, Japan) and assembled together to enclose the flow channels. After placing in an oven at 75 °C for 2 h to form an irreversible bond, the PDMS block containing flow and control layers was removed from the control layer silicon template, through-holes were punched at the end of all control channels, and the assembly was finally bonded to an unpatterned PDMS piece to enclose the control channels using oxygen plasma treatment. The final PDMS microdevice was cured in an oven at 120 °C for at least 48 h for complete curing and hydrophobic recovery. Figure 1D shows the droplet extraction interface portion of the PDMS microfluidic device.

2.3 Selective modification of the aqueous channel surface

The aqueous stream flow channel surface was selectively modified to be hydrophilic using corona discharge, while the oil stream flow channel surface retained the hydrophobicity of native PDMS. Deionized water was first gently introduced into the oil flow channel using a syringe through the oil outlet. Due to the hydrophobic PDMS surface and high flow resistance through the small apertures, water was confined to the oil flow channel and a water/air interface was produced at the junction (Figure 1E), leaving the aqueous channel open while protecting the oil flow channel. A corona discharge unit (BD-20AC; Electro-Technic Products, Chicago, IL) with a fine tip electrode was then placed close to the aqueous buffer inlet and actuated for 1–2s. The generated plasma was dispersed into the channel and then water immediately entered into the aqueous channel through the interface, indicating that the aqueous channel surface was oxidized to be hydrophilic. Finally, the water in the oil flow channel was replaced by oil, but the modified aqueous channel surface was kept in the water environment to maintain the modified surface.

2.4 Device operation

Droplet generation in this work was controlled by integrated pneumatic valves, which were actuated by a computer-controlled external solenoid valve as described previously (Sun, 2011a). The valving control channels in the PDMS device were filled with water to avoid introduction of air bubbles into the flow channels. The continuous oil (Fluorinert FC-40, 3M) flow in the channel was controlled by a syringe pump (PHD 2000; Harvard Apparatus, Holliston, MA) through a fused-silica capillary (75 µm i.d., 360 µm o.d.; Polymicro Technologies, Phoenix, AZ). One end of the capillary was connected with a 50 µL syringe (Hamilton, Reno, NV) and the other end was inserted into a ~2 mm long section of Tygon tubing (TGY-101-5C; Small Parts, Miramar, FL) to create a pressure fitting at the oil inlet on the microchip. The dispersed aqueous solutions were stored in sealed vials with a nitrogen gas inlet to pressurize the liquid and an outlet to allow liquid to be transferred into the microchannels via a capillary. Before operation, any air bubbles trapped in the transfer lines and microchannels were removed. The pneumatic valves were actuated to disperse aqueous solutions into the oil flow channel to generate droplets. The continuous aqueous/ESI buffer solution was infused into the aqueous channel from a syringe through a capillary transport line. The infusion rate was controlled by a syringe pump. For experiments in which electroosmotic flow was employed in the aqueous channel, a high voltage was applied over that channel using a PS-350 high-voltage power supply (Stanford Research Systems, Sunnyvale, CA, USA) via platinum electrodes placed in the reservoirs.

2.5 Detection

A fluorescence detection system was employed to monitor fluorescein-containing droplets (Sun, 2011a). The UV light generated by a mercury lamp (Olympus, Tokyo, Japan) was passed into an inverted optical microscope (Olympus, Tokyo, Japan), and fluorescence was collected with a digital camera (Nikon, Tokyo, Japan). For MS detection, an ion funnel-modified (Kelly, 2010) orthogonal time-of-flight MS instrument (G1969A LC/MSD TOF, Agilent Technologies, Santa Clara, CA) was used. Before coupling with MS, an integrated ESI emitter was created by making two vertical cuts through the PDMS device based on patterned guide lines such that the aqueous channel terminated at the apex (Kelly et al., 2008; Sun et al., 2010). The microfluidic emitter was positioned 3 mm in front of the MS inlet capillary, which was heated to 120 °C. The electrospray potential was applied on the ESI buffer-delivering syringe needle by a high-voltage power supply.

3. Results and discussion

3.1 Droplet extraction interface

Achieving reliable and highly efficient droplet extraction is of great importance for many droplet-based microfluidic systems integrated with further analytical functions such as chemical separations or MS detection. Ideally, each aqueous droplet should be entirely transferred to the aqueous flow channel to avoid losing any important quantitative information. On the other hand, oil should be prevented from entering the aqueous flow channel to avoid negatively impacting the downstream analysis (e.g., blocking flow, interrupting the electrospray circuit or adding chemical background to MS signal). The present droplet extraction interface retains the previously developed geometry of an array of cylindrical posts with narrow apertures in between, but we further strengthened the interface by modifying the aqueous channel to be hydrophilic while the oil flow channel surface remained hydrophobic. Corona discharge was used to treat the aqueous flow channel surface, and the hydrophobic surface in the oil flow channel was prevented from corona exposure by filling the channel with water beforehand. A similar method utilizing corona treatment was reported recently to extract droplets entering the device from a capillary for electrochemical detection (Filla et al., 2011). In the work described here, after actuating the corona discharge for 1–2s, the water entered into the modified channel quickly through the apertures and wetted the oxidized surface, which protected the PDMS surface from recovering its hydrophobicity. The treated device could usually survive for 2–3 days, and repeated surface treatments could likely be applied thereafter (although not explored here). Thus, the channel surfaces on both sides of the interface had different surface energies, which resulted in a very stable aqueous/oil interface established at the junction (Figure 1F).

3.2 Droplet extraction without hydrodynamic aqueous flow

In many of the previously reported systems, an appropriate aqueous stream flow was required to balance the pressure at the interface. In order to test the interface stability of our present device, we first investigated the droplet extraction performance without any hydrodynamic flow in the aqueous channel. Figure 2 shows the fluorescein droplet transfer at the interface when the oil flow rate was 500 nL/min and there was no flow in the aqueous channel. The entire droplet (~650 pL) was extracted in approximately 100 ms. In contrast to our previous report (Kelly et al., 2009), the droplet transfer occurred through all the apertures. Based on the fluorescence intensity measurement immediately before and after extraction, the droplet content was not significantly diluted during the extraction process. The extracted droplet displaced the aqueous buffer toward the aqueous inlet due to lower resistance, where it remained due to an absence of flow in that channel. To transport the extracted droplet contents, a 500 V/cm electric field was applied over the aqueous channel. The fluorescein molecules migrated at a rate of ~1 mm/s driven by a combination of electrophoresis and electroosmotic flow, which indicated that the effective mobility of fluorecein was approximately 2×10^{-4} $cm^2V^{-1}s^{-1}$ in this case. This arrangement suggests that the droplet transfer event could potentially constitute a novel injection method for an electrophoretic separation. While the possibility is enticing, in the present form the droplet volumes were too large to serve as an efficient CE injection plug. However, utilizing EOF for post-extraction transport is attractive in that the plug-shaped flow profile can help to avoid the band broadening Taylor dispersion associated with pressure-driven laminar flow.

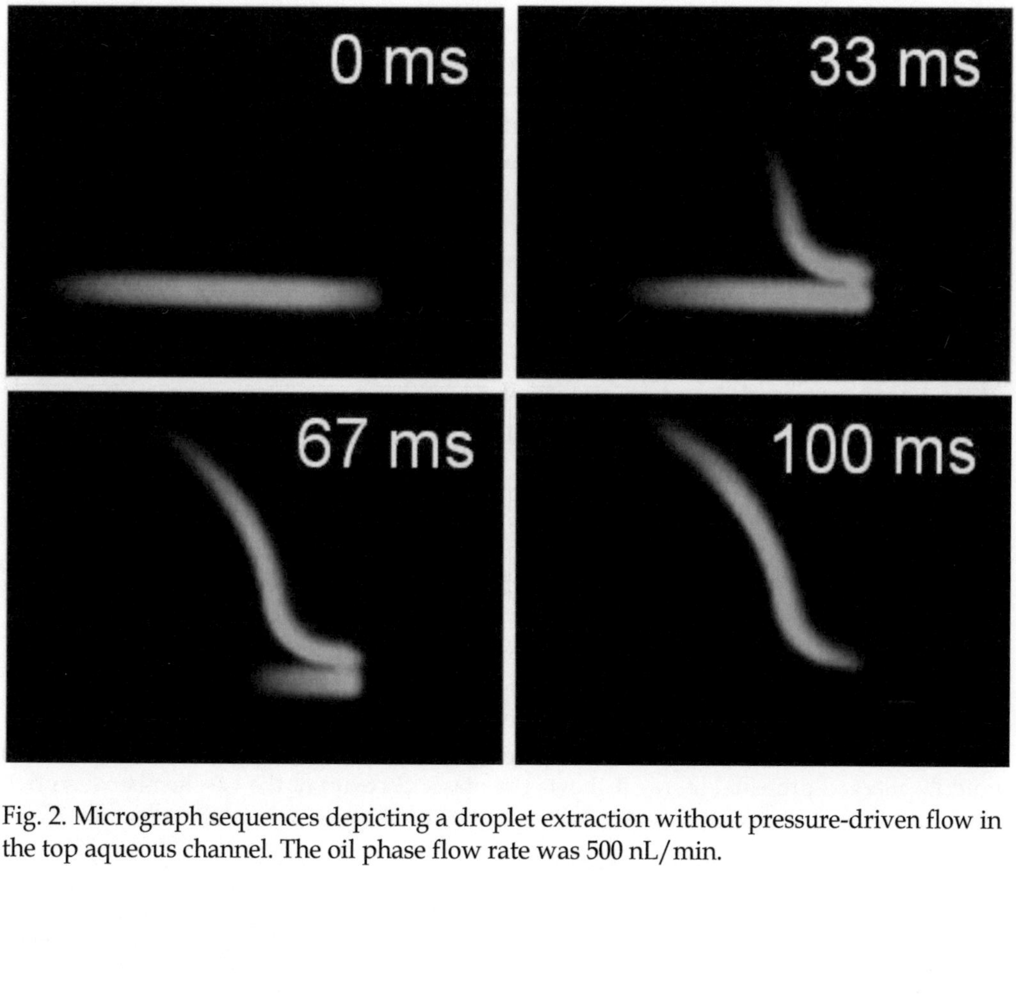

Fig. 2. Micrograph sequences depicting a droplet extraction without pressure-driven flow in the top aqueous channel. The oil phase flow rate was 500 nL/min.

3.3 Impact of oil and aqueous flow rates on droplet extraction

Where EOF-based transport is not practical, hydrodynamic flow is utilized for transport of the extracted contents in the aqueous channel. We showed previously that by positioning the detector sufficiently close to the extraction region, the droplet contents could be analyzed with essentially no dilution (Kelly et al., 2009). However, the flow rates in the oil and aqueous channels needed to be carefully balanced to maintain a stable extraction interface as even a small variance of the aqueous flow rate could disrupt the droplet extraction. The current interface maintained much more robust and effective droplet extraction, enabling the effect of large changes in the aqueous flow rate to be explored. As a baseline, Figure 3A presents a micrograph sequence depicting the droplet extraction procedure in which both the oil and aqueous flow rates were 400 nL/min. When the aqueous plug arrived at the interface, the entire droplet rapidly transferred to the aqueous flow channel. The extracted droplet content was forced to flow in the direction of the hydrodynamic aqueous flow immediately, and the sample stream filled the entire channel cross-section with minimal dilution. Figure 3B shows the droplet transfer behavior at different aqueous flow rates. The oil flow rate was kept constant at 200 nL/min. In each case, the interface was kept stable, and effective droplet transfer was obtained because of the low surface energy and strong capillary force applied on the droplet. When the aqueous flow rate was larger than the oil flow rate, the high pressure originating from the aqueous buffer flow resulted in large resistance to droplet transfer and confined the extracted sample plug into a narrow stream in the aqueous flow channel. Further increasing the aqueous flow rate led to an increase in the droplet transfer (Figure 3C) and a narrowing of the extracted sample stream (Figure 3B), which indicated that the sample would be diluted much more after extraction. When the aqueous flow rate was lower than that in the oil channel, the extraction event essentially disrupted the aqueous flow and the extracted sample inserted itself into the aqueous stream. The excellent stability of the interface enables the oil and aqueous flow rates to be selected independently based upon the desired properties; where dilution should be minimized, it is important that the aqueous flow rate does not exceed that of the oil stream. Varying the oil stream flow rate while holding the aqueous stream constant produced analogous results as shown in Figure 4.

3.4 NanoESI-MS analysis of extracted droplets

The extracted droplet contents could be delivered by the aqueous flow to a monolithically integrated ESI emitter (Kelly et al., 2009; Zhu & Fang, 2010) for MS detection. Figure 5 shows the MS detection of droplets containing 1 μg/μL apomyoglobin in DI water. The oil and ESI buffer flow rates were 100 nL/min and 400 nL/min, respectively. The droplet generation frequency was 0.1 Hz. For a continuous analysis of 60 droplets in 10 min (shown in Figure 5A), the relative standard deviation (RSD) of the peak heights was 3.2%, and RSD of the peak widths was 5.7%. Figure 5B is a zoomed-in view of the MS detection of apomyoglobin droplets in 2 min from the same experiment as shown in Figure 5A. The mass spectrum of peak (a) indicated from Fig. 5B is shown in Fig. 6A with the characteristic signal for the multiply charged protein. Figure 6B shows the mass spectrum of the baseline between two droplets. No protein peaks were observed, which demonstrates there was no cross-contamination between extracted droplets.

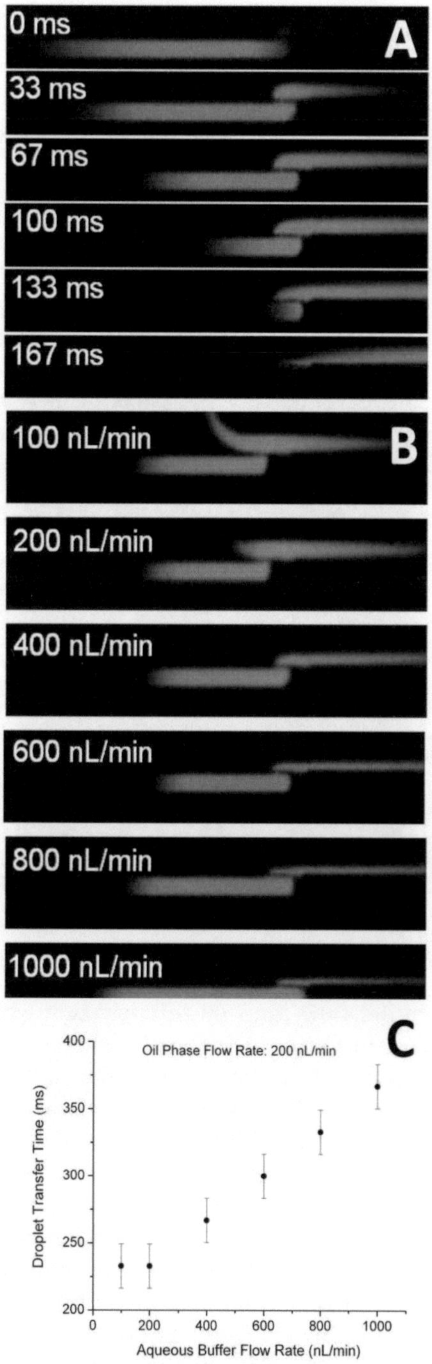

Fig. 3. (A) Micrograph sequence depicting the extraction of an individual fluorescein droplet. The flow rate in both channels was 400 nL/min. (B) Photographs of droplet extraction at different aqueous buffer flow rates. The flow rate of the continuous oil phase was 200 nL/min. (C) Plot of droplet transfer time versus aqueous buffer flow rate.

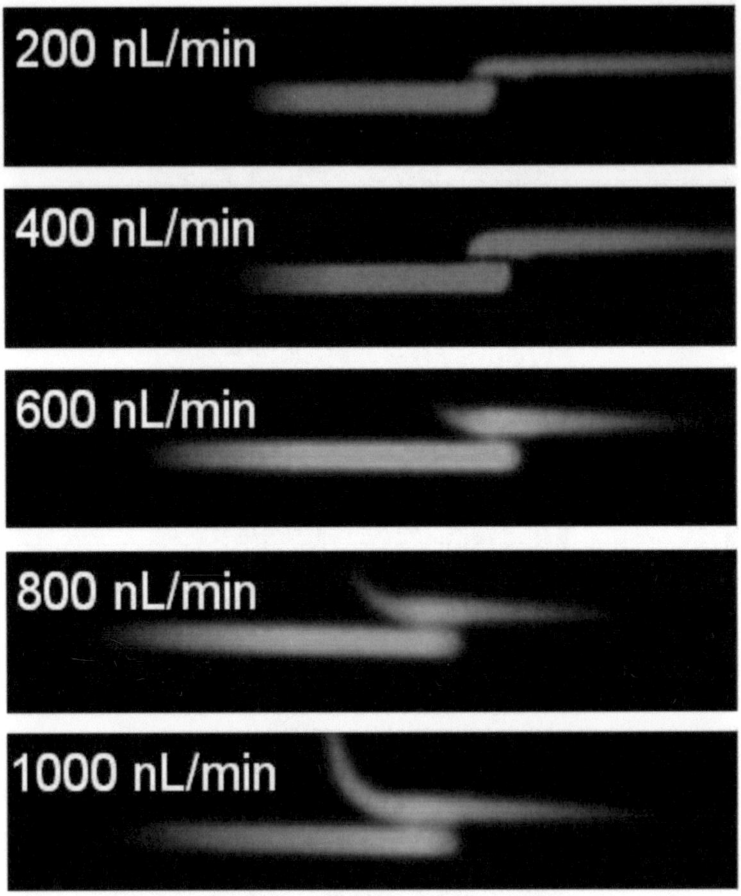

Fig. 4. Photographs of droplet extraction at different oil phase flow rates. The flow rate of the aqueous buffer was 400 nL/min.

The impact of oil and ESI buffer flow rates on MS detection was investigated using 1 μM leucine enkephalin as the aqueous sample. Figure 7A shows the variance of the MS peak intensity and width with ESI buffer flow rate, in which the oil flow rate was kept constant at 100 nL/min. With an increase in the ESI buffer flow rate, both the MS peak intensity and width decreased. This result is consistent with the discussion above, i.e., that the increase in the ESI buffer flow rate would dilute the extracted droplet content much more and broaden the actual sample band. The decrease in peak width is due to the increased flow rate as shown in Figure 3B. In addition, the electrospray ionization efficiency was reduced when increasing the ESI buffer flow rate, which would also influence the MS peak intensity negatively. Figure 7B shows the effect of oil flow rate on MS detection. The ESI buffer flow rate was held constant at 500 nL/min. At lower oil flow rates (< ESI buffer flow rate), the MS peak intensity increased and the peak width decreased with increase in the oil phase flow rate, which resulted from the minimized dilution during droplet extraction. When the oil flow rate increased further (≥ ESI buffer flow rate), the MS peak intensity and width changed less, which was due to the droplet transfer with minimal dilution in these situations.

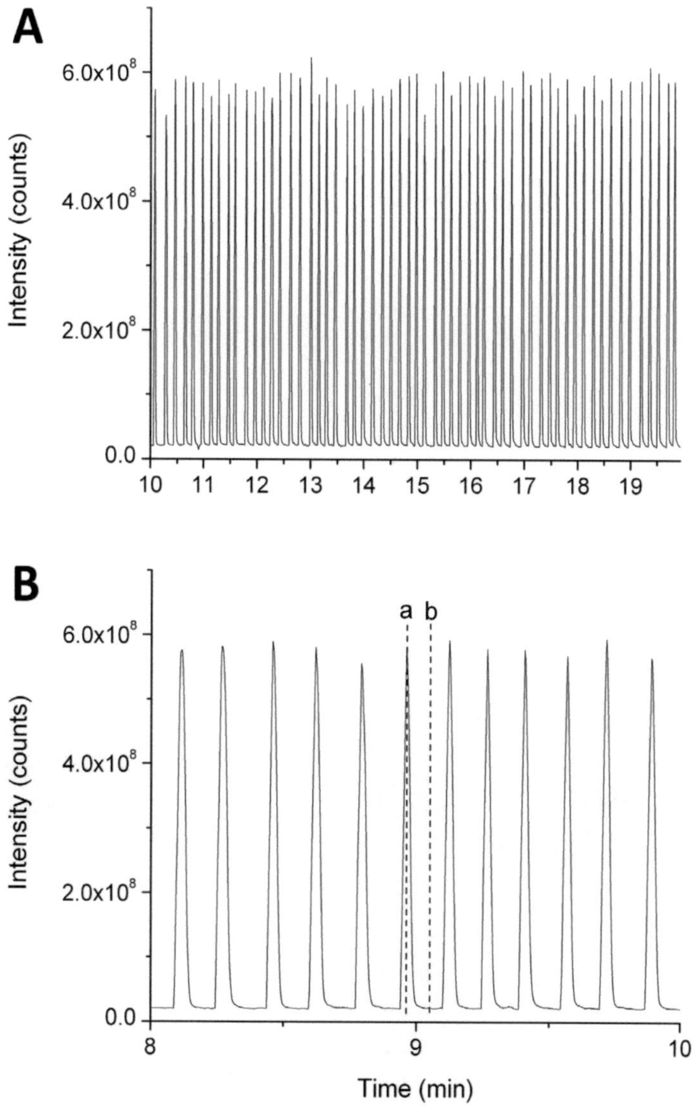

Fig. 5. MS signal intensity (total ion count) vs. time. (A) Detection of the extracted 1 µg/µL apomyoglobin droplets. Oil phase flow rate was 100 nL/min, ESI buffer flow rate was 400 nL/min, and droplet generation frequency was 0.1 Hz. (B) Detailed view of the MS-detected extracted apomyoglobin droplets. (a) and (b) correspond to mass spectra in Fig. 6.

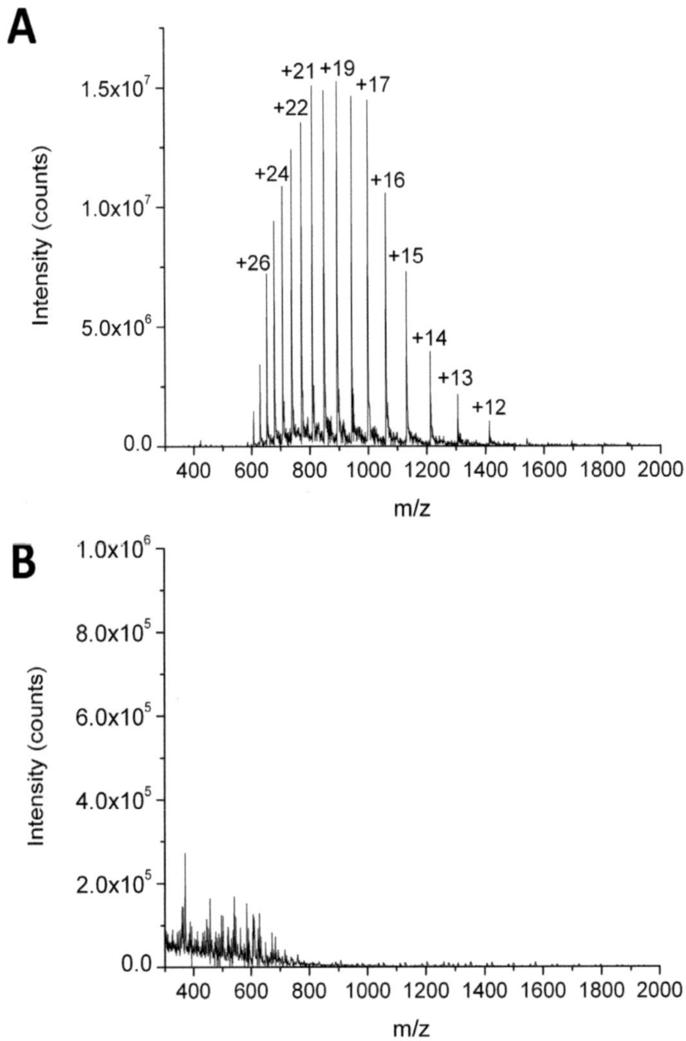

Fig. 6. (A) and (B) Mass spectra obtained from the peak and baseline indicated as (a) and (b) in Fig. 5B, respectively.

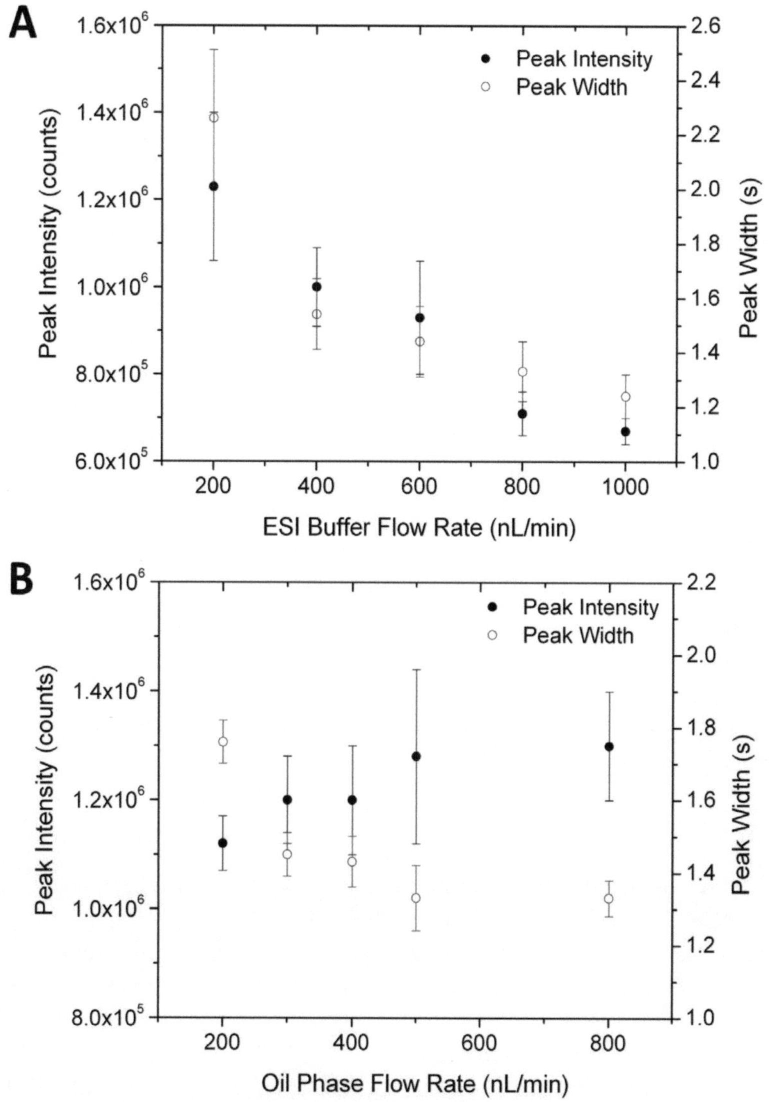

Fig. 7. (A) Plot of MS peak intensity and width vs. ESI buffer flow rate. The sample was 1 µM leucine enkephalin. Oil phase flow rate was 100 nL/min. (B) Plot of the MS peak intensity and width vs. oil phase flow rate. ESI buffer flow rate was 500 nL/min.

4. Conclusion

We developed a robust and efficient droplet extraction interface integrated in a droplet-based PDMS microfluidic device. The interface was constructed with an array of cylindrical posts with small apertures in between. On one side of the interface, the aqueous phase flow channel surface was selectively modified to be hydrophilic by corona discharge. The other side, i.e. the continuous oil phase flow channel surface, remained hydrophobic by protecting the surface with water. The combination of the narrow apertures and surface modification changed the surface tension at the interface. The lower surface energy of the hydrophilic surface for aqueous droplet and capillary force originated from the small structure facilitated the aqueous droplet transfer while preventing the oil from entering the aqueous flow channel. The resulting surface tension not only helped to establish a stable interface at the junction, but also enabled large pressure imbalances resulting from different oil and aqueous flow rates to be tolerated. All droplets could be transferred through the interface entirely within broad ranges of aqueous flow rates (0 - 1 μL/min, i.e. 0 - 1.33 cm/s) and oil flow rates (100 nL/min - 1 μL/min, i.e. 0.13 - 1.33 cm/s). When the oil flow rate was smaller than the aqueous flow rate, the higher pressure from the aqueous flow slowed the droplet transfer and squeezed the extracted sample stream, which induced dilution of the droplet content after extraction. If the aqueous flow rate was lower, the droplets were extracted rapidly and without dilution. The extracted droplet contents could be analyzed by MS through a monolithically integrated ESI emitter and the droplet extraction efficiency was 100%. The interface described here will form a key component in integrated platforms for reactions and analyses at the picoliter scale.

5. Acknowledgements

Portions of this research were supported by the U.S. Department of Energy (DOE) Office of Biological and Environmental Research, the NIH National Center for Research Resources (RR018522) and the Environmental Molecular Sciences Laboratory (EMSL), a U.S. DOE national scientific user facility located at the Pacific Northwest National Laboratory (PNNL) in Richland, WA. PNNL is a multiprogram national laboratory operated by Battelle for the DOE under Contract No. DE-AC05-76RLO 1830.

6. References

Aebersold, R. & Mann, M. (2003). Mass spectrometry-based proteomics. *Nature*, Vol.422, No.6928, (March 2003), pp. 198-207, ISSN 0028-0836

Chiu, D. T. & Lorenz, R. M. (2009). Chemistry and Biology in Femtoliter and Picoliter Volume Droplets. *Accounts of Chemical Research*, Vol.42, No.5, (May 2009), pp. 649-658, ISSN 0001-4842

Chiu, D. T.; Lorenz, R. M. & Jeffries, G. D. M. (2009). Droplets for ultrasmall-volume analysis. *Analytical Chemistry*, Vol.81, No.13, (July 2009), pp. 5111-5118, ISSN 0003-2700

Fidalgo, L. M.; Whyte, G.; Bratton, D.; Kaminski, C. F.; Abell, C. & Huck, W. T. S. (2008). From microdroplets to microfluidics: selective emulsion separation in microfluidic devices. *Angewandte Chemie International Edition*, Vol.47, No.11, (February 2008), pp. 2072-2075, ISSN 1433-7851

Fidalgo, L. M.; Whyte, G.; Ruotolo, B. T.; Benesch, J. L. P.; Stengel, F.; Abell, C.; Robinson, C. V. & Huck, W. T. S. (2009). Coupling Microdroplet Microreactors with Mass Spectrometry: Reading the Contents of Single Droplets Online. *Angewandte Chemie International Edition*, Vol.48, No.20, (May 2009), pp. 3665-3668, ISSN 1433-7851

Filla, L. A.; Kirkpatrick, D. C. & Martin, R. S. (2011). Use of a corona discharge to selectively pattern a hydrophilic/hydrophobic interface for integrating segmented flow with microchip electrophoresis and electrochemical detection. *Analytical Chemistry*, Vol.83, No.15, (June 2011), pp. 5996-6003, ISSN 0003-2700

Galas, J. C.; Bartolo, D. & Studer, V. (2009). Active connectors for microfluidic drops on demand. *New Journal of Physics*, Vol.11, (July 2009), pp. 075027, ISSN 1367-2630

Huebner, A.; Sharma, S.; Srisa-Art, M.; Hollfelder, F.; Edel, J. B. & deMello, A. J. (2008), Microdroplets: a sea of applications? *Lab on a Chip*, Vol.8, No.8 (August 2008), pp. 1244-1254, ISSN 1473-0197

Kelly, R. T.; Tang, K.; Irimia, D.; Toner, M. & Smith, R. D. (2008). Elastomeric microchip electrospray emitter for stable cone-jet mode operation in the nanoflow regime. *Analytical Chemistry*, Vol.80, No.10, (May 2008), pp. 3824-3831, ISSN 0003-2700

Kelly, R. T.; Page, J. S.; Marginean, I.; Tang, K. & Smith, R. D. (2009). Dilution-Free Analysis from Picoliter Droplets by Nano-Electrospray Ionization Mass Spectrometry. *Angewandte Chemie International Edition*, Vol.48, No.37, (September 2009), pp. 6832-6835, ISSN 1433-7851

Kelly, R. T.; Tolmachev, A. V.; Page, J. S.; Tang, K. & Smith, R. D. (2010). The ion funnel: Theory, implementations, and applications. *Mass Spectrometry Reviews*, Vol.29, No.2, (March/April 2010), pp. 294-312, ISSN 1098-2787

Koster, S. & Verpoorte, E. (2007). A decade of microfluidic analysis coupled with electrospray mass spectrometry: An overview. *Lab on a Chip*, Vol.7, No.11, (November 2007), pp. 1394-1412, ISSN 1473-0197

Liu, T.; Belov, M. E.; Jaitly, N.; Qian, W.-J. & Smith, R. D. (2007). Accurate mass measurements in proteomics. *Chemical Reviews*, Vol.107, No.8, (August 2007), pp. 3621-3653, ISSN 0009-2665

Mellors, J. S.; Gorbounov, V.; Ramsey, R. S. & Ramsey, J. M. (2008). Fully integrated glass microfluidic device for performing high-efficiency capillary electrophoresis and electrospray ionization mass spectrometry. *Analytical Chemistry*, Vol.80, No.18, (September 2008), pp. 6881-6887, ISSN 0003-2700

Pei, J.; Li, Q.; Lee, M. S.; Valaskovic, G. A. & Kennedy, R. T. (2009). Analysis of samples stored as individual plugs in a capillary by electrospray ionization mass spectrometry. *Analytical Chemistry*, Vol.81, No.15, (August 2009), pp. 6558-6561, ISSN 0003-2700

Pei, J.; Nie, J. & Kennedy, R. T. (2010). Parallel electrophoretic analysis of segmented samples on chip for high-throughput determination of enzyme activities. *Analytical Chemistry*, Vol.82, No.22, (November 2010), pp. 9261-9267, ISSN 0003-2700

Roman, G. T.; Wang, M.; Shultz, K. N.; Jennings, C. & Kennedy, R. T. (2008). Sampling and electrophoretic analysis of segmented flow streams using virtual walls in a microfluidic device. *Analytical Chemistry*, Vol.80, No.21, (November 2008), pp. 8231-8238, ISSN 0003-2700

Song, H.; Chen, D. L. & Ismagilov, R. F. (2006). Reactions in droplets in microfluidic channels. *Angewandte Chemie International Edition*, Vol.45, No.44, (November 2006), pp. 7336-7356, ISSN 1433-7851

Sun, X.; Kelly, R. T.; Tang, K. & Smith, R. D. (2010). Ultrasensitive nanoelectrospray ionization-mass spectrometry using poly(dimethylsiloxane) microchips with monolithically integrated emitters. *Analyst*, Vol.135, No.9, (September 2010), pp. 2296-2302, ISSN 0003-2654

Sun, X.; Kelly, R. T.; Danielson III, W. F.; Agrawal, N.; Tang, K. & Smith, R. D. (2011a). Hydrodynamic injection with pneumatic valving for microchip electrophoresis with total analyte utilization. *Electrophoresis*, Vol.32, No.13, (June 2011), pp. 1610-1618, ISSN 0173-0835

Sun, X.; Kelly, R. T.; Tang, K. & Smith, R. D. (2011b). Membrane-based emitter for coupling microfluidics with ultrasensitive nanoelectrospray ionization-mass spectrometry. *Analytical Chemistry*, Vol.83, No.14, (July 2011), pp. 5797-5803, ISSN 0003-2700

Teh, S.-Y.; Lin, R.; Hung, L.-H. & Lee, A. P. (2008). Droplet microfluidics. *Lab on a Chip*, Vol.8, No.2, (February 2008), pp. 198-220, ISSN 1473-0197

Theberge, A. B.; Courtois, F.; Schaerli, Y.; Fischlechner, M.; Abell, C.; Hollfelder, F. & Huck, W. T. S. (2010). Microdroplets in microfluidics: An evolving platform for discoveries in chemistry and biology. *Angewandte Chemie International Edition*, Vol.49, No.34, (August 2010), pp. 5846-5868, ISSN 1433-7851

Unger, M. A.; Chou, H.-P.; Thorsen, T.; Scherer, A. & Quake, S. R. (2000). Monolithic Microfabricated Valves and Pumps by Multilayer Soft Lithography. *Science*, Vol.288, No.5463, (April 2000), pp. 113-116, ISSN 0036-8075

Wang, M.; Roman, G. T.; Perry, M. L. & Kennedy, R. T. (2009). Microfluidic chip for high efficiency electrophoretic analysis of segmented flow from a microdialysis probe and in vivo chemical monitoring. *Analytical Chemistry*, Vol.81, No.21, (November 2009), pp. 9072-9078, ISSN 0003-2700

Yang, C.-G.; Xu, Z.-R. & Wang, J.-H. (2010). Manipulation of droplets in microfluidic systems. *Trends in Analytical Chemistry*, Vol.29, No.2, (February 2010), pp. 141-157, ISSN 0165-9936

Zeng, S.; Li, B.; Su, X.; Qin, J. & Lin, B. (2009). Microvalve-actuated precise control of individual droplets in microfluidic devices. *Lab on a Chip*, Vol.9, No.10, (May 2009), pp. 1340-1343, ISSN 1473-0197

Zeng, S.; Pan, X.; Zhang, Q.; Lin, B. & Qin, J. (2011). Electric control of individual droplet breaking and droplet contents extraction. *Analytical Chemistry*, Vol.83, No.6, (March 2011), pp. 2083-2089, ISSN 0003-2700

Zhu, Y. & Fang, Q. (2010). Integrated droplet analysis system with electrospray ionization-mass spectrometry using a hydrophilic tongue-based droplet extraction interface. *Analytical Chemistry*, Vol.82, No.19, (October 2010), pp. 8361-8366, ISSN 0003-2700.

Part 3

Applications

8

Microfluidics in Single Cell Analysis

Caroline Beck and Mattias Goksör
University of Gothenburg
Sweden

1. Introduction

Microfluidics is a research field that evolved during the early 1980's and is still under constant and rapid progress. As indicated by the name, microfluidics concerns the behaviour of fluids in micro length scale environments and originally has its roots in four different disciplines: molecular biology and analysis, microelectronics and, maybe a bit surprisingly, biodefence (Whitesides 2006). The applications of microfluidic systems are usually called μTAS (micro total analysis systems) or LOC (lab-on-a-chip) devices, and refer to the possibility to scale down analyses and reactions to a miniaturized format, something that has opened up new exciting applications. Analytical systems for DNA sequencing, separation and manipulation, enzymatic assays, protein analysis, cell sorting, counting and culturing etc. are applications readily applied (Beebe & Paguirigan 2008; Jensen et al 2006; Manz et al 2002a; Manz et al 2002b; Tudos et al 2001; van den Berg & Andersson 2003; van den Berg & Lammerink 1998; Weibel et al 2007). Even in the fast growing research field of systems biology, microfluidic systems have come to play an important role. Systems biology aims to reduce and describe complex biological systems with mechanistic models such as chemical kinetics and control theory. The models are in turn used to simulate and predict the behaviour of a biological system (Kitano 2002). In order to refine these models, traditional population based data from e.g. Western blot and microarrays, needs to be complemented with data acquired from single cells (Ryley & Pereira-Smith 2006). Since a microfluidic device offers the possibility to spatially and temporally control the fluid within, it has proven to be a powerful tool when following biological responses to perturbations on the level of individual cells (Bennett et al 2008). In traditional methodology, the removal of a specific substance or drug from cells to investigate a certain response is, if not impossible, difficult without changing the equilibrium of the whole cell culture. It is also possible that, depending on the method of investigation, the time required to change the cellular environment actually takes longer than the event you are trying to capture. Hence, the temporal resolution of such an experiment would be less than required and likely impossible to improve by traditional methods. However, these obstacles may be overcome by the use of microfluidics. Other benefits of microfluidics are vastly reduced sample and reagent volumes, lower costs and power consumption and the design versatility (Whitesides & Sia 2003).

2. Single cell analysis

Basic biological processes are often studied using a reductionist approach. Thus, only a few isolated events are monitored and any interference from external factors is kept to a minimum. The use of model organisms has gained much attraction and has contributed to the understanding of other higher species. Since many fundamental properties are conserved throughout the evolution, the investigation of simpler life forms can thus contribute to the understanding of human biological mechanisms and diseases. Within medical research, model organisms such as rats and mice designed to express a specific set of genes are often used. In molecular biology as well as in genetics, model organisms include bacteria, yeast cells, banana flies and plants. Often *in vivo* (within the living) experiments are separated from *in vitro* (in glass) experiments, where the latter indicate that the biochemical process is removed from its natural environment and instead takes place in an artificial environment. However, a cell biologist would instead use this term when referring to cells cultured outside the specific organism from which they are retrieved. *In vitro* experiments allow manipulations and measurements that would be impossible *in vivo*. In addition, an *in vitro* experimentalist may take advantage of the possibility to use genetically homogenous cell populations. However, also in populations with low cell-to-cell variation, traditional averaging techniques only provide the mean response for the cell population as a whole.However, also in populations with low cell-to-cell variation, traditional averaging techniques only provide the mean response for the cell population as a whole. Any information on the heterogeneity within such a cell culture, although genetically homogenous, will thus be lost. The remaining cell-to-cell variation depends on intrinsic and extrinsic noise (O'Shea & Raser 2004) and concerns differences in the concentration of regulatory proteins or bio-molecules, cell state (cell cycle stage and cell age) as well as involved reactions and in which order these occur within the cell. These parameters play a crucial role in the biological processes and will of course mirror the experimental results.

A schematic example of the response from a cell culture measured with a bulk measurement technique that could be misinterpreted is illustrated in figure 1 and 2. In the first example (Figure 1) the cells respond to an external stimulus either (left) as a homogenous population (all cells react at the same time and approximately to the same extent); (center) in a bimodal distribution (only a fraction of the cells react, but the reaction is to the fullest extent); or (right) the cells show a response as a combination of the two. Hence, both the kinetics and measured mean values might not correspond to the actual scenario taking place. In the second example (Figure 2) the single cell data reveal a much faster response to outer stimuli than the averaged response. Since the response is temporally distributed an averaged result will thus be falsely interpreted. Here is where single cell analysis becomes a vital and complementary tool to traditional techniques when interpreting biological events.

There are techniques available that to some extent can separate a cell population into sub-populations. FACS (Fluorescent Activated Cell Sorting) utilises fluorescence and light-scattering properties of individual cells that pass through a laser beam (Dean & Hoffman 2007). This results in high throughput data and the cells can even be separated into sub-populations and collected for further analyses.

However, since the cells are in constant motion through the laser beam they are therefore impossible to track. Hence, there is no possibility to do time series experiments or to re-analyse a specific cell. Re-analyses of specific cells are, however, possible with the technique, laser scanning cytometry. Here, labeled cells are located on a glass slide and their emitted fluorescence is recorded with a microscope. This technique will give detailed morphological information but has limited possibilities to sort the cells into sub-populations. Even though the throughput is lower compared to FACS, this limitation could be overcome by a high content screening set up in which a large number of cells can be imaged over time (Deptala et al 2001). One should remember that high quality single cell analysis should be accompanied by relevant statistics. If possible, the techniques should combine environmental control with cell manipulation in an attempt to limit the extracellular influences that otherwise would broaden the population distribution. With the progressions within μTAS/LOC development and automated analyses approaches, it is only a question of time before single cell analysis and high throughput studies will be established (Pepperkok & Ellenberg 2006).

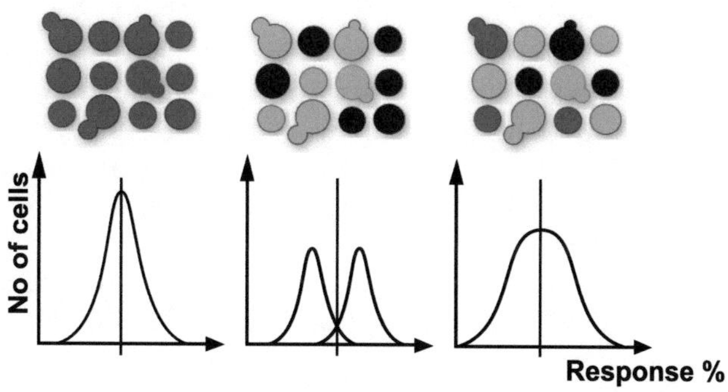

Fig. 1. Heterogenous responses may be obscured in population techniques. Population data of cells represents the mean response value from all cells studied. Hence, it is impossible to distinguish from the mean response value if the cells within the population react in the same way; as a homogenous population where all cells react at the same time and approximately to the same extent (left); in a bimodal distribution where only a fraction of the cells react to the fullest (centre); or the cells show a response as a combination of the two (right). The mean values correspond to the vertical black lines.

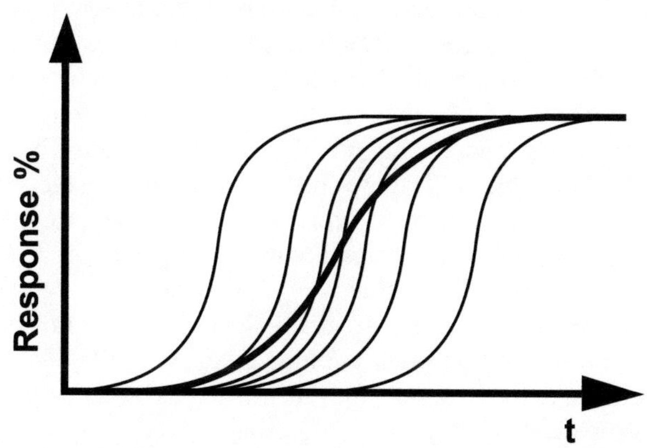

Fig. 2. Differences in measured response time of single cells versus population. Averaging techniques could be misleading when interpreting the cellular kinetic response. The mean response value of the population (thick line) shows a slower response over time (t) than the individual cell responses (thin lines).

2.1 Yeast cells as model organism

Saccharomyces cerevisiae (*S. cerevisiae*) is often known in popular parlance as baker's yeast or budding yeast, and belongs to the fungi species. The organism commonly inhabits sugar containing substrates (fruits and flowers) and can adapt to a wide range of environmental conditions. It can tolerate temperatures between freezing and 55°C (although proliferation occurs between 12 and 40°C), grows in both acidic and moderately alkaline milieus (pH 2.8-8.0) and tolerates both high sucrose (3 M) and ethanol (20%) concentrations. Not surprisingly, *S. cerevisiae* is the main organism used in both wine and beer production, but is also used for drug screening and functional analyses as well as a safe production organism of commercially important proteins (Bayne et al 1988; Mcaleer et al 1984). Many eukaryotic characteristics were first discovered in yeast and due to powerful genetics and molecular biology, yeast is today used not only to study the function of living cells, but also to improve or generate new biotechnological approaches (Stagljar et al 2006). *S. cerevisiae* is a unicellular organism, approximately 4-7 µm in size, and represents the simplest form of eukaryotic cell, i.e. complex internal structures are enclosed by membranes. Several processes, such as chromosome replication and cell cycle control, intracellular signal transductions, as well as transcription and translation events, are similar between *S. cerevisiae* and higher eukaryotic cells and therefore frequently studied in yeast. However, *S. cerevisiae* is protected from the extracellular milieu by a cell wall (Klis et al 2006), a feature distinguishing them from mammalian cells but shared with bacteria, algae and plants. In 1996 the complete genome of *S. cerevisiae* was successfully sequenced (Goffeau et al 1996) and the 16 linear chromosomes were shown to consist of approximately 13 million base pairs (including the ribosomal DNA) encoding in total roughly 6000 open reading frames (ORFs). Even though it was the first eukaryote genome ever sequenced, molecular geneticists had transformed *S. cerevisiae* with foreign DNA already in 1978 (Beggs 1978; Hinnen et al 1978). Nearly 31% of the possible ORFs show statistically robust homologues in the mammalian genomes and many of the known genes in *S. cerevisiae* have mammalian orthologs (Ploger et al 2000). This makes yeast an ideal model organism in which biological events that could have future impact on disease prevention, easily can be studied (Hartwell 2004). The knowledge of yeast genetics is today vast and summarised at the public Saccharomyces Genome Database (www.yeastgenome.org), where various search options and analysis tools for homology and gene comparisons are also available. The knowledge of each gene function is a joint goal for the yeast community. This is facilitated by the yeast GFP clone collection, containing approximately 75% of all yeast genes tagged to green fluorescent protein (GFP), enabling protein detection and localization (http:// http://yeastgfp.yeastgenome.org/). Yeast is overall a very robust model organism to work with; it is easy to cultivate, has a short generation time (about 90 minute), is non-pathogenic and possesses a well known genome that easily can be manipulated. *S. cerevisiae* are grown on solid agar plates or in liquid nutrient broth and can survive storage in glycerol stocks at -70°C for long periods of time. The most common liquid nutrient is yeast peptone dextrose (YPD). However, since YPD is highly auto-fluorescent, a synthetic minimum yeast nitrogen base (YNB) medium is used for fluorescent imaging applications.

In nature *S. cerevisiae* is found as diploids, meaning that they have two copies of their chromosomes, probably since the extra gene copy increases their chances of surviving essential gene mutations. They divide by budding off new cells (hence the name budding yeast) which results in two genetically identical cells of unequal size, a larger one (the mother) and a smaller one (the daughter). After budding, every daughter cell leaves a bud scar on the cell wall of the mother cell. Mother cells may bud as many as 30-50 times and hence have the same amount of bud scars (Mortimer & Johnston 1959). Under certain circumstances, such as e.g. nitrogen starvation, the diploid cells sporulate and form four haploid (one copy of each chromosome) spores, of which usually only a single spore survives. This spore will in turn germinate when the nutrient access is sufficient and reproduce as a haploid cell. Two haploid cells can via a sexual phase, however, fuse to a diploid cell, provided they express the opposite mating types, a and α (Burgess 2003). This mating can even happen between mother and daughter cells, since the mother cell can switch mating type. The life cycle of *S. cerevisiae* has many advantages. The spores can survive harsh environments and the sporulation event "cleans" the genome from any possible accumulated mutations and gives rise to genetic diversity by allele crossover. Furthermore, mating partners from other spores might result in possible genetic advantages. In a laboratory environment the yeast cells are kept as stable haploid cell cultures genetically modified so that no mating type switches are possible.

2.1.1 HOG - A MAPK pathway

External perturbations of a cell are sensed and conveyed into its interior by processes referred to as signal transduction. Even though the perturbation can differ from stress conditions and growth factors to specific cytokines and hormones, the signal transduction triggers an appropriate response to counteract or adapt the cell to the new environment. The principles of the intracellular signal transductions of *S. cerevisiae* are considered to be well conserved, especially those belonging to the so called mitogen activated protein (MAP) kinase pathways (Shields et al 1999). These pathways all share a common architecture where the most downstream component, the MAP kinase (MAPK), becomes phosphorylated and thereby activated by its upstream MAPK kinase (MAPKK) (Figure 3). This MAPKK is in turn phosphorylated by a further upstream MAPKK kinase (MAPKKK).

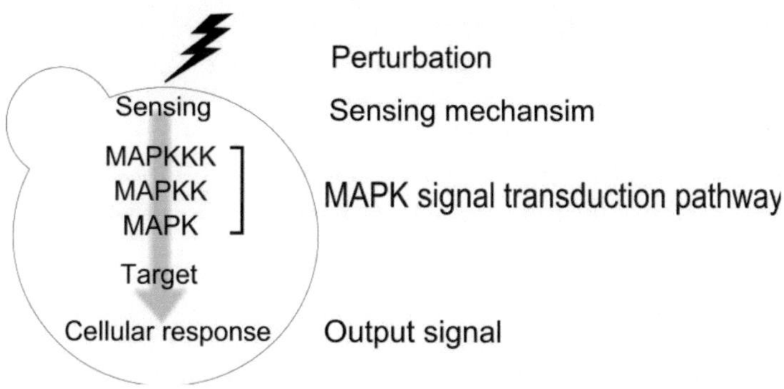

Fig. 3. The common architecture of a typical MAPK signal transduction pathway in yeast. Outer stimulus is sensed and transduced into the interior of the cell via the phosphorylation events of the pathway. When reaching the targets, the output signal corresponds to a cellular response to counteract and/or adapt the cell to the perturbation.

An external stimulus is responsible for activating this most upstream kinase of the signal transduction pathway that leads to the cellular response. The High Osmolarity Glycerol

(HOG) response pathway is one of four MAPK pathways present in *S. cerevisiae* (Figure 4) (Gustin et al 1998). The perturbation is constituted by hyper osmotic stress and results in glycerol production to prevent dehydration (Thorner et al 2004). The signal transduction is mediated by activated Hog1 (Hog1-PP) which migrates from the cytosol to the cell nucleus via the nuclear importer Nmd5 (Silver et al 1998). Inside the nucleus the transcription of up to 600 genes is up regulated (Thorner et al 2004). The HOG pathway has two separate branches of activation, the Sln1 and the Sho1 branch. In total, three plasma transmembrane sensors activate the pathway; Sln1, Msb2 and Hkr1, where the latter two belong to the Sho1 branch (Posas et al 1996; Tatebayashi et al 2007). Together with two other components, Ypd1 and Ssk1, Sln1 constitutes what is called a phospho-relay module. Sln1 is active under normal osmotic conditions and becomes inactivated during hyper osmotic stress. When active, Sln1 is auto-phosphorylated and the phospho-group is mediated via Ypd1 to Ssk1. Ssk1 acts as a response regulator and becomes inactivated when receiving the phospho-group, i.e. does not activate the downstream transduction components (Saito & Posas 1998).

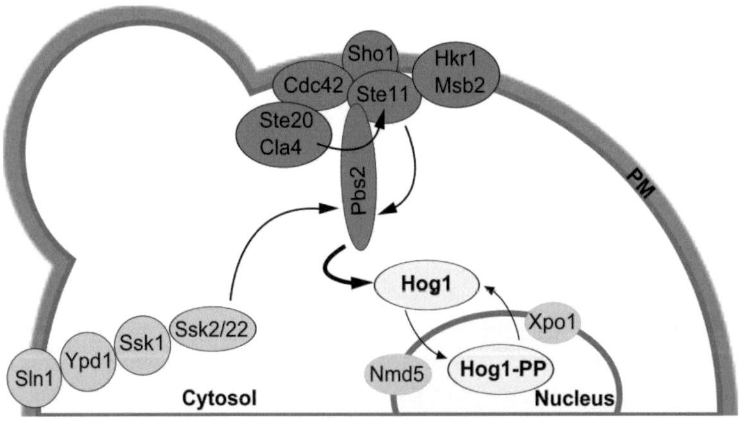

Fig. 4. A schematic overview of the HOG pathway in *S. cerevisiae*. The pathway sensors (Sln1, Msb2 and Hkr1) are situated in the plasma membrane (PM) of the cell. Upon high osmolarity the cytosolic MAPK Hog1 is phosphorylated (Hog1-PP) by the upstream MAPKK Pbs2. Pbs2 in turn can be activated by two separate sensing branches, the Sln1 and Sho1 branch. Upon activation, Hog1 migrates into the nucleus via the nuclear importer Nmd5 where it binds to specific DNA regions and together with cofactors up regulates gene transcription. Upon dephosphorylation and inactivation, Hog1 is rerouted into the cytosol via the exporter protein Xpo1.

However, during hyperosmotic stress Sln1 is dephosphorylated and consequently the amount of unphosphorylated Ssk1 accumulates. Ssk1 binds to the components Ssk2 and Ssk22, which becomes auto-phosphorylated and activated, and in turn can activate the MAPKK Pbs2 by yet another phosphorylation. The activation mechanism of the Sho1 branch is still partly unknown. Since Sho1 is a transmembrane protein it was first thought to act as a sensor of this branch (Maeda et al 1995). Instead, it seems to be important as a scaffold protein, recruiting other components important for active cell growth and remodeling of the cell membrane (Ammerer et al 2000). The stimulation of the Sho1 branch sensors probably recruits the MAPKK Pbs2 to the cell membrane, which will then act as a scaffold protein for the other components of the Sho1 branch. The kinases Ste11, Ste20 and Cla4 will in association with the G-protein Cdc42 all contribute to the phosphorylation of Pbs2 (Hohmann 2009).

As previously described, the two branches merge by the activation of the MAPKK Pbs2 that in turn phosphorylates and activates the transcription factor and MAPK, Hog1. Although functionally redundant, the Sln1 branch is more sensitive and shows gradual activation characteristics, while the Sho1 branch requires a higher degree of osmosis to react in an "all-or-nothing" fashion (O'Rourke & Herskowitz 2004). The cellular response to the increased osmolarity involves transcription of both Hog1 specific and general stress-dependent genes, which are thought of as the cells' long-term adaptation to the new environment (Ferreira et al 2005; Rep et al 2000; Rep et al 1999). However, since cells are still able to adapt to increased osmolarity in the absence of nuclearized Hog1 (Thorner et al 2008), mechanisms extending beyond gene transcription seem to be of higher importance when instant adaption is required (Dihazi et al 2004; Tamas et al 2006). Since a constitutively active Hog1 can be detrimental to the cell, the amplitude and period of Hog1 activation needs to be precisely controlled. This is achieved by dephosphorylation of Hog1. Several different phosphatases can act on Hog1 which allows precise temporal and spatial modulation of its activity (Molina et al 2005; Warmka et al 2001). When Hog1 is dephosphorylated it returns to the cell cytosol via the nuclear exportin, Xpo1, where it awaits a new perturbation (Silver et al 1998). The shuttling between cytosol and nucleus is thought to correspond to the activity of Hog1. The mammalian Hog1 ortholog, MAPK p38, is involved in chronic inflammatory diseases and its four isoforms are activated upon different environmental stresses such as e.g. inflammatory cytokines, hypoxia and oxidative stress (Ono & Han 2000).

2.1.2 Inhibitors

The alterations of gene expression for most model organisms are today possible through molecular biology and cloning techniques. The transient transformation (or transfection) of cells with a plasmid will usually result in an over expression of the gene of interest. Mutations can also be introduced by transiently or permanently altering the gene expression by changing the genetic code. Some of these mutations will lead to non-viable changes and hence kill the cell while other mutations will not affect the cell at all. Some mutations will, however, change the targeted gene expression as intended. Unfortunately, also, other gene expressions than the one targeted can change as a secondary effect. This is referred to as compensatory mechanisms (Specht & Shokat 2002). This "adaption mechanism" for the model organism to the new environmental parameters may interfere with the initial purpose of the cloning and consequently make the data more difficult to interpret. Compensatory mechanisms are especially common in deletion mutants, resulting in interrupted pathways, and as a consequence other intertwined pathways or similar proteins could compensate for the loss.

An alternative approach is to instead use a substance that can be administrated to the cells that will inhibit or disable a specific protein or group of proteins. This will allow studies of response dynamics without any compensatory mechanisms. Since protein kinases share a common sequence and structural homology in their ATP-binding site, many synthesized kinase inhibitors lack selectivity (Johnson et al 2004). Approaches have been conveyed where a protein inhibitor is synthesized to recognize a mutated version of the protein kinase of interest, referred to as the ASKA approach (Shokat et al 2000). It is, however, more convenient to use the wild type protein kinase not only in order to abolish the need of molecular genetics work, but also to limit possible negative effects (such as changed activity and/or instability of the protein kinase) caused by the protein kinase mutation itself. Nevertheless, the ASKA approach has been used in recent reports on HOG pathway control and target identification (Thorner & Westfall 2006). In 2010, a selective inhibitor targeting the MAPK Hog1 was reported which undoubtedly will be a valuable tool investigating the HOG transduction pathway (Diner et al 2010).

3. Microfluidic devices

Depending on the size and application of the microfluidic devices, both the materials and developmental process used to manufacture the devices vary greatly. For biological applications the analyte is of course of crucial importance and different microfluidic designs and qualities are thus required if an organism, a cell, a metabolite or intracellular structure will be studied. In any case, the construction of the microfluidic device should be validated both theoretically and practically before use. Theoretical simulations of flows through the microfluidic channels are a valuable tool to evaluate new microfluidic designs and also provides optimum dimensions of the microfluidic device.

3.1 Flow simulations

The flows within microfluidic devices are usually laminar and characterised by a high surface area to volume ratio. This is an advantage when designing the system since it provides a possibility to beforehand estimate the flow profile at different locations within the device when changing the channel dimensions. The flows are expressed in terms of the Reynolds numbers (Re); a dimensionless number that depend on the fluid's density (ρ), velocity (U) and viscosity (η) as well as on the typical length scale (L) of the device

$$\mathrm{Re} = \frac{\rho U L}{\eta} \tag{1}$$

Generally, flows with a Reynolds number smaller than 2300 are considered laminar, while a larger Reynolds number represents flows of turbulence (Beebe et al 2002). When working with microfluidic devices, where at least one of the dimensions is on the microscale, the Reynolds numbers are usually well below 1. Consequently, two adjacent fluids within a microfluidic device are mixed only by diffusion. The diffusion is driven by Brownian motions and is characterised by the diffusion coefficient, D. The diffusion coefficient is inversely proportional to the size of the diffusing particle, where the denominator is constituted by the so called Stokes drag coefficient

$$D = \frac{k_B T}{6\pi\eta\alpha} \tag{2}$$

The Stokes drag coefficient involves the viscosity of the fluid (η), and the radius of the particle (a), while the nominator contains Boltzmann's constant (k_B) and the absolute temperature (T). Even though several theoretical diffusion coefficients are available in the literature, specific substances used should if possible be experimentally measured by e.g. diffusion NMR. The effect of diffusion on a relevant time scale is of great importance on the microscale and the distance (d) a molecule diffuses over the time (t) can be calculated by

$$d = \sqrt{2D_t} \tag{3}$$

Another important aspect is the flow profile of the fluids within the microfluidic device. Pressure driven forces, the geometry of the channels as well as surface tensions and surface to volume ratios affect the net flow through the device and need to be taken into consideration when simulated. Hence, theoretical simulations of specific substance within the flows are required for correct experimental setup. Figure 5 shows a simulation of 500 mM sorbitol through a three-channel microfludic chamber at a flow speed of 1000 nl/min while the flows in the two other channels are 0 nl/min. This specific chamber is 27 μm high, the wider inlet channel is 200 μm wide, the thinner channels are 100 μm wide and the outlet channel is 400 μm wide.

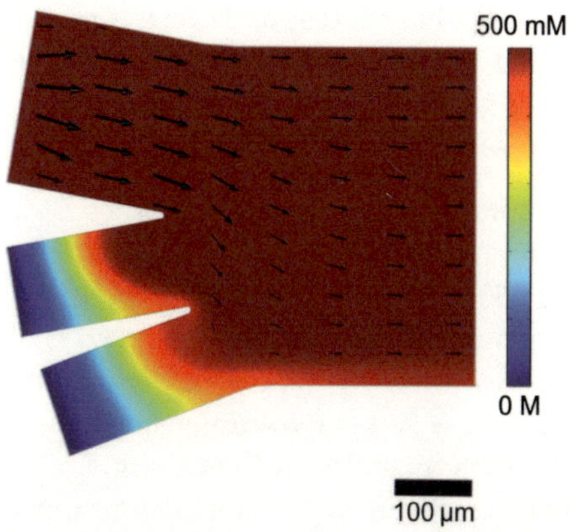

Fig. 5. Simulated concentration and flow profile at the surface of a three-channel microfluidic chamber during a static stress experiment. The lengths of the arrows are proportional to the flow velocities in the different inlet channels. The colours represent the concentration gradient of the perturbation agent sorbitol; blue corresponds to the lowest concentration of 0 mM sorbitol and red the highest concentration (500 mM sorbitol). This simulation is valid for 1000 nl/min velocity flow in the sorbitol inlet channel and no flow in the lower two inlet channels.

3.2 Construction of microfluidic devices

Microfluidic devices can with advantage be constructed using a soft lithography approach. Usually a mould representing the microfluidic system is created, onto which a plastic polymer is added. After hardening, the plastic polymer will thus have the form of the wanted microfluidic device. The mould is made from a light sensitive photoresist that is dispensed on top of a planar silicon wafer. The liquid resist is spun out in a vacuum centrifuge to create a uniformly thick layer of photoresist. Different density properties of the resist will thus result in different layer thicknesses at specific centrifugation speeds. One commonly used resist is SU-8, which has good properties for achieving nearly vertical walls even when relatively thick resist layers are required. Typically, spin speed between 1000 and 3000 rpm for approximately 30-50 seconds results in 1.5-250 μm thick resist layers, depending on which type of SU-8 is used. The SU-8 film needs to be densified and its solvents evaporated. This is done by heating steps; first at 65°C and a subsequent incubation at 95°C. This gradual heating is recommended and the incubation times are closely related to the SU-8 thickness. A near UV-exposure is required to start the polymerisation process. This is preferably done in a so called mask aligner (i-line at 365 nm is recommended) where parallel rays from a UV light source illuminate the exposed surface of SU-8. To achieve selective cross-linkage of the resist, a post exposure bake step must be performed. Again, a stepwise heating (first at 65°C and then at 95°C) is recommended to minimise wafer stress and/or cracking. Non-cross-linked resist is removed in a developer solution wash, for instance in ethyl lactate or diacetone alcohol, after which the master is rinsed with isopropanol and blown dry with nitrogen. Should the surface of the wafer contain a cloudy film after this treatment, a longer developer incubation time is necessary. There are both positive and negative photoresists available. A positive resist will become soluble in the developer after UV light illumination, while a negative photoresist will not. Hence, using a negative photoresist and a mask that corresponds to the microfluidic channel pattern will result in ridges of polymerised resist on top of the silicon wafer. This is referred to as the master and can also be reused several times. The inclusion of a final hard bake step at 150-200°C further increases the cross linking process and hence the lifetime of the master.

The most commonly used soft polymer in microfluidics today is polydimethylsiloxane (PDMS). This polymeric organo-silicon is considered to be inert and non-toxic, has optic transparency (lower limit of 280 nm) and is rather inexpensive. Even though the surface is highly hydrophobic after curing, the surface properties can easily be modified (Whitesides et al 2000). The PDMS is constituted by repeats of the monomer $[SiO(CH_3)_2]$ that after addition of the curing agent (1:10 w/w) is two-dimensionally cross-linked. When moulding the microfluidic system, PDMS is mixed with the curing agent and degassed in a vacuum dessicator. After approximately 10 min in the desiccators the vacuum is released and the PDMS mixture is dispensed on top of the master and incubated at 95°C for one hour to polymerise. Access holes to the microfluidic channels can either be drilled afterwards or made already during the moulding phase by the use of flattened needles centred to the ridges of the master. The polymerised PDMS can be peeled off and the master reused for a new moulding procedure. To tightly seal the PDMS system onto a glass slide, oxygen plasma treatment of the two is preferable. This can for instance be achieved in a plasma cleaner where only 30 seconds of air plasma treatment is necessary for the methyl groups in the PDMS to be replaced for hydroxyl groups (-OH). When put in contact with the plasma treated glass, irreversible, covalent bonds (-Si-O-Si-) are spontaneously formed.

Figure 6 shows a thick PDMS device, connected via teflon tubing to small volume syringes with a total volume of 250 μl that in turn can be placed in a syringe pump that pump the syringe fluids into the microfluidic device.

4. The applications of microfluidics

Microfluidics have complemented biological and biochemical research in many ways and the development of many variants of LOC and μTAS indicates the vast variety of applications that could be performed (Chiu & Kuo 2011; Cho et al 2010; Levchenko et al

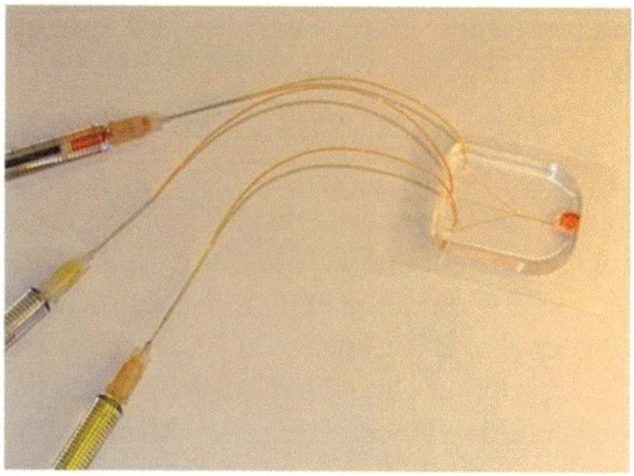

Fig. 6. A three-inlet microfluidic PDMS chamber. Syringes fitted to disposable needles and subsequent tubing attached. To visualise the channels the syringes contain highly concentrated fluorescein.

2010; Lu et al 2010; Park & Hwang 2011; Veres et al 2010). However, cell biology would benefit even more if the level of control could be extended not only to include the microenvironment around the cells, but also the cells themselves. Optical manipulation and optical tweezers (OT) are optical techniques that provide the opportunity to manipulate the cells inside the microfluidic device in a non-invasive and non-intrusive manner.

The following part will give an example of how a microfluidic setup can be used in a single cell analysis project focusing on perturbations of the HOG MAPK pathway.

4.1 The microscope and optical tweezers setup

OT can easily be integrated into most optical microscopes setup and combined with a range of fluorescence imaging techniques. The setup used for the single cell experiments this chapter is based upon was built around an inverted epi-fluorescence microscope (Leica DMI6000B) containing a single stationary OT. In epi-fluorescence microscopy the microscope objective not only collects the emitted fluorescence light but also illuminates the sample with the excitation light. The equipment used for imaging the sample was a mercury metal halide bulb in combination with a filter cube containing excitation and emission filters for imaging green fluorescent protein (GFP) (excitation peak at 396 nm) and mCherry (excitation peak at 587 nm) as well as a dichroic mirror. The images were acquired by a 14 bit dynamic electron multiplying charged coupled device (EMCCD) that can detect very weak intensities emitted by the sample. The microscope was also equipped with a motorized stage and a motorized objective that allowed for optical sectioning of the sample.

The OT consisted of a laser beam (λ=1070 nm) with a Gaussian intensity profile that was focused to a diffraction limited beam waist by the same microscope objective that was used to image the sample. The laser beam was introduced into the microscope via a dichroic mirror that was situated between the microscope objective and the filter cube cassette. This enabled the imaging filters to be changed without affecting the laser trapping (Eriksson et al 2010). At the laser focus the power was estimated to be approximately 240 mW and allowed the trapping of yeast cells from one of the inlet channels within the microfluidic device. The theory behind the optical trapping mechanisms and important experimental considerations is beyond the scope of this chapter and can be found elsewhere (e.g. (Neuman & Block 2004)).

4.2 The imaging of the cells

The development of the green fluorescent protein (GFP) has really come to revolutionise the field of biological imaging (Chalfie et al 1994). Since its discovery, GFP and other fluorescent proteins have been applied in a wide range of applications (Chudakov et al; Ckurshumova et al; Palmer et al 2011; Shaner et al 2007; Tsien et al 2005). By PCR, the gene sequence of the fluorescent protein can be inserted at almost any position in the gene of interest. The cells are either transiently or constitutively expressing one or several fluorescent protein-fusion proteins (i.e. reporter proteins), that can thus be monitored both spatially and temporally on a sub-cellular level. As a consequence, cellular events such as dynamic intracellular signalling, transport and localisation, protein-protein interactions and gene expression can easily be monitored using an ordinary fluorescence microscope.

When elucidating the mechanisms of the HOG transduction pathway, the cells are engineered to express GFP coupled Hog1 (Hog1-GFP). This enables the localisation of Hog1 and the protein can be imaged during the performed time lapse studies while the cell is perturbed by rapid environmental changes. Cells expressing the general stress protein Msn2 coupled to GFP (Msn2-GFP) were used as control in the experiments. Msn2 is also cytosolic during normal conditions, but migrates into the nucleus upon osmotic stress (Gorner et al 1998). Both cell types also express the nucleus resident reporter protein, Nrd1-mCherry, which facilitates cell nuclei localisation during imaging. The cells were imaged in 3D by seven optical sections (using the filter for GFP and mCherry) and followed for 45 minutes. Since the cellular response to the environmental change occurs within minutes following the perturbation, images was captured every 30 seconds in the beginning of the time lapse study. After approximately 5 minutes, the images were captured more and more sparsely (see Figure 11).

4.3 Microfluidics in single cell analysis

To improve cell adhesion, the microfluidic device is flushed with the lectin protein Concanavalin A (ConA). ConA specifically binds to the glucose aminoglycans residues on the cell wall of yeast cells, and distributed to the microfluidic device at a concentration of 1 mg/ml for 30-60 min prior to experiment is enough to acquire proper cell binding to the surface (Eriksson et al 2010). The yeast cells as well as the other substances or mediums needed for the experiment are introduced into the microfluidic channels via tubing coupled to glass syringes. The syringes are in turn placed in syringe pumps that allow separate control of the flow speeds in each channel. When a stable flow from all inlet channels is established, the optical tweezers are used to trap single cells flowing past. In our three-channel chamber, this is done while the pump speed settings are 80 nl/min in the two thinner inlet channels and 40 nl/min in the wider one. In figure 7, yeast cells are introduced in the lower channel while the middle channel is "the neutral channel", containing the same solutions as the cells have been cultured in. The third, wider channel contains the perturbation solution.

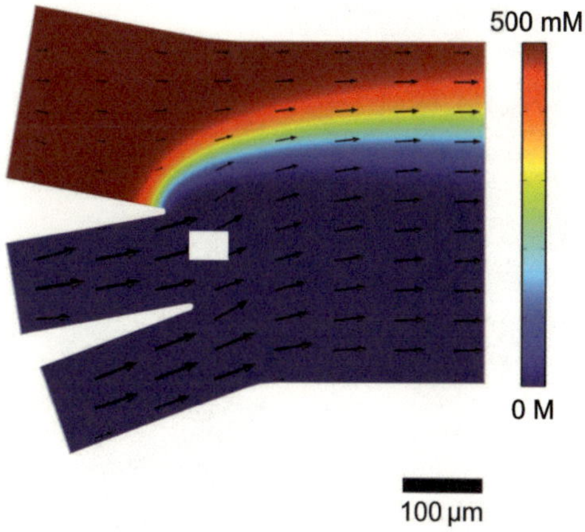

Fig. 7. Simulated concentration and flow profile at the surface of a three-channel microfluidic chamber during loading of the cell array using the OT. The lengths of the arrows are proportional to the flow velocity which is also printed in each of the inlet channels (nl/min). The colours represent the concentration gradient of the perturbation agent sorbitol; blue corresponds to the lowest concentration of 0 mM sorbitol and red the highest concentration (500 mM sorbitol). This simulation is valid for 40 nl/min velocity flow in the sorbitol inlet channel and 80 nl/min velocity flow in the lower two inlet channels. The cells are introduced in the lower inlet channel and the white area shows where the cell array is placed.

The cells are individually trapped by the OT and attached to the bottom surface of the device. The trapping to the release of the cell into the array usually takes less than 5 seconds, which leave the cells reproduction ability unaffected by the laser light (Eriksson et al 2010). The effective area of the cell array depends both the width of the channels as well as on the diffusion coefficient of the substances used in the experiment. When using for instance the combination of a three-channel microfluidic device with sorbitol as the perturbing agent, up to 25 cells can be placed in the array that constitutes the area of interest. The environment of the cells attached to the bottom of the system is changed by changing the individual pump speeds. As visualised in figure 8 the 25 cells are placed within the area of the white rectangle in a three-channel chamber. Fluorescein is introduced in the upper channel to visualise the borders of fluids as would be the case during yeast cell trapping and placement. During this process, no fluorescein (i.e. the perturbation) reaches the cell array, leaving the cells unaffected (left panel). The environment is then changed by increasing the pump speed of the pump controlling the fluorescein flow while decreasing the pump speed controlling the other two flows (right panel). By using this procedure, a change in the environment can be successfully accomplished within 2 seconds. After the environmental switch the cell array is covered by the fluorescein and in the case of introducing a drug, the cellular response to the drug exposure can be imaged by the microscope. By automating and synchronising the pump speed and the image acquisition, the experimental precision can be increased. The removal of a substance or even multiple treatment pulses is thus possible by simply changing the speed of the individual flows within the device.

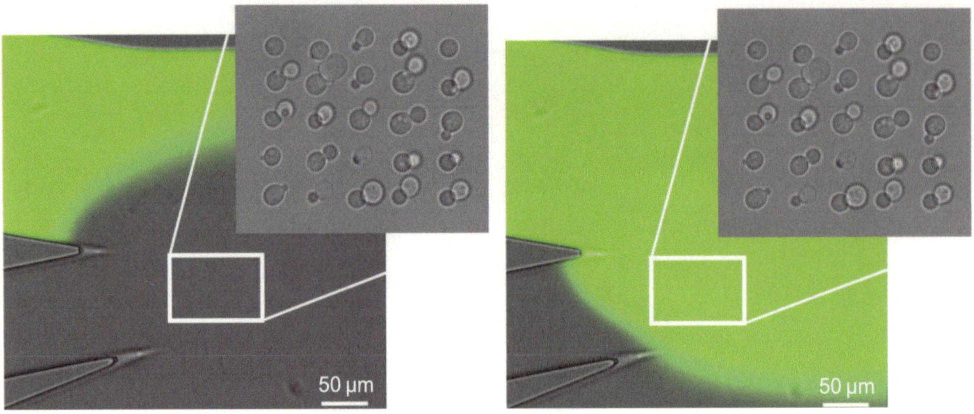

Fig. 8. Practical evaluations of the flow speeds in a three-inlet microfluidic chamber. The images show the field of view through a 20X objective. The upper inlet channel contains fluorescein and the white square corresponds to the placement of the yeast cell array (shown in enlargement with yeast cells in both images). The left image corresponds to the flow situations when capturing the cells by the OT and array placement. No perturbation (here represented by fluorescein) is affecting the cells. In the right image the pump speeds have been changed and the fluorescein now covers the cells in the array that consequently will be perturbed and monitored while responding. This environmental change is completed within 2 seconds.

A three-channel microfluidic chamber is a convenient setup for treating cells with *one* additional external stimulus. However, if yet another media or reagent needs to be introduced, a fourth channel would be required. This turned out to be necessary when the inhibitor against the MAPK Hog1 in the HOG pathway was to be evaluated. The inhibitor was designed to selectively bind to Hog1 and its effect on Hog1 nuclear migration upon osmotic stress treatment was thus to be investigated. There was, however, no other way of performing this experiment using a three-channel system but to pre-treat the cells with the inhibitor and then introduce the cells into the microfluidic device. Thus it was impossible to study cells that had been exposed to the inhibitor for less than 15 min, since this was the minimum time from the inhibitor pre-treatment to the environmental change of the cells within the microfluidic device. It also turned out to be difficult to compare data from different experimental runs, since cells in different experiments had been exposed to the inhibition for different periods of time before the experiment could start. The four-channel microfluidic device shown in figure 9 enabled us not only to treat the yeast cells with the perturbing agent sorbitol, but also to perform the pre-treatment of the cells with the inhibitor for a predetermined time within the microfluidic device.

The workflow is as follows; (i) cells are optically trapped and placed in the array by OT. (ii) The cells are exposed to the inhibitor by increasing the flow speed in the inhibitor inlet channel, before (iii) the cells are perturbed by exposure to sorbitol by yet another speed change of the flow in the sorbitol inlet channel. Meanwhile the yeast cells were imaged and the Hog1 localisation monitored on a single cell basis. Another benefit of using an extra inlet channel was the possibility to also introduce control cells in the yeast cell array. The inhibitor was designed to bind to the ATP-binding site of Hog1 and hence inhibit

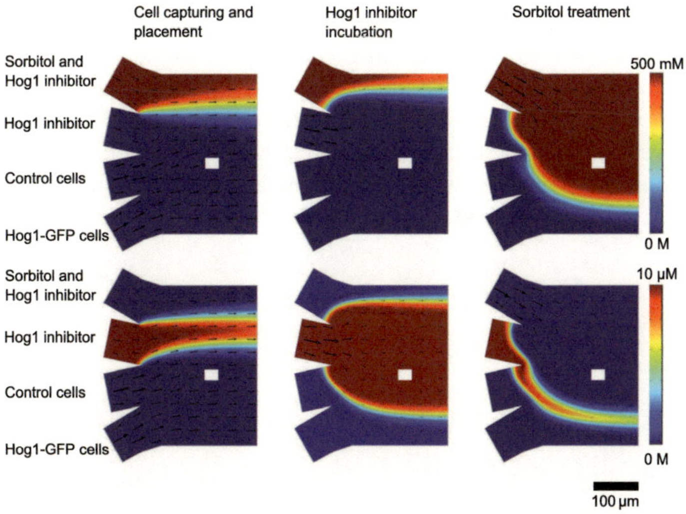

Fig. 9. Simulations of the flow profile used in the four-channel microfluidic chamber setup. In the upper channel, the perturbation of the MAPK pathway is introduced (sorbitol) in combination with the inhibitor; the second inlet channel from the top contains the inhibitor only; the third and fourth inlet channel from the top contains the control cells and Hog1-GFP cells respectively. The three simulation scenarios (from left to right) represent cell capturing and placement, Hog1 inhibitor incubation and sorbitol treatment. The red colour in the upper row simulations correspond to 500 mM sorbitol, while in the lower row simulations red colour corresponds to 10 μM Hog1 inhibitor.

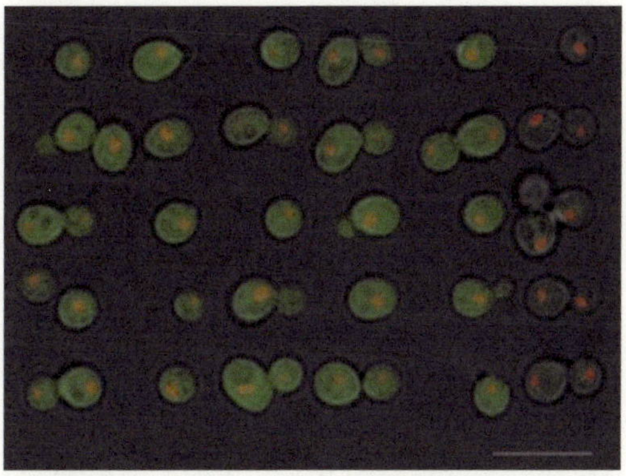

Fig. 10. Imaged response of the yeast cell array in the four-channel microfluidic chamber. The image shows the cell array at t=-30 seconds, i.e. before the perturbation is introduced. The last column of cells constitutes the control cells whose expressed reporter protein (Msn2-GFP) migration is not affected by the Hog1 inhibitor. All cells express a nuclear resident reporter protein (Nrd1-mCherry) that enables visual localisation of the nuclei at all times. The scale bar represents 10 μm.

downstream phosphorylation events in the HOG signal transduction pathway. Since it is suggested that the ATP-binding site of Hog1 must be active for nuclear envelope translocation (Thorner & Westfall 2006), a complete and proper inhibitor-binding to Hog1 would also inhibit Hog1 nuclear migration. Therefore, instead of imaging a "non-response" of the Hog1-inhibitor treated cells only, also cells expressing a reporter protein unaffected by the inhibitor but responsive to the osmotic stress was introduced. Such cells would also constitute a propriety control of the microfluidic chamber itself, ensuring that all the cells in the array indeed experience the osmotic stress (Figure 10). The cell response due to osmotic stress can be presented as the ratio (R) of nuclear to cytosolic Hog1-GFP and thus corresponds to the GFP intensity in the nucleus and cytosol respectively.

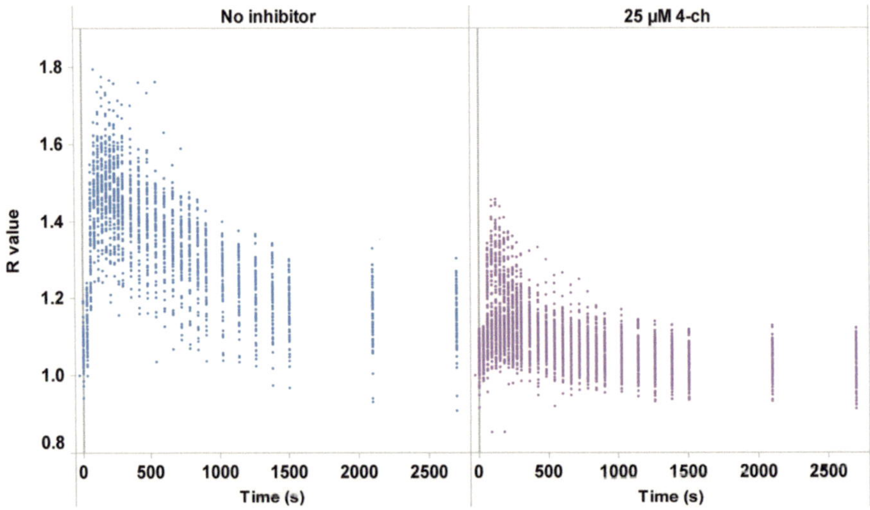

Fig. 11. Single cell data of the Hog1-GFP intensities. The left panel shows the normalised R-values of yeast cells experiencing sorbitol perturbation only (two separate experiments). The right panel shows the yeast cells that have been pre-treated with the Hog1 inhibitor for 20 minutes before sorbitol perturbation in the four-channel microfluidic chamber (four separate experiments). The sorbitol perturbation is initiated at time=0.

A high R-value is therefore attained during sorbitol perturbation where Hog1 has migrated into the nuclei. Pre-treating the cells with the Hog1 inhibitor would therefore impair the R-value increase upon sorbitol perturbation compared with cells experiencing sorbitol perturbation only. When treating the yeast cell array with 25 µM of the Hog1 inhibitor for 20 minutes before sorbitol perturbation, the R-value did indeed decrease compared with cells that had experienced sorbitol perturbation only, as can be seen in figure 11. The graphs show the time lapse single cell data of two (no inhibitor treatment) and four (25 µM Hog1 inhibitor treatment) separate experiments respectively. The Msn2-GFP in the control cells did migrate into the nucleus upon sorbitol perturbation and hence proved that sorbitol perturbation was indeed induced (data not shown).

5. Conclusions and future aspects

The experimental setup and the results obtained indicate how mechanisms of the MAPK HOG pathway can be investigated using a microfluidic approach. The four-channel microfluidic chamber demonstrated allows the removal of not only the sorbitol perturbation, but also the inhibitor, thus enabling inhibitor turnover rate investigations. The controlled exposure to stress substances as well as inhibitors allows kinetic and dynamic events to be studied with great temporal and spatial resolution, something that usually is very difficult to achieve with traditional techniques. The single cell analysis is also an excellent complement to ensemble averaging techniques, since the cellular heterogeneity within the genetically identical population is easily revealed. We are confident that similar microfluidic approaches will be a valuable tool in the future when delineating, not only the HOG pathway, but similar pathways in which dynamics of fluorescent reporter proteins can be followed in single cells. Furthermore, the inclusion of two different cell types in the same cell array enables a direct comparison between, for instance, cells with different reporter proteins or mutants versus wild type cells.

6. Acknowledgments

The authors thank Charlotte Hamngren for the development and evaluation of the four-channel microfluidic chamber and Morten Grötli for the Hog1 inhibitor. The project was funded by the FP7 programme UNICELLSYS.

7. References

Ammerer G, Reiser V, Salah SM. 2000. Polarized localization of yeast Pbs2 depends on osmostress, the membrane protein Sho1 and Cdc42. *Nat Cell Biol* 2:620-7

Bayne ML, Applebaum J, Chicchi GG, Hayes NS, Green BG, Cascieri MA. 1988. Expression, Purification and Characterization of Recombinant Human Insulin-Like Growth Factor-I in Yeast. *Gene* 66:235-44

Beebe DJ, Mensing GA, Walker GM. 2002. Physics and applications of microfluidics in biology. *Annu Rev Biomed Eng* 4:261-86

Beebe DJ, Paguirigan AL. 2008. Microfluidics meet cell biology: bridging the gap by validation and application of microscale techniques for cell biological assays. *Bioessays* 30:811-21

Beggs JD. 1978. Transformation of Yeast by a Replicating Hybrid Plasmid. *Nature* 275:104-9

Bennett MR, Pang WL, Ostroff NA, Baumgartner BL, Nayak S, et al. 2008. Metabolic gene regulation in a dynamically changing environment. *Nature* 454:1119-22

Burgess JCMaSM. 2003. Yeast as a model organism. *In Encyclopedia of Life Sciences*

Chalfie M, Tu Y, Euskirchen G, Ward WW, Prasher DC. 1994. Green Fluorescent Protein as a Marker for Gene-Expression. *Science* 263:802-5

Chiu DT, Kuo JS. 2011. Disposable microfluidic substrates: Transitioning from the research laboratory into the clinic. *Lab on a Chip* 11:2656-65

Cho YK, Gorkin R, Park J, Siegrist J, Amasia M, et al. 2010. Centrifugal microfluidics for biomedical applications. *Lab on a Chip* 10:1758-73

Chudakov DM, Matz MV, Lukyanov S, Lukyanov KA. Fluorescent proteins and their applications in imaging living cells and tissues. *Physiol Rev* 90:1103-63

Ckurshumova W, Caragea AE, Goldstein RS, Berleth T. Glow in the Dark: Fluorescent Proteins as Cell and Tissue-Specific Markers in Plants. *Mol Plant*

Dean PN, Hoffman RA. 2007. Overview of flow cytometry instrumentation. *Curr Protoc Cytom* Chapter 1:Unit1

Deptala A, Bedner E, Darzynkiewicz Z. 2001. Unique analytical capabilities of laser scanning cytometry (LSC) that complement flow cytometry. *Folia Histochem Cyto* 39:87-9

Dihazi H, Kessler R, Eschrich K. 2004. High osmolarity glycerol (HOG) pathway-induced phosphorylation and activation of 6-phosphofructo-2-kinase are essential for glycerol accumulation and yeast cell proliferation under hyperosmotic stress. *J Biol Chem* 279:23961-8

Diner P, Veide Vilg J, Kjellen J, Migdal I, Andersson T, et al. Design, synthesis, and characterization of a highly effective Hog1 inhibitor: a powerful tool for analyzing MAP kinase signaling in yeast. *PLoS One* 6:e20012

Eriksson E, Sott K, Lundqvist F, Sveningsson M, Scrimgeour J, et al. 2010. A microfluidic device for reversible environmental changes around single cells using optical tweezers for cell selection and positioning. *Lab on a Chip* 10:617-25

Ferreira C, van Voorst F, Martins A, Neves L, Oliveira R, et al. 2005. A member of the sugar transporter family, Stl1p is the glycerol/H+ symporter in Saccharomyces cerevisiae. *Mol Biol Cell* 16:2068-76

Goffeau A, Barrell BG, Bussey H, Davis RW, Dujon B, et al. 1996. Life with 6000 genes. *Science* 274:546-&

Gorner W, Durchschlag E, Martinez-Pastor MT, Estruch F, Ammerer G, et al. 1998. Nuclear localization of the C2H2 zinc finger protein Msn2p is regulated by stress and protein kinase A activity. *Genes Dev* 12:586-97

Gustin MC, Albertyn J, Alexander M, Davenport K. 1998. MAP kinase pathways in the yeast Saccharomyces cerevisiae. *Microbiol Mol Biol Rev* 62:1264-300

Hartwell LH. 2004. Yeast and cancer. Bioscience Rep 24:523-44

Hinnen A, Hicks JB, Fink GR. 1978. Transformation of Yeast. *P Natl Acad Sci USA* 75:1929-33

Hohmann S. 2009. Control of high osmolarity signalling in the yeast Saccharomyces cerevisiae. *Febs Lett* 583:4025-9

Jensen KF, El-Ali J, Sorger PK. 2006. Cells on chips. *Nature* 442:403-11

Johnson LN, Noble MEM, Endicott JA. 2004. Protein kinase inhibitors: Insights into drug design from structure. *Science* 303:1800-5

Kitano H. 2002. Computational systems biology. *Nature* 420:206-10

Klis FM, Boorsma A, De Groot PW. 2006. Cell wall construction in Saccharomyces cerevisiae. *Yeast* 23:185-202

Levchenko A, Gupta K, Kim DH, Ellison D, Smith C, et al. 2010. Lab-on-a-chip devices as an emerging platform for stem cell biology. *Lab on a Chip* 10:2019-31

Lu H, Crane MM, Chung K, Stirman J. 2010. Microfluidics-enabled phenotyping, imaging, and screening of multicellular organisms. *Lab on a Chip* 10:1509-17

Maeda T, Takekawa M, Saito H. 1995. Activation of Yeast Pbs2 Mapkk by Mapkkks or by Binding of an Sh3-Containing Osmosensor. *Science* 269:554-8

Manz A, Auroux PA, Iossifidis D, Reyes DR. 2002a. Micro total analysis systems. 2. Analytical standard operations and applications. *Analytical Chemistry* 74:2637-52

Manz A, Reyes DR, Iossifidis D, Auroux PA. 2002b. Micro total analysis systems. 1. Introduction, theory, and technology. *Analytical Chemistry* 74:2623-36

Mcaleer WJ, Buynak EB, Maigetter RZ, Wampler DE, Miller WJ, Hilleman MR. 1984. Human Hepatitis-B Vaccine from Recombinant Yeast. *Nature* 307:178-80

Molina M, Martin H, Flandez M, Nombela C. 2005. Protein phosphatases in MAPK signalling: we keep learning from yeast. *Mol Microbiol* 58:6-16

Mortimer RK, Johnston JR. 1959. Life Span of Individual Yeast Cells. *Nature* 183:1751-2

Neuman KC, Block SM. 2004. Optical trapping. *Rev Sci Instrum* 75:2787-809

O'Rourke SM, Herskowitz I. 2004. Unique and redundant roles for HOG MAPK pathway components as revealed by whole-genome expression analysis. *Mol Biol Cell* 15:532-42

O'Shea EK, Raser JM. 2004. Control of stochasticity in eukaryotic gene expression. *Science* 304:1811-4

Ono K, Han J. 2000. The p38 signal transduction pathway: activation and function. *Cell Signal* 12:1-13

Palmer AE, Qin Y, Park JG, McCombs JE. 2011. Design and application of genetically encoded biosensors. *Trends Biotechnol* 29:144-52

Park JK, Hwang H. 2011. Optoelectrofluidic platforms for chemistry and biology. *Lab on a Chip* 11:33-47

Pepperkok R, Ellenberg J. 2006. Innovation - High-throughput fluorescence microscopy for systems biology. *Nat Rev Mol Cell Bio* 7:690-6

Ploger R, Zhang J, Bassett D, Reeves R, Hieter P, et al. 2000. XREFdb: cross-referencing the genetics and genes of mammals and model organisms. *Nucleic Acids Res* 28:120-2

Posas F, WurglerMurphy SM, Maeda T, Witten EA, Thai TC, Saito H. 1996. Yeast HOG1 MAP kinase cascade is regulated by a multistep phosphorelay mechanism in the SLN1-YPD1-SSK1 "two-component" osmosensor. *Cell* 86:865-75

Rep M, Krantz M, Thevelein JM, Hohmann S. 2000. The transcriptional response of Saccharomyces cerevisiae to osmotic shock. Hot1p and Msn2p/Msn4p are required for the induction of subsets of high osmolarity glycerol pathway-dependent genes. *J Biol Chem* 275:8290-300

Rep M, Reiser V, Gartner U, Thevelein JM, Hohmann S, et al. 1999. Osmotic stress-induced gene expression in Saccharomyces cerevisiae requires Msn1p and the novel nuclear factor Hot1p. *Mol Cell Biol* 19:5474-85

Ryley J, Pereira-Smith OM. 2006. Microfluidics device for single cell gene expression analysis in Saccharomyces cerevisiae. *Yeast* 23:1065-73

Saito H, Posas F. 1998. Activation of the yeast SSK2 MAP kinase kinase kinase by the SSK1 two-component response regulator. *Embo J* 17:1385-94

Shaner NC, Patterson GH, Davidson MW. 2007. Advances in fluorescent protein technology. *J Cell Sci* 120:4247-60

Shields DC, Caffrey DR, O'Neill LAJ. 1999. The evolution of the MAP kinase pathways: Coduplication of interacting proteins leads to new signaling cascades. *J Mol Evol* 49:567-82

Shokat KM, Bishop AC, Ubersax JA, Petsch DT, Matheos DP, et al. 2000. A chemical switch for inhibitor-sensitive alleles of any protein kinase. *Nature* 407:395-401

Silver PA, Ferrigno P, Posas F, Koepp D, Saito H. 1998. Regulated nucleo/cytoplasmic exchange of HOG1 MAPK requires the importin beta homologs NMD5 and XPO1. *Embo J* 17:5606-14

Specht KM, Shokat KM. 2002. The emerging power of chemical genetics. *Curr Opin Cell Biol* 14:155-9

Stagljar I, Suter B, Auerbach D. 2006. Yeast-based functional genomics and proteomics technologies: the first 15 years and beyond. *Biotechniques* 40:625-44

Tamas MJ, Thorsen M, Di YJ, Tangemo C, Morillas M, et al. 2006. The MAPK Hog1p modulates Fps1p-dependent arsenite uptake and tolerance in yeast. *Mol Biol Cell* 17:4400-10

Tatebayashi K, Tanaka K, Yang HY, Yamamoto K, Matsushita Y, et al. 2007. Transmembrane mucins Hkr1 and Msb2 are putative osmosensors in the SHO1 branch of yeast HOG pathway. *Embo J* 26:3521-33

Thorner J, Westfall PJ. 2006. Analysis of mitogen-activated protein kinase signaling specificity in response to hyperosmotic stress: Use of an analog-sensitive HOG1 allele. *Eukaryot Cell* 5:1215-28

Thorner J, Westfall PJ, Ballon DR. 2004. When the stress of your environment makes you go HOG wild. *Science* 306:1511-2

Thorner J, Westfall PJ, Patterson JC, Chen RE. 2008. Stress resistance and signal fidelity independent of nuclear MAPK function. *P Natl Acad Sci USA* 105:12212-7

Tsien RY, Shaner NC, Steinbach PA. 2005. A guide to choosing fluorescent proteins. *Nat Methods* 2:905-9

Tudos AJ, Besselink GAJ, Schasfoort RBM. 2001. Trends in miniaturized total analysis systems for point-of-care testing in clinical chemistry. *Lab on a Chip* 1:83-95

van den Berg A, Andersson H. 2003. Microfluidic devices for cellomics: a review. *Sensor Actuat B-Chem* 92:315-25

van den Berg A, Lammerink TSJ. 1998. Micro total analysis systems: Microfluidic aspects, integration concept and applications. *Top Curr Chem* 194:21-49

Warmka J, Hanneman J, Lee J, Amin D, Ota I. 2001. Ptc1, a type 2C Ser/Thr phosphatase, inactivates the HOG pathway by dephosphorylating the mitogen-activated protein kinase Hog1. *Mol Cell Biol* 21:51-60

Weibel DB, Diluzio WR, Whitesides GM. 2007. Microfabrication meets microbiology. *Nat Rev Microbiol* 5:209-18

Veres T, Malic L, Brassard D, Tabrizian M. 2010. Integration and detection of biochemical assays in digital microfluidic LOC devices. *Lab on a Chip* 10:418-31

Whitesides GM. 2006. The origins and the future of microfluidics. *Nature* 442:368-73

Whitesides GM, McDonald JC, Duffy DC, Anderson JR, Chiu DT, et al. 2000. Fabrication of microfluidic systems in poly(dimethylsiloxane). *Electrophoresis* 21:27-40

Whitesides GM, Sia SK. 2003. Microfluidic devices fabricated in poly(dimethylsiloxane) for biological studies. *Electrophoresis* 24:3563-76

9

A Tunable Microfluidic Device for Drug Delivery

Tayloria Adams*, Chungja Yang*, John Gress,
Nick Wimmer and Adrienne R. Minerick
Michigan Technological University
USA

1. Introduction

The field of microfluidics, small-scale tests from nanoscale to microscale, has grown dramatically over the past two decades as evidenced by greater than 30,000 papers published over the last 10 years on the topic [Web of Knowledge search using 'microfluidic' terms October 2011]. Microfluidic platforms, also known as lab-on-a-chip (LOC), include a set of miniaturized integrated unit operations that are touted to lead to fast, easy, precise control in biological and chemical systems. LOCs include the development of point-of-care (POC) medical diagnostic devices with the advantages of increased sensitivity, lower sample volumes, lower reagent volumes, low energy, low cost, low labor need, and less likelihood of human error (Xiao & Young, 2011). Due to these advantages, LOCs have substantial potential to be widely utilized in medicine for analytical and diagnostic assays, biosensors, and drug delivery.

Microfluidic technology has been used for a wide variety of applications such as forensics, cell phone facilitated micro-imaging, and analytical testing. In 2006 Bienvenue at al., compared the use of microfluidic technology with a commercial kit that utilized dithiothreitol to extract and purify DNA from sperm samples. The sample volume was less than 10 μL and the resulting electropherograms were very similar for both techniques (Bienvenue et al., 2006). DNA separation has also been studied by Aboud et al. Pentameric short tandem repeat (STR) markers were tested in a microfluidic device on single-stranded DNA. Coupling microfluidics with pentameric STRs improved allele resolution by 3.7 times (Aboud et al., 2010). In these cases, microfluidics can be used as a rapid screening tool for forensic DNA analysis to help resolve the backlog of DNA casework (Aboud et al., 2010; Bienvenue et al., 2006). Zhu et al., combined optofluidics with cell phone technology. A cell phone was converted to a microscope analysis tool by integrating optofluidic fluorescent cytometry with compact optical attachments. The cell phone optical attachment included a lens, plastic color filter, two light emitting diodes, and batteries, which altogether weighed less than 1 lb. To test the effectiveness of this new imaging system, the density of white blood cells were measured using the cell phone-based fluorescent image cytometry and compared with the white blood cell density found with a commercial hematology analyzer.

---

* These authors contributed equally

The blood sample was injected into the microfluidic chamber using a syringe pump and the cell phone recorded the fluorescent emission. This study demonstrated that the densities found by both systems were a good match with less than 5% error and that cell phone optofluidic fluorescent imaging cytometry was useful for rapid blood cell counts or screening of water quality (Zhu et al., 2011).

Research into microfluidic devices tailored for the medical field is extensive. Weng et al., developed a suction type microfluidic device to detect the dengue virus. This three-layer device used pneumatics, mixing, and transport to detect the virus in 30 minutes (Weng et al., 2011). Digital microfluidic devices transport biochemical materials in the form of miniature discrete droplets (Xiao & Young, 2011) and have been used for immunosensing, proteomics, DNA, and cell based assays (Vergauwe et al., 2011). Dielectrophoresis (DEP) has been incorporated into microfluidic devices for transport, separation, and blood typing (Minerick et al., 2008; Srivastava et al., 2011; C. Wang et al., 2011). DEP phenomena is the movement of cells from an external applied electric field and has been used to continuously separate breast cancer cells from normal blood cells (Alazzam et al., 2011). The device developed by Alazzam et al. can potentially be used as an early detection method for cancer.

Professor Robert Langer and other researchers at MIT investigated the idea of a "pharmacy on a chip". They performed controlled release studies to determine if a microfluidic platform could act as a pulsatile release drug delivery system. Pulsatile release is a common controlled release method used to treat people with disorders that require drugs to be delivered at varying rates over time. A prototype microchip made from silicon was developed. The microchip had multiple reservoirs for drug storage and the reservoirs were covered with gold membranes. The reservoirs were filled with sodium fluorescein and calcium chloride using ink jet printing. To release the drugs an electric potential of approximately 1V was applied and the gold membranes were dissolved in 10 to 20 seconds. The results from this study revealed that storage and on-demand delivery of drugs can be achieved from microfluidic LOC technology. One major advantage of using microfluidic platforms for drug delivery is that small microchips can be implanted inside the body to locally treat diseases (Santini et al., 1999, 2000). Farokhzad et al. gave a possible application of microfluidic technology in the field of urology (Farokhzad et al., 2006). Other researchers have implemented the proof-of-concept that Langer demonstrated for ambulatory emergency care treatment. A plethora of drug delivery systems that can be embedded in the body have been researched for use in chronic and non-chronic diseases. When treating chronic and non-chronic diseases drugs are delivered over long periods of time. These systems are now modified to rapidly deliver drugs in emergency situations (Elman et al., 2009). Elman et al. developed a smart microchip implant to deliver a drug bolus when disease symptoms are detected. The device is composed of three layers: reservoir layer where drug solution is stored, membrane layer where reservoir is sealed and location of drug is released, and actuation layer where bubbles are formed to trigger the release of the stored drugs. The actuation layer triggers the operation of the device. Micro-resistors heat the drug to generate bubbles, pressure is produced, and the membranes burst delivering the drug ($\approx$ 20 μL) rapidly from the device to its target area in 45 seconds. In this work vasopressin was used as the drug and it was found that 92.5% of the solution loaded into the device was released. Devices of this nature have the potential to accompany cardiac devices such as defibrillators and pacemakers (Elman et al., 2009). Langer's findings have even been extended to nanotechnology. Brammer et al., has shown that silicon nanowires are a viable drug delivery system for antibiotics. It was shown that silicon nanowires sustained drug release levels for 42 days (Brammer et al., 2009).

Despite this wide breadth of research success, commercial implementation of POC devices for diagnostics assays, biosensors, and drug delivery have been much slower than originally predicted. A feature article in Time magazine in 2001 exemplified this dream touting safer and more effective drug delivery techniques (Bjerklie & Jaroff, 2001). However, only a few notable LOC platforms have come to market and are most advanced in the areas of bioassays (pregnancy/ovulation tests, etc.) and gene chips. Bioassay companies include eBioscience (http://www.ebioscience.com/), and Chembio Diagnostic Systems, Inc. (http://www.chembio.com/). Notable gene profiling chips include those by Affymetrix (http://www.affymetrix.com/), Fluidigm (http://www.fluidigm.com/snp-genotyping.html), Gyros (http://www.gyros.com/en/home/index.html), and Sage (http://www.sagescience. com/). Blood chemical analyzers are marketed by PiccoloXpress (http://www.piccoloxpress.com/), while versatile analytical LOCs are marketed by Caliper (http://www.caliperls.com/products/labchip-systems/) and Dolomite Microfluidics (http://www.dolomite-microfluidics.com/). Commercialization is more advanced in the diagnostics arena than in the drug delivery area due to the complexity of sensing the concentration of the drug and controlling the release of new drug. However, as demonstrated by the growth in foundational research, popular news source stories, and commercialization of products, new innovations in this area are being sought.

Cancer is a disease that touches everyone in the world; people are either directly affected by cancer or know someone suffering from the disease. Globally, cancer is responsible for $1/8^{th}$ of all deaths, which is more than HIV/AIDS, malaria, and tuberculosis combined (American Cancer Society, 2011). It is estimated that 1.5 million new cases of cancer will be diagnosed in the U.S. in 2011. Cancer is growing at an increasingly high rate and it is expected that there will be 21.4 million new cases of cancer in 2030 and 13.2 million cancer deaths (American Cancer Society, 2011). Gastric cancer is malignant cell growth originating in the gastro-intestinal tissue lining and kills 650,000 people with 870,000 new cases diagnosed annually (Balcer-Kubiczek & Garofalo, 2009). It is the second most fatal disease in the world (Balcer-Kubiczek & Garofalo, 2009; National Cancer Institute, 2010) and has a poor prognosis due in part to late stage development of any symptoms. People diagnosed with gastric cancer often do not experience symptoms until the disease is metastatic and spreading elsewhere in the body. This then dictates systemic chemotherapy treatment, which traditionally is conducted with regular injections or an embedded catheter. These methods add suffering and additional pain beyond the discomforts of chemotherapy. Further, these methods of drug delivery have large variations in patient exposure concentrations over the course of the treatment; survival rates for gastric cancer suggest this approach is not entirely effective. Therefore, there is a great need for development of new technology to treat cancer patients. The new technology should have two goals **(1)** *effectively treat cancer patients to eradicate disease* and **(2)** *make cancer treatments as painless and noninvasive as possible*. Here we wish to combine four unique technologies into a microfluidic device to provide novel nanoscale drug delivery for cancer patients via a wrist device resembling a watch. Figure 1 shows the global view of our chemotherapy drug delivery system and Figure 2 shows how these four technologies fit together on the drug-delivery microfluidic device and are then discussed separately in the following sections.

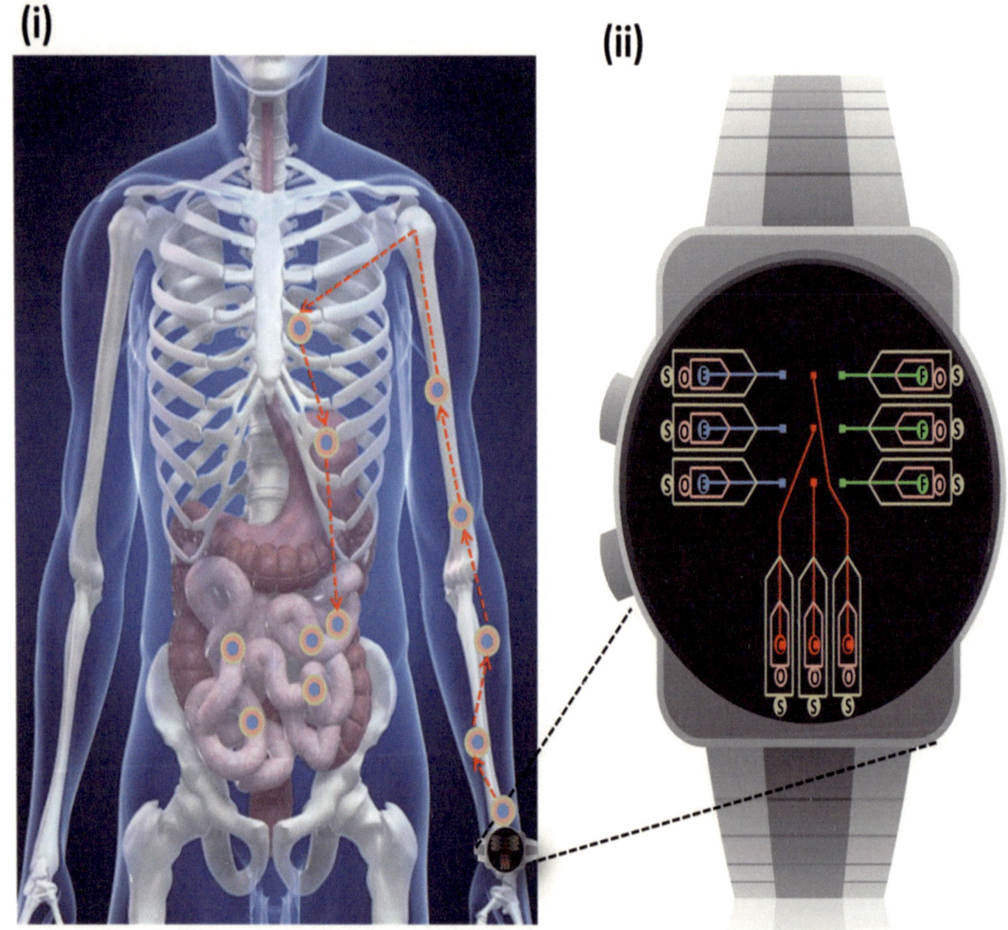

Fig. 1. Global view of chemotherapy drug delivery system, (i) path of emulsified drug from the wrist microdevice through the human body and (ii) enlarged view of wrist device depicting the chemotherapy drug delivery system. The encapsulated chemotherapy drug droplets travel from the wrist device to the intestines contacting circulating tumor cells (CTCs) to treat the gastric cancer.

In Figure 2(i), the reservoirs for each drug are centralized into larger chambers above the layers shown in Figure 2(iv). There is a primary and three secondary reservoirs for oil (one for each drug), and the same for saline. The primary reservoir allows the flow rate of each drug to be independently controlled. This device uses microchannels and tunable electrodispersion to form in-line emulsions of the chemotherapy drug, which are then delivered to the patient using adjustable dielectrophoretic pumping and painless microneedles that penetrate the dermis of the skin. The focus of this new technology has been to specifically treat gastric cancer, but can be adapted to treat many other types of cancer and possibly other diseases.

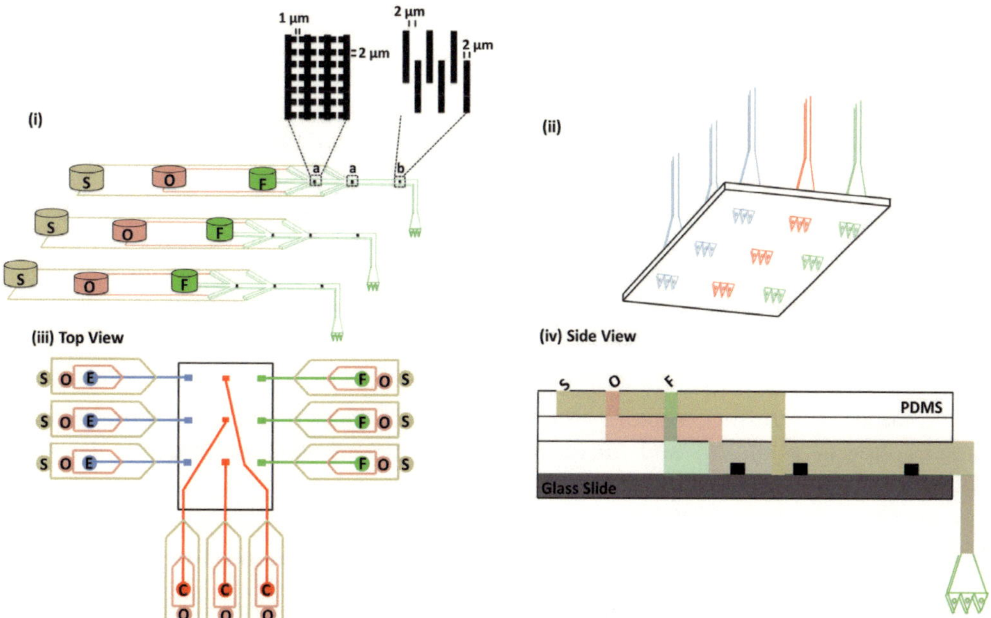

Fig. 2. Microfluidic drug delivery device for cancer treatment. (i) Overview of fluorouracil drug system including drug, oil and saline storage including (a) electrodispersion electrodes and (b) dielectrophoretic micropumping electrodes. Microchannel dimensions are 25 μm (width) x 25 μm (height). (ii) Termination of microchannels into the microneedle system. (iii) Top view of ECF droplet microdevice. And (iv) Side view of multilayered system for droplet dispersion and flow to the microneedles. The abbreviations are S = saline, O = poppy seed oil, F = fluorouracil, E = epirubicin, and C = cisplatin.

Chemotherapy is a common treatment option for gastric cancer. Several single chemotherapy drugs have been used to treat gastric cancer including 5-fluorouracil, mitomycin, doxorubicin, cisplatin, etoposide, docetexal, and methotrexate. Efficacy of these drugs are typically measured via clinical response rates, which is the percentage of patients that respond to cancer treatment such that cancer cells are no longer detected. The response rates to these drugs were poor ranging from 15-35% (Cleveland Clinic Foundation, 2010; Hershock, 2006; Levi et al., 1979). More effective treatments use a combination of two, three or more chemotherapy drugs. Combining two chemotherapy drugs has been examined by Levi et al. and response rates for drug cocktails increased to 40-50% (Levi et al., 1979). McDonald et al. combined three chemotherapy drugs fluorouracil, doxorubicin, and mitomycin and results showed a 55% response rate (Levi et al., 1979). Rivera et al. studied docetexal, a newer chemotherapy drug, in combination with cisplatin and 5-fluorouracil (DCF). DCF was compared with docetexal/cisplatin (DC) and cisplatin/5-fluorouracil (CF), and the objective response rates for DCF were 37-43%, 26% for DC, and 25% for CF. Based on these results it can be concluded that a combination of three chemotherapy drugs are more effective than two chemotherapy drugs (Rivera et al., 2007). Other combination chemotherapy drugs have been studied, and their response rates were: epirubicin, cisplatin and, 5-fluorouracil (ECF) 71%; 5-fluorouracil, adriamycin, and mitomycin 50% and 9%; 5-

fluorouracil, leucovorin, and cisplatin 44%; 5-fluorouracil, adriamycin, mitomycin, and methotrexate 42%; cisplatin, epirubicin, leucovorin, and 5-fluorouracil 43% docetexal, cisplatin, and 5-fluorouracil 37-43%; (Hershock, 2006; Power et al., 2010). Epirubicin, cisplatin, and 5-fluorouracil (ECF) had the highest response rate of 71%. In summary, response rate data suggests that combination chemotherapy drug treatment is the superior treatment option. Therefore, the microfluidic device described here will utilize the combination of ECF. In the device depicted in Figures 1 and 2 each individual drug in the ECF drug system is stored in separate reservoirs so their dosage can be independently controlled via feedback electronics. There is also redundancy in the microchannels and microneedles for backup in case any of the channels become clogged over time. Within the microchannels each drug is sheathed in a biocompatible oil in order to protect the integrity and enhance drug efficacy over the dosage cycle.

This chapter will further explore the integration of microchannels, electrodispersion, dielectrophoretic pumping, and microneedles in a dynamically controllable microfluidic platform to deliver ECF to gastric cancer patients.

2. Technologies utilized in the drug delivery microfluidic device

2.1 Microchannels and electrodispersion

Emulsions are mixed dispersions of more than two immiscible fluids via encapsulation of one layer by the other layer. These emulsion droplets are useful in areas such as foods, cosmetics, pharmaceutical drug delivery, and chemical synthesis. Examples of foods include milk, yogurt, sauce, butter, etc. and cosmetics of lotion (oil-in-water, O/W), cream (water-in-oil, W/O), hair, shaving, and bath products that are predominantly viscous liquids (Mezzenga, 2005; S.H. Kim et al., 2011). In chemical synthesis, droplets are being used as a new reaction platform due to their ability to function as a batch reactor such as antimicrobial agent and preservatives (Hamouda et al., 1999; Jensen & Lee, 2004; Mejia et al., 2009). The forms of emulsion droplets to contain various physical and chemical compositions are effective in delivering drugs and cosmetics in human body (Wibowo & Ng, 2001; Kiss et al., 2011).

In this drug delivery microfluidic device, flow focusing (FF) hydrodynamics and electrodispersion technology are combined to dynamically generate oil-sheathed drug droplets on the order of 100 nm outer diameter dispersed in saline. Poppy seed oil is used to decrease the toxicity of the chemotherapy drug (Pai et al., 2003) while maintaining its potency and efficacy once it reaches the target malignant cells, and saline is used to carry the droplets into the tissue during injection. Our drug delivery microdevice will combine both FF and electrodispersion technologies in order to achieve narrower size distribution of particles by preventing droplet interactions and coalescence. Electric fields are added for chemotherapy drug droplet formation to decrease the size of droplets, improve robustness of continuity of the droplet thread formed, and increase velocity as droplets travel downstream in the microchannel. Thus, the main focus of our study is developing FF geometry with electrodispersion and adequate use of surfactants to generate submicron droplets (~100 nm) with highly uniform sizes to promote a quick transport into cells.

Emulsion droplets of very narrow size distributions can be strategically generated by harnessing hydrodynamic behaviors within microfluidic systems (Anna et al., 2003; Martin-

Banderas et al., 2005; W. Lee et al., 2009). FF geometries are used in mixing immiscible phases or encapsulating one phase within a second sheathing phase. Typically, hydrophobic drops are dispersed in a hydrophilic fluid or vice versa. The drug delivery microdevice utilizes this technology in order to protect the inner fluid (drug) by an outer fluid (oil) which is then dispersed in a continuous saline stream. The inner phase is traditionally termed the dispersed phase while the outer phase is termed the continuous phase when they meet at an orifice. Most microfluidic emulsions involve a single droplet dispersed in a continuous phase such as water-in-oil (W/O) and oil-in-water (O/W) (Ha and Yang, 1999; Anna et al., 2003; W. Lee et al., 2009; Kiss et al., 2011). This concept can also be expanded to droplets that include more than one internal droplet such as double emulsions (W/O/W or O/W/O) (Utada et al., 2005; Seo et al; 2007; Liao and Su, 2010; S.H. Kim et al., 2011) and are the foundation for the drug delivery microdevice described in this chapter.

During small droplet synthesis, mechanical shear stress was utilized in order to achieve small and highly stable emulsion droplets, but this approach yielded tens of nano- to hundreds of micro-scale droplets with large size distributions, which was problematic (Pacek et al., 1999; Abismaïl et al., 1999). The key advantage of FF technique is precise control in producing droplets into the range of hundreds of nanometers (Anna et al., 2003; Thiele et al., 2010), but ambiguity remains regarding the lower limit of droplet sizes that can be achieved, the size distribution and continuity of the droplet threads still remain unreported due in part to the difficulty of in-line droplet size analysis and the length of the droplet thread (Anna et al., 2003; W. Lee et al., 2009). The goal in this drug delivery microdevice is to generate the smallest droplets possible because larger droplets are less stable and more likely to come in contact with each other which leads to droplet deformation and coalescence. Other studies have determined that 50-150 nm droplets ensure an optimal intake in cells for drug delivery applications (Thiele et al., 2010). In addition, larger droplets are less stable and more likely to come in contact with each other which leads to droplet deformation and coalescence.

Because the drug delivery microfluidic system dimensions are designed to fit within a wristwatch-like system on a human wrist, multilayered FF and electrodispersion are proposed for ECF chemotherapy drug emulsion and delivery, Figure 2(ii-iv). The total dimensions of multilayered ECF drug system is small enough to be non-obtrusive, so that it can be worn for continuous drug delivery with minimal discomfort as shown in Figure 1. In the drug delivery microdevice, two FF orifice/junction geometries are utilized in series as shown in Figure 3. The two junctions whereby sheathing flows of poppy seed oil and saline are added to the main channel are 3mm long, 20 μm wide, and 20 μm deep in the microchannel, and the continuous and dispersed phases are injected through pressure regulated membrane deflection into the fluid reservoirs which operate as micropumps. Such designs are also commercially available and meet the volume and portability limitations of the proposed wristwatch system, as well as the energy demand limitations (Lima et al., 2004). The flow rate inputs are 0.01 μL/min of drug, 0.1 μL/min for poppy seed oil, and 1 μL/min for saline based on the reduction of the optimal flow conditions achieved by Zagnoni et al., 2009 & 2010 to meet our channel dimensions in which the ratio of the continuous and dispersed phase flow rate are held at 10 to achieve submicron droplets in the downstream microchannels.

Achieving drug droplets with diameters at approximately 500 nm is feasible via the hydrodynamic flow focusing achieved with the 4 μm wide orifice and combined with the strategic use of surfactant chemistry (W. Lee et al., 2009). Droplet size is determined initially by the orifice geometry; however, a surfactant mixed with either the continuous or dispersed phase balances interfacial tension and enables droplet sizes to be orders of magnitude smaller in comparison with those without a surfactant (W. Lee et al., 2009). That is, adding a proper surfactant can improve stability of the droplets and decrease the size of droplets because the surfactant molecules reduce the interfacial tension between different fluids in a droplet, thus avoiding the undesirable coalescence among droplets. For this chemotherapy drug delivery application, proper surfactant selection requires biocompatibility which must be considered as well as long-term drug-surfactant interactions. To avoid the later, surfactants will be dispersed in the oil phase and in the saline phase.

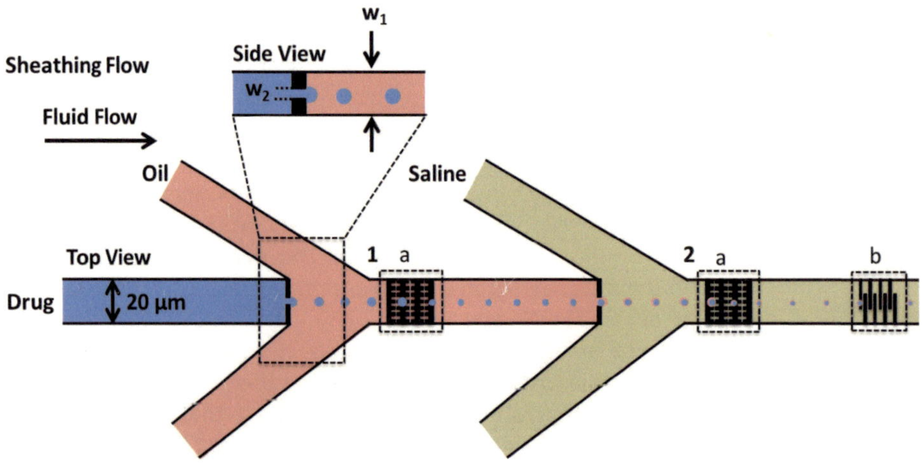

Fig. 3. Microchannel design for the chemotherapy drug emulsion formation with flow focusing and sheathing flow; appropriate scale not reflected. Dashed boxes (a) and (b) correspond to the placement of the electrodes for electrodispersion and dielectrophoretic micropumping, respectively. w_2 is the orifice depth 4 μm, and w_1 is the depth of the junction 20 μm . The aqueous drug solution flows through the microchannel until it is dispersed into the continuous oil phase at junction 1. At junction 1 orifice, drug-in-oil droplets are formed and flow until they are then dispersed into the saline phase at junction 2 to achieve an oil sheathing of the drug droplet in the continuous saline phase. Double emulsion chemotherapy droplet is produced via the second orifice. See Figure 4 for the COMSOL simulation of this fluid flow.

The droplet formation behavior is mainly determined by capillary number (Ca), which represents the balance between the viscous forces and interfacial tension at the surface of two fluids as defined in Equation 1. In Equation 1, μ is the viscosity of the continuous phase, V is the characteristic velocity, and γ is the surface tension. Capillary number is the most appropriate dimensionless number to describe droplet formation behavior because Reynolds number and Weber number are less significant in FF (W. Lee et al., 2009). In addition to Ca, the flow rate ratio (φ), the viscosity ratio (λ), and the expansion ratio (A) in Equations 2-4 (W. Lee et al., 2009) are also parameters to govern droplet formation because they balance viscous stresses and exerted shear stresses which result in droplet thread trajectory and velocity profile. In Equation 2, Q_d and Q_c are flow rate of dispersed phase and continuous phase, respectively. In Equation 3, μ_d and μ_c are viscosity of dispersed phase and continuous phase and w_1 and w_2 are depth of outside and inside orifice.

$$Ca = \frac{\mu V}{\gamma} \quad (1)$$

$$\varphi = \frac{Q_d}{Q_c} \quad (2)$$

$$\lambda = \frac{\mu_d}{\mu_c} \quad (3)$$

$$\mathrm{A} = \frac{w_1}{w_2} \quad (4)$$

Accordingly, the size of droplets changes by capillary number due to viscous force and the interfacial tension. While flow rate ratio affects the droplet formation behavior significantly, the viscosity ratio has a relatively weak impact on the droplet diameter compared to capillary number (Ca) since the flow rate is associated with characteristic velocity of Ca. In addition, as the expansion ratio between the orifice and the junction depth decreases, a longer thread of smaller droplets is achieved, which is desired in the drug delivery microdevice. Much remains to be learned in regards to the underlying physics in the system as well as the stability of the droplet threads over long operation times. Therefore, a few assumptions were made in the FF designs of the microdevice. These assumptions include the combination of the FF droplet formation with the electrodispersion design to decrease droplet size, increases velocity of droplets, as well as with regards to the continuous robustness of producing droplets over long operation times (days) of the device.

Research has also been conducted in the related field of electrically manipulated emulsifications (H. Kim et al., 2007; Zagnoni et al., 2009; Zagnoni et al., 2010). Electric fields with a resonance frequency between 10Hz and 10MHz have been used as a separation method to remove water dispersed in oil for applications in the petroleum industry. This data is applicable to droplets in the drug delivery microdevice since the frequency-modulated electric field utilized can be used as a deformation tool for emulsion droplets (Zagnoni et al., 2009; Zagnoni et al., 2010). Electric fields have also been used to focus and space nano and micro emulsions/particles in microchannels of 50 μm (width)* 50 μm (length)* 61 μm (height) orifice dimensions in DC fields (H. Kim et al., 2007). H. Kim et al. studied electrospray emulsification and produced emulsion < 1 μm in diameter with ~2% of size distribution at a field of $1.4*10^2$ - $5.5*10^3$ V/m, and also Arya et al. successfully synthesized hundreds of nm chitosan micro/nano spheres for drug delivery application in $2.3*10^5$- $4.7*10^5$ V/m. Furthermore, Mejia et al. formed wax emulsions with high uniformity for water-proof painting, cosmetics, and adhesives that supports the idea of utilizing electric field (2.6-2.9 kV in 500 ml wax mixture) in the production of fine emulsions. Compared to these literature values of the applied electric fields, the device operating condition of $5*10^6$-10^7 V/m is several orders are higher.The electric energy acting on the particles cause the water-oil interface to charge such that it behaves as a capacitor, which leads to a tip of Taylor cone which enables tiny droplets and narrow size distribution ~2% (H. Kim et al., 2007).

Castellation configuration electrodes (Zagnoni et al., 2009; Zagnoni et al., 2010) were designed for electrodispersion because this design resulted in the highest localized electric field in the z-dimension which has the potential to most efficiently manipulate the droplets and minimize deformation in the x- and y- directions. These are placed 200 μm downstream from both junction 1 and junction 2 as shown in Figure 3 (a). This electric field energy does two things: a) it breaks apart the droplets from hundreds of nanometers in diameter into the more effectively adsorbed size of < 100 nm and b) it spatially disperses the droplets in the continuous phase in order to minimize coalescence as droplets travel forward in microchannel.

In order to simulate behaviors in the drug delivery microdevice, COMSOL 4.2 was used to simulate the 3D electric field gradient of the electrodispersion design in Figure 4, which is castellated gold electrodes of gap width 1 μm (x-direction), width 2 μm (y-direction), and thickness 20 nm (z-direction). The electrostatics module with the electrostatic potential (Equation 5), a relationship between electric displacement and the electric field (Equation 6), and Gauss's law (Equation 7) were used to simulate the electric field gradient in a fluid medium of saline.

$$\vec{E} = -\vec{\nabla} V \tag{5}$$

$$\vec{D} = \varepsilon_0 \vec{E} + \vec{P} \tag{6}$$

$$\vec{\nabla} \cdot \vec{E} = \frac{\rho_V}{\varepsilon_0} \tag{7}$$

In Equations 5-7, $\vec{E}$ is the electric field, V is the electric potential, $\vec{D}$ is the electric displacement, ε_0 is the permittivity of a vacuum, $\vec{P}$ is the electric polarization and was assumed to be zero in the fluid medium, and ρ_V is the space charge density. Combining Equations 5-7 with zero electric polarization gives

$$\vec{\nabla} \cdot \left(\varepsilon_0 \varepsilon_r \vec{E}\right) = \rho_V \tag{8}$$

Equation 8 is the governing equation used in to simulate the electric field gradient of the electrodispersion electrode design (Figure 3). Equation 8 is modified slightly and employed in COMSOL as

$$\nabla E = \sqrt{\frac{\partial^2 E}{\partial x^2} + \frac{\partial^2 E}{\partial y^2} + \frac{\partial^2 E}{\partial z^2}} \tag{9}$$

In COMSOL, the physical properties for water were altered slightly to simulate the saline such as 8.9 x 10^{-4} Pa·s dynamic viscosity and 80 relative permittivity. The physical properties used for poppy seed oil were dynamic viscosity 5.58 x 10^{-2} Pa·s and relative permittivity of 4. Poppy seed oil is used to decrease the immediate toxicity of the chemotherapy drug (Pai et al., 2003) as it enters the tissue, and saline is used to carry the sheathed droplets and match tissue isotonicity during injection. Gold was the material used for the electrodes, and PDMS was the material used to form the orifices at the microchannel junctions (see Figure 2 for orifice design) as reported in the literature (W. Lee et al., 2009). The initial flow rates in the microchannels were chosen to be 0.01 μL/min of drug, 0.1 μL/min for poppy seed oil, and 1 μL/min for saline. These initial flow rate values were chosen based on Zagnoni et al. 2009 and 2010.

The fluid flow velocity in the FF microchannel was modeled with the laminar flow module employing the Navier Stokes equation and the continuity equation, which were simplified by assuming a steady-state system and an incompressible fluid as follows:

$$\rho(v\cdot\nabla)v=-\nabla p+\nabla\left(\mu\left(\nabla v+\nabla^{T}v\right)\right)+F_S \tag{10}$$

$$\nabla\cdot v=0 \tag{11}$$

In Equations 10 and 11, ρ is the fluid density, v is the fluid velocity, t is time, p is the pressure, μ is the viscosity, and F_S is the volumetric force on the fluid resulting from surface tension.

Two types of fluidic conditions of water (to represent drug and saline) and oil were employed, and the velocity of the formed emulsion droplets were calculated from the summation of pressure driven flow velocity and electro-osmotic velocity. The electro-osmotic flow velocity was calculated from Smoluchowski slip velocity equation via a wall boundary condition on the microchannel and added to simulate the velocity of the chemotherapy droplets in the field created by the electrodispersion, Figure 3. The zeta potential for PDMS was assumed at -0.1V (Kirby and Hasselbrink Jr., 2004). The boundary condition for electrical potential was an applied DC field of 10V and ground across each pair of electrodes. The velocities without electrodes were calculated from pressure driven flow in the x-, y-, and z-direction and expressed with u_i. The governing equations for electro-osmotic flow used for this simulation are Equation 12-16 below,

$$\mu_E=\frac{\varepsilon_r\varepsilon_0\zeta}{\eta} \tag{12}$$

$$v_{EOF,i}=\mu_E\cdot\vec{E}_i \quad \text{where i = x,y,z} \tag{13}$$

In Equations 12 and 13, μE is electrophoretic force, ε_r is the relative permittivity of the fluid, ε_0 is the permittivity of a vacuum, ζ is the zeta potential, η is the dynamic viscosity, $v_{EOF,i}$ is the velocity due to electro-osmotic flow (EOF), and $\vec{E}$ is the electric field. The normal EOF velocity and the total velocity of the droplets are given as

$$v_{norm}=\sqrt{v_{EOF,x}^2+v_{EOF,y}^2+v_{EOF,z}^2} \tag{14}$$

$$v_{total}=\sqrt{v_{x,total}^2+v_{y,total}^2+v_{z,total}^2} \tag{15}$$

The EOF velocity is related to the total velocity by

$$v_{i,total}=u_i+v_{EOF,i}^2 \quad \text{where i = x,y,z} \tag{16}$$

The total velocity is displayed in the simulation scale bar next to each COMSOL diagram in Figures 4 and 5.

Figure 4 shows the electric field gradient with and without the droplets. The maximum electric field gradient without droplets is 1.1 x 10^7 V/m and the maximum electric field gradient with droplets is 2.3 x 10^7 V/m. This difference is because when the droplets pass in between the two electrodes, the droplets have an induced field which influences the applied field gradient by reducing the gap over which the potential acts. The electric field is tuned to the resonant frequency of the droplets in order to break apart the ~500 nm droplets into < 100 nm droplets as well as distribute the droplets spatially within the continuous fluid phase.

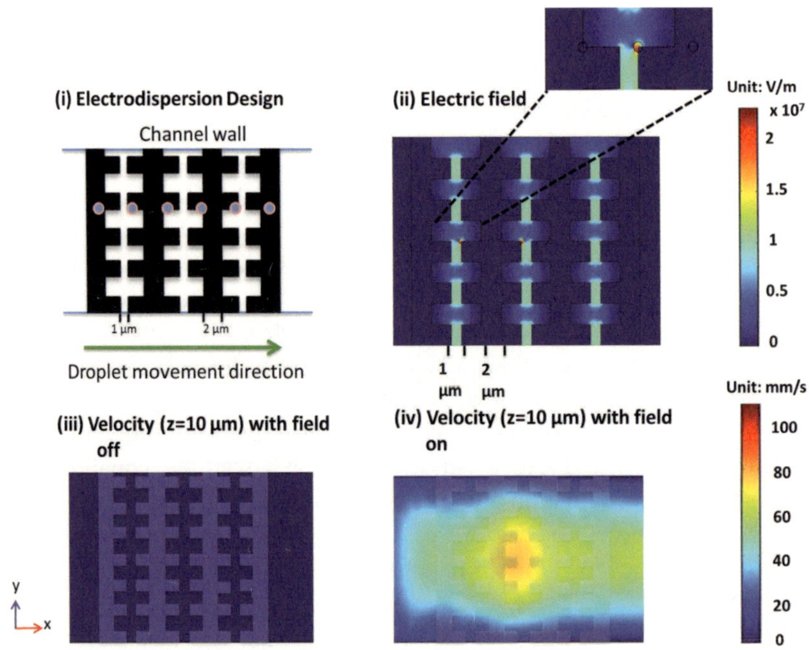

Fig. 4. COMSOL simulations of the electric field gradient as well as the velocity profiles (without and with the E field active) at the midpoint (x, y plane at z = 10 µm height) of the microchannel. (i) Channel level view of electrodes with cartooned drug droplets, (ii) electric field magnitude, (iii) fluid velocity above the inactivated electrodispersion electrodes at junction 1 (roughly 0.056 mm/s), and (iv) fluid velocity once the electric field is activated (maximum velocity roughly 73.0 mm/s). The maximum observed electric field strength during fluid flow is 2.3 x 10^7 V/m.

The drug delivery microdevice is designed so that it can be easily fabricated with standard UV-photolithographic methods in a Class 100 or greater cleanroom. The FF geometry integrated within the microchannel design is fabricated in poly (dimethylsiloxane) (PDMS) using standard printed masks, UV soft photolithography techniques, and multilayer alignment of the channels from one layer to the next. The electrodes can be fabricated on a silicon or glass support via photoresist masking followed by deposition of a 10 nm titanium layer then a 10 nm gold layer by electron-beam evaporation.

Each layer of the molded channels can be sealed via oxygen plasma bonding procedures. PDMS is hydrophobic so that oil wets the walls, but surface treatment defines the required hydrophobic and hydrophilic patterns depending on fluid phases used. The dispersed drug phases will be hydrophilic (viscosity, μ_c= 6 mPa·s) and a biocompatible surfactant will be mixed with either the continuous phase or the dispersed phase at a concentration approximately 2.5 times greater than the critical micelle concentration (CMC) which is a moderate surfactant concentration and is favorable to test a wide flow rate range (W. Lee et al., 2009). The presence of the surfactant does not substantially change the viscosity as evidenced by W. Lee et al.

A positive displacement chamber pump connected across a membrane to a pressurized canister will be used to drive the fluids from each inlet into the FF microchannels. Separate micropumps will connect to each solution reservoir such that flow in each channel can be separately controlled to achieve a feedback controlled fluidic system that can be worn on a wrist (Lima et al., 2004). Pressure drops for flows between 0.01 and 1 μL/min are not expected to cause deformation of microchannels due to either high-pressure injection or the PDMS elasticity used in our drug microdevice (Soller et al., 2011). If necessary, Thermoset Polyester (TPE) would be the best alternative material of PDMS due to its high rigidity and suitability with droplet microfluidics (Soller et al., 2011).

In order for a stable, efficient, and continuous small dosage of drug delivery, optimized selection of a surfactant, microfabrication condition, and pumping system are discussed. The unique combination of FF and electrodispersion to generate drug droplets protected by a sheathing layer of biocompatible poppy seed oil is described. Further, the electric field and fluid flow conditions were simulated and results are used to optimize the design. The droplets exiting the FF and electrodispersion region must then be accelerated in the drug delivery microdevice channel in order to generate a high enough pressure difference for the fluid to exit the microneedles into the dermis of the skin to achieve drug delivery.

2.2 Dielectrophoretic pumping

Traveling wave DEP (twDEP) is incorporated into the drug delivery microdevice in order to accelerate the chemotherapy droplets as they travel to an array of microneedles for painless injection into the body. DEP is an efficient nondestructive way to manipulate bioparticles (Cheng et al., 2011), and twDEP is being investigated as a possible drug delivery technique (Bunthawin et al., 2010). The electrode configuration consists of an array of parallel rectangular electrodes configured in an intercalated pattern as shown in Figure 1(ib). The intercalated configuration of the electrodes facilitates horizontal movement of particles when a non-uniform AC field is applied that is offset by 90° with each successive electrode. This causes the field maxima to travel in waves down the array of electrodes thus driving the particle forward. Typically the spacing between the electrodes is fixed to the width of one single electrode with the optimal width of an electrode being close to the diameter of the target particle. The spacing between the electrodes is usually 10 μm to 50 μm and remains constant (Lin & Yeow, 2007).

Parallel electrodes are used for collecting, transporting, and/or separating particles. For the drug delivery microdevice, we will be using the parallel electrodes for transporting the oil-sheathed chemotherapy droplets. To help facilitate transportation the frequency and conductivity of medium is chosen specifically to induce the largest positive DEP force on the dielectric particles. Positive DEP is the movement of particles up the electric field gradient. When the AC field is applied to the electrodes a dipole moment is induced in the particles and the time-average DEP force, $\langle F_{DEP} \rangle$, is given as (Pethig, 2010)

$$\langle F_{DEP} \rangle = 2\pi\varepsilon_m R^3 \left\{ \mathrm{Re}[f_{cm}] \nabla E^2 + \mathrm{Im}[f_{cm}] \sum E^2 \nabla \phi \right\} \tag{17}$$

In Equation 17 ε_m is the absolute permittivity of the medium, R is the radius of the particle, f_{cm} is the Clausis-Mossotti factor, E is the amplitude of the electric field, and ϕ is the phase. The Clausius-Mossotti factor is related to the polarizability of the particle and $\sum E^2 \nabla \phi$, is the summation of the magnitude and phase of each field component. The Clausis-Mossotti factor, f_{cm}, for a spherical, homogeneous particle is given as (Pethig, 2010)

$$f_{cm} = \frac{\underline{\varepsilon}_p - \underline{\varepsilon}_m}{\underline{\varepsilon}_p + 2\underline{\varepsilon}_m} \tag{18}$$

In Equation 18 $\underline{\varepsilon}_p$ is the complex permittivity of the particle and $\underline{\varepsilon}_m$ is the complex permittivity of the medium defined as (Pethig, 2010)

$$\underline{\varepsilon}_p = \varepsilon_p + \frac{\sigma_p}{j\omega} \tag{19}$$

$$\underline{\varepsilon}_m = \varepsilon_m + \frac{\sigma_m}{j\omega} \tag{20}$$

In Equation 19 and 20 ε_p is the absolute permittivity of the particle, σ_p is the conductivity of the particle, σ_m is the conductivity of the medium, ω is the angular frequency and j is the imaginary number. It is important to note that the DEP force for twDEP is dependent on both the real and imaginary part of the Clausius-Mossotti factor which causes particles to experience both an in-phase force (real part) and the out of phase force (imaginary part). Classical DEP force is only dependent on the real portion of the Clausius-Mossotti factor, $\langle F_{DEP} \rangle = 2\pi\varepsilon_m R^3 \mathrm{Re}[f_{cm}] \nabla |E|^2$.

Further, the assumption that the oil-sheathed droplets of aqueous drug can be represented as a first approximation homogeneous particle is acceptable because the conductivity of the oil layer is substantially different ($\approx$ 1 nS/m) from the supporting saline medium that the Clausius-Mossotti factor is nearly 0.50 – 3.36*10^{-8}i and varies very little (< 0.01%) over the anticipated frequency range of interest from 1 Hz to 10 MHz (Durr et al., 2003; Felten et al., 2008). Since twDEP is being used to transport chemotherapy droplets, it's important to have an expression that relates the particle electrophoretic mobility with twDEP force. The DEP force can also be written utilizing the zeta potential ζ_p (Kang & Li, 2009),

$$\vec{F}_{DEP} = 6\pi\zeta_p \varepsilon_r R E \tag{21}$$

The DEP force written in this form allows it to be related to the electrophoretic mobility of the particle given as (Kang & Li, 2009)

$$\mu_E = \frac{\varepsilon_r \varepsilon_0 \zeta_p}{\eta} \tag{22}$$

$$v_{DEP} = \frac{R^2 \varepsilon_r \varepsilon_0 \operatorname{Re}\left[f_{cm}\right]}{3\eta} \nabla E \tag{23}$$

In equations 21-23 ζ_p is the zeta potential, ε_r is the relative permittivity of the medium, R is the radius of the particle, E is the magnitude of the electric potential, ε_0 is the permittivity of free space, $\operatorname{Re}[f_{cm}]$ is the real-part of the Clausis-Mossotti factor, and η is the dynamic viscosity of the medium.

Parallel electrodes were added to the end of the microchannels, Figure 2(ib), to increase the velocity of the chemotherapy droplets to ensure the droplets reach the microneedles at the end of the device with sufficient velocity to penetrate the dermis. The desired velocity at the microneedle tips was 10 μm/s based on modeling of transdermal delivery (Lv et al., 2006). COMSOL was used to simulate the electric field gradient via the conservation of electrical potential between the insulating channel walls and the velocity of the droplets via the Navier-Stokes (Equation 12) and conservation relationships (Equation 13). Equations 5-9 developed in section 2.1 were used to obtain the electric field gradient with and without droplets in the channel as shown in Figure 5. Droplets were defined as separate spherical

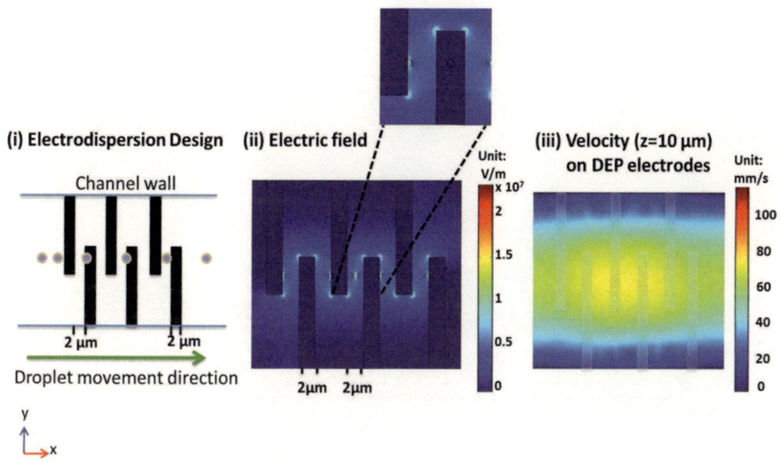

Fig. 5. COMSOL simulation of traveling wave dielectrophoretic pumping electrodes before and during fluid flow in the microchannel. The combined total length of the electrodes including gaps is 22 μm (i) channel level view of electrodes, (ii) normalized electric field strength from twDEP electrodes, (iii) velocity distribution in the channel view. In (i) when the twDEP electrodes are energized at an instantaneous half cycle with max potential of 10Vpp to ground at subsequent electrodes. The maximum observed electric field strength before fluid flow is 7.2×10^6 V/m, and the maximum electric field strength during fluid flow is 1.1×10^7 V/m. The droplet fluid maximum velocity was 73 mm/s in this DEP pumping region at z = 10 μm.

fluids with the laminar flow module in COMSOL. The material selected in COMSOL for the particles was oil as previously described in section 2.1 and modified water to represent saline. It can be seen in Figure 5 that the electric field gradient decreases by 65% when the droplets are present because of the change in the electric charge of the dielectric oil layer. Further, the droplets align along the centerline of the electrode configuration because the DEP force is directed toward the symmetric centerline. This phenomenon is observed because electric field interact with droplets and induce force due to the dielectric property difference between immiscible phases (water and oil) since free charges can accumulate on the interface between inner and outer fluid. The behavior of the droplet velocity was examined in z-dimension revealing that the electric field primarily affects the velocity of the droplets near the electrodes (z < 3-4μm). The droplet equilibrium height is 10 microns above the bottom of the channel. DEP forces are predicted to be on the order of 152 pN for the 10 Vpp field over the 22 μm electrodes.

The electrophoretic mobility was used to simulate the velocity of the chemotherapy droplets in the field created by the dielectrophoretic pumping electrodes. In this drug delivery microdevice system, the droplet is a chemotherapy drug coated with poppy seed oil in a continuous phase of saline. In order to look at electric field effect, electrophoretic mobility was an added term.

In Figure 5, the local fluid velocities increased by over 5000 times from 0.021mm/s to 109 mm/s by implementing dielectrophoretic pumping at 10 Vpp via the electrode designs in the COMSOL simulation. Comparing without and with dielectrophoretic pumping (Figure 4(iv) and Figure 5(iii), respectively), the velocities with electrodispersion and DEP pumping electrodes remarkably changed as demonstrated in Figure 6. These increases in velocity are necessary to move the drug droplets forward to the microneedles for injection.

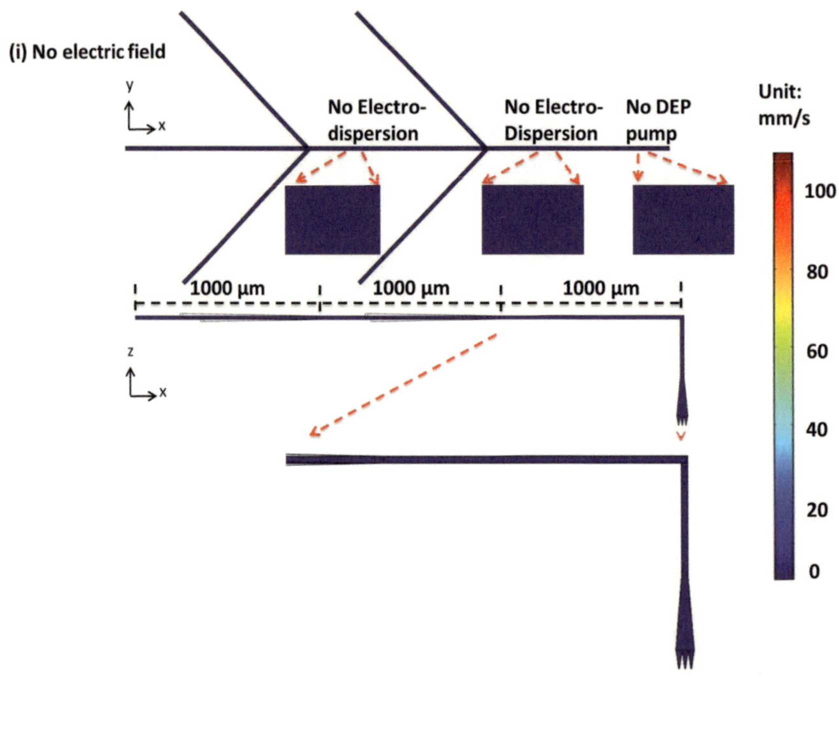

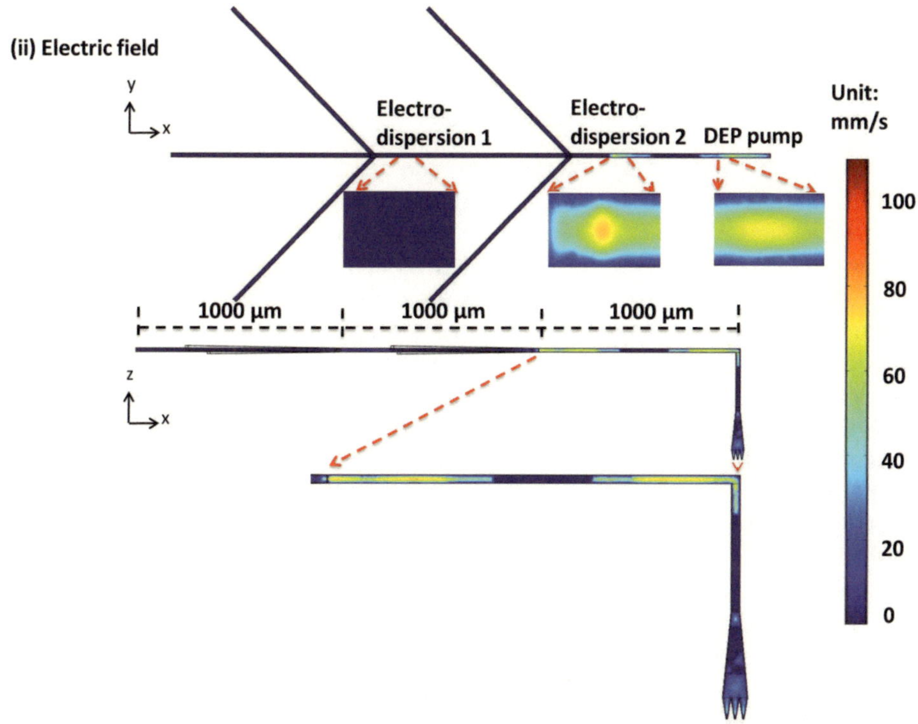

Fig. 6. The top of each simulation shows x-y plane, top view, and bottom of each simulations shows z-x plane to display overall velocity profile. COMSOL simulation of fluid velocity without (i) and with (ii) dielectrophoretic pumping electrodes. Interdigitated dielectrophoretic pumping electrodes were added to the end of the microchannel to increase the velocity of the chemotherapy drug droplets to ensure sufficient velocity for microneedle injection. (i) The maximum fluid velocity inside the microchannel without electrodes was 0.025 mm/s reached at junction 2, (ii) the maximum fluid velocity with electrodes was 109 mm/s also at junction 2. The magnified sections for (i) and (ii) show that the fluid velocity substantially increases from 0.0056 mm/s to 35 mm/s at the end of the microchannel once the pumping electrodes are added.

2.3 Microneedles

The skin is a common area for drug delivery, and offers advantages over other non-invasive drug delivery techniques. Drug delivery through the skin avoids drug metabolism by enzymatic reactions and the gastro-intestinal, has the potential for continuous drug delivery (Migalska et al., 2011), and facilitates reaching circulating tumor cells (CTCs) that metastacized from their origin. Hypodermic needles are commonly used as the dominant method for transdermal drug delivery for gastric cancer treatment. Needles are painful, inconvenient and require professional administration for each dose (P.M. Wang et al., 2006). Another problem is that drugs are delivered all at once which causes an immediate spike in drug concentration profiles which then rapidly diminishes. This can cause physiological instabilities, other dosage-related side-effects and can potentially fail to completely eradicate the tumor and/or CTCs (Zahn et al., 2004). To overcome these drawbacks, this work proposes a novel drug delivery microdevice amenable to feedback control; the device integrates microchannel hydrodynamic flow focusing, electrodispersion to decrease drug droplet size, dielectrophoretic pumping, and microneedles together in one microdevice to deliver chemotherapy drugs in the form of droplets to a gastric cancer patient.

The skin has four layers: (1) the stratum corneum, principal barrier composed of corneocytes, (2) epidermis, (3) dermis, and (4) subcutaneous tissue (Escobar-Chavez et al., 2011). The four layers of the skin are barriers to transdermal drug delivery and microneedles are used as physical enhancers for transdermal drug delivery. They are designed to increase the permeability of the skin up to four orders of magnitude, so that drug passage through the stratum corneum and outer most layer of the epidermis becomes simple (Escobar-Chavez et al., 2011; K. Lee et al., 2011). Microneedles allow for drugs to be delivered across the skin in four ways: (1) "poke-with-patch", this method uses a solid microneedle array to penetrate the skin creating micropores, the microneedle array is removed and drugs are delivered through the micropores via a transdermal patch, gel or solution; (2) "coat-and-poke", this method coats an array of microneedles with a drug and inserts the coated microneedles into the skin; (3) "poke-and-release" embeds the drug molecules into the structure of polymer biodegradable microneedles and inserts them into the skin; and (4) "poke-and-flow", uses hollow microneedles to insert liquid drugs into the skin (Migalska et al., 2011). There are advantages and disadvantages to each approach and these vary with the application. Micropores created by microneedle penetration last for more than a day when left covered and they last for less than 2 hours when left uncovered. Microneedles that dissolve under the skin are perceived as the safest with the least chance of prolonged irritation (K. Lee et al., 2011). Methods 1, 2, and 3 are best suited for applications where a one time dose or daily dose of drug is desired. However, for the chemotherapy drug delivery microdevice, method 4 is optimal. The microneedles would remain inserted in the skin and held in place via a wristband as shown in Figure 1.

Microneedles can be inserted into the skin easier than hypodermic needles because stress on the skin is inversely proportional to the area of the top (Zahn et al., 2004). Microneedles require sharpness to overcome stress forces on the skin surface as well as strength against fracturing, bending, and buckling, sufficient flow rate, and biocompatibility of the needle material (Nguyen & Wereley, 2006). The insertion force can be lowered by utilizing kinetic energy such as vibration which can reduce the force by as much as 30% (Nguyen & Wereley, 2006), and provides an increase in infusion flow rate (P.M. Wang et al., 2006). Retracting the microneedles after insertion by approximately 100-300 μm also achieves a much greater flow rate (P.M. Wang et al., 2006).

Microneedles have been shown to increase the transdermal delivery of many molecules including aminovulinic acid, anthrax, bovine serum albumin, desmospressin, erythropoirtin, meso-tera(N-methyl-4-pyridyl)porphine tetra tosylate, ovalbumin, plasmid DNA, low molecular weight tracers to proteins, and nanoparticles. Insulin is the most widely studied drug with microneedles and enhanced skin permeability has been reported in vivo and in vitro (Donnelly et al., 2011; Escobar-Chavez et al., 2011). Recently researchers have looked into using microneedles fabricated from maltose to deliver methotrexate to rats via iontophoresis to treat cancer. The result of this study was a synergistic 25-fold increase of drug delivery (Escobar-Chavez et al., 2011).

Microneedles are a painless drug delivery method and create larger transport pathways for larger molecules; our device extends this to nanometer droplets. The painless characteristics of microneedles allow them to overcome the limitation of hypodermic needles. Advantages of microneedle technology are that the transport mechanisms are not dependent on the diffusion into the tissue, placement in the epidermis allows for drugs to reach target areas more readily, while only penetrating the stratum corneum without piercing nerve endings thus reducing pain, infection, or other injury. The microneedles are nontoxic, minimally invasive, can be mass-produced for a range of materials such as silicon dioxide and polymers, are easily disposable/interchangeable, and can be made from biodegradable materials. Some disadvantages to using microneedles are local inflammation and skin irritation. Another disadvantage is that the microneedles may break and be left under the skin; to avoid this, the diameter of the microneedle should be smaller than the diameter of a hair, ≤ 50 μm (Escobar-Chavez et al., 2011). In comparison hypodermic needles are inconvenient, not easily self-administered, and have poor targeted delivery because they have to be manually injected (P.M. Wang et al., 2006).

Our microdevice uses an array of microneedles, Figure 7, to deliver drug-in-oil microdroplets, applying the "poke-and-flow" method. Figure 7(i) shows the design of a single microneedle wherein the diameter of hole in the microneedles is approximately 40 μm which is sufficiently large to allow delivery of the oil sheathed drug droplets without disruptive shearing effects while simultaneously being large enough to avoid breakage in the skin. Figure 7(ii) is color coded to show the array of microneedles for each of the chemotherapy drugs with blue representing epirubicin, red representing cisplatin, and green representing fluorouracil. Each microneedle is connected to its own FF channel as shown in Figure 2. Figure 7(iii) shows a photograph of an array of microneedles (Baek et al., 2011) fabricated from polylactic acid. The fluid emulsion velocities are approximately 35 mm/s as they leave the DEP pumping electrode region. Once the emulsions enter their final descent, to the microneedles the midchannel linear velocity decreases to approximately 30 to 40 mm/s as the channel expands followed by velocity increases within the microneedle tip, as constrained by the continuity equation.

Several methods are used to fabricate hollow microneedles and most are made from silicon or silicon-based materials. This drug delivery microdevice requires a relatively straight channel with minimal bends so that the microdroplets are not sheared and are delivered to the dermis intact. The pyramid-shaped microneedles were chosen because fabrication is simple and thus it has the best shape to achieve tip sharpness and strength (Moon & Lee, 2003). The pyramid microneedles can be fabricated by an inclined LIGA process combining lithography, electroplating, and molding techniques. This process utilizes X-ray's directed towards a protective electroplated gold mask over a poly(methyl methacrylate) (PMMA) substrate on a silicon wafer (Moon & Lee, 2003).

As discussed in this section, optimal material, design, and operating conditions were gleaned from the literature. This information was combined to simulate electric field and hydrodynamic flow behaviors in each section of the microdevice. The design was iteratively optimized based on these simulation results and then integrated together into the drug delivery microdevice.

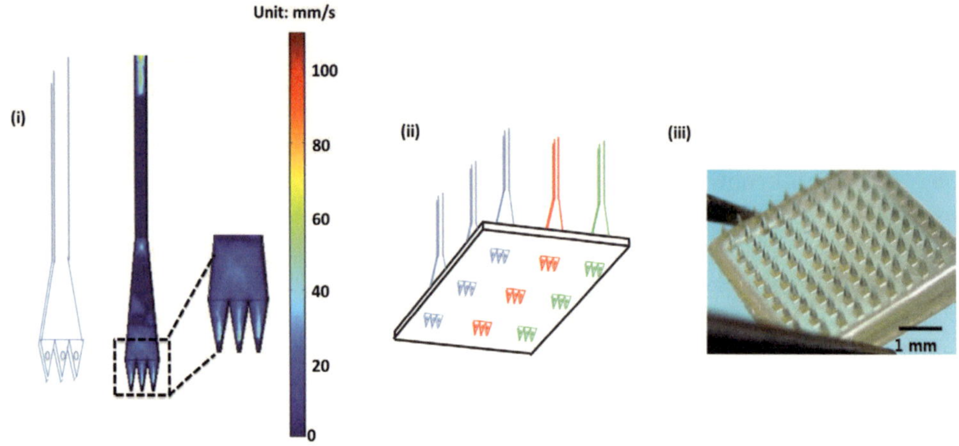

Fig. 7. Microneedle design to deliver chemotherapy emulsion to cancer patient, (i) pyramid-shaped hollow microneedle for delivery of 3 chemotherapy droplets (epirubicin, cisplatin, and fluorouracil) and the COMSOL simulation of the fluid velocity in the microneedle, (ii) microneedle array, and (iii) real image of a microneedle array. The initial velocities of the fluids in the microchannel were $4.2*10^{-4}$ m/s drug, $4.2*10^{-3}$ m/s poppy seed oil, and $4.2*10^{-2}$ m/s saline, and the exiting velocities from the microneedles is approximately 40mm/s. In (iii) microneedles are made out of poly-lactic acid (Baek et al., 2011).

3. Integration of technology into fully conceived device

This microfluidic drug delivery device will operate in sequence as described in sections 2.1 through 2.3. First, three different chemotherapy drugs, Epirubicin, Cisplatin, and Fluorouracil, will be separately dispersed into oil and then subsequently into saline by proven flow-focusing microchannel technology. Current state of the art results in this field suggest droplets formed will be approximately 500 nm in diameter. This design utilizes two stages of interdigitated electrodes on the bottom surface of the microchannel leaving the flow focusing junction in order to electrodisperse the droplets into <100nm droplets and to evenly disperse them spatially as they flow downstream to reduce coalescence. The oil-sheathed drug droplets are then pumped down the microchannel using traveling wave DEP technologies before flowing into an array of microneedles inserted into the dermis layer of the skin. This entire system is packaged inside of a small unit that can be worn on the wrist. The reservoirs for each drug, poppy seed oil, saline, and pressurized air to pump from the reservoirs can be individually replaced based on usage. Further, integrated electronic feedback control and monitoring (not described here) can be utilized to monitor chemotherapy drug delivery into the dermis and subsequently the blood stream of a cancer patient. The main components of this microfluidic device were optimized from literature data and COMSOL simulations.

4. Conclusions and perspectives for future integrated microdevice technologies

There is a great need for new technologies to effectively treat all forms of metastasized cancer. Gastric cancer provides a poignant example because patient symptoms typically do not arise until the cancer has progressed to stage IV. Any new technologies developed should increase the comfort level of patients as well as concurrently improve treatment efficacy or even eradicate the disease. The goal of this work was to develop a novel drug delivery system to effectively treat gastric cancer patients with minimal pain or lifestyle interruptions while undergoing treatment.

In addition, this work links together technologies that have been progressing in isolation from each other. For example, electrodispersion integrated with flow focusing and surfactant stabilization is a novel technique with the potential to produce droplets less than 100 nm in diameter. Smaller droplets are desired in diverse applications such as nanoparticle synthesis or pharmaceutical packaging. One key advantage of this combined technique is that it can be integrated into lab-on-a-chip devices provided the capillary number and volumetric flow rate ratio which are optimized and the surfactant required for optimizing interfacial tension increases the long-term stability of droplets.

Traveling wave DEP is an advantageous technique for transport of droplets in microchannels, which has minimal power requirements and thus is ideal for portable microdevices operating on batteries. Incorporating traveling wave DEP electrodes into the drug delivery microdevice described increased the velocity of the droplets for optimal microneedle injection rates. Further, this technique has proven to be minimally disruptive to a particle which is a key advantage with this adaption of the technology in the drug delivery microdevice.

Microneedles have been explored in many forms as physical enhancers for drug delivery. Within the drug delivery microdevice, the microneedle array was adopted to reduce pain and facilitate continuous delivery of the Epirubicin, Cisplatin, and Fluorouracil chemotherapy drug cocktail into the dermis. Based on evidence from previous studies, the pain level can be greatly reduced and the chemotherapy droplets reach their target areas in a more efficient manner thus reducing side effects.

For future work this drug delivery microdevice wrist system could be improved by incorporating a biosensor, in-line feedback control, and wireless reporting to measure the concentration and metabolites of the chemotherapy droplets in the blood stream, dynamically adjust dosage, and keep the primary care physician informed of progress. This will allow for real-time drug and treatment monitoring. The advantages of adding in this technology would be decreases in patient drug side effects, uniform maintenance of the critical drug concentration delivered to the gastric tumor and CTCs, increased treatment effectiveness, and increased patient comfort.

5. References

Abismaïl, B., Canselier, J.P., Wilheml, A.M., Delmas, H., & Gourdon, C. (1999). Emulsification by ultrasound: drop size distribution and stability. *Ultrasonics Sonochemistry*, 6, 1-2, (March 1999), pp.75–83, ISSN 1350-4177.

Aboud, M.J., Gassmann, M., & McCord, B.R. (2010). The development of mini pentameric STR loci for rapid analysis of forensic DNA samples on microfluidic system. *Electrophoresis*, 31, 15, (August 2010), pp. 2672-2679, ISSN 0173-0835.

Alazzam, A., Stiharu, I., Bhat, R., & Meguerditchian, A. (2011). Interdigitated comb-like electrodes for continuous separation of malignant cells from blood using dielectrophoresis. *Electrophoresis*, 32, 11, (June 2011), pp. 1327-1336, ISSN 0173-0835.

American Cancer Society. (n.d.). Cancer facts & Figures 2012, In: *Cancer.Org*, August 2011, Available from: <http://www.cancer.org/Research/CancerFactsFigures/CancerFactsFigures/cancer-facts-figures-2011>.

Anna, S.L., Bontoux, N., & Stone, H.A. (2003). Formation of dispersions using "flow focusing" in microchannels. *Applied Physics Letters*, 82, 3, (January 2003), pp. 364-366, ISSN 0003-6951.

Arya, N., Chakraborty, S., Dube, N., & Katti, D.S. (2009). Electrospraying: A Facile Technique for Synthesis of Chitosan-Based Micro/Nanopsheres for Drug Delivery Applications. *Journal of Biomedical Materials Research Part B: Applied Biomaterials*, 88B, 1, (January 2009), pp. 17-31, ISSN: 1552-4973.

Baek, C., Han, M., Min, J., Prausnitz, M.R., Park, J.H., & Park, J.H. (2011). Local transdermal delivery of phenylephrine to the anal sphincter muscle using microneedles. *Journal of Controlled Release*, 154, 2, (September 2011), pp. 138-147, ISSN 0168-3659.

Reprinted from Journal of Controlled Release, 154, Baek, C., Han, M., Min, J., Prausnitz, M.R., Park, J.H., & Park, J.H., Local transdermal delivery of phenylephrine to the anal sphincter muscle using microneedles, pp. 138-147, Copyright (2011), with permission from Elsevier.

Balcer-Kubiczek, E.K., & Garofalo, M.C. (2009). Molecular targets in gastric cancer and apoptosis, In: Apoptosis in Carcinogenesis and Chemotherapy, Chen, C.G. and Lai, P.B.S., pp. 157- 192, Springer Science + Business Media, Retrieved from <http://www.crcnetbase.com/search/advanced>.

Bienvenue, J.M., Duncalf, N., Marchiarullo, D., Ferrance, J.P., & Landers, J.P. (2006). Microchip-based cell lysis and DNA extraction from sperm cells for application to forensic analysis. *Journal of Forensic Sciences*, 51, 2, (March 2006), pp. 266-273, ISSN 0022-1198.

Bjerklie, D. & Jaroff, L. (January 15, 2001). Beyond needles and pills, In: Time Magazine U.S., October 2011, Available from: < http://www.time.com/time/magazine/article/0,9171,998968,00.html>.

Brammer, K.S., Choi, C., Oh, S., Cobb, C.J., Connelly, L.S., Loya, M., Kong, S.D., & Jin, S. et al. 2009. Antibiofouling, sustained antibiotic release by Si nanowire templates. Nano Letters, 9, 10, (October2009), pp. 3570-3574, ISSN 1530-6984.

Bunthawin, S., Wanichapichart, P., Tuantranont, A., & Coster, H.G.L. (2010). Dielectrophoretic spectra of translational velocity and critical frequency for a spheroid in traveling electric field. *Biomicrofluidics*, 4, 1, (March 2010), ISSN 1932-1058.

Cheng, I.F., Chung, C.C., & Chang, H.C. (2011). High-throughput electrokinetic bioparticle focusing based on a travelling-wave dielectrophoretic field. *Microfluidics and Nanofluidics*, 10, 3, (March 2011), pp. 649-660, ISSN 1613-4982.

Cleveland Clinic Foundation. (n.d.).What is Chemotherapy, In: Chemocare.com, September 2010, Available from: <http://www.chemocare.com/whatis/how_do_the_doctors_decide_which.asp>.

Donnelly, R.F., Majithiya, R., Singh, T.R.R., Morrow, D.I.J., Garlad, M.J., Demir, Y.K., Migalska, K., Ryan, E., Gillen, D., Scott, C.J., & Woolfson, A.D. (2011). Design, optimization and characterisation of polymeric microneedle arrays prepared by a novel laser-based micromoulding technique. *Pharmaceutical Research*, 28, 1, (January 2011), pp. 41-57, ISSN 0724-8741.

Durr, M., Kentsch, J., Muller, T., Schnelle, T., & Stelzle, M. (2003). Microdevices for manipulation and accumulation of micro- and nanoparticles by dielectrophoresis. *Electrophoresis*, 24, 4, (February 2003), pp. 722-731, ISSN 0173-0835.

Elman, N.M., Duc, H.L.H., & Cima, M.J. (2009). An implantable MEMS drug delivery device for rapid delivery in ambulatory emergency care. *Biomedical Microdevices*, 11, 3, (June 2009), pp. 625-631, ISSN 1387-2176.

Escobar-Chavez, J.J., Bonilla-Martinez, D., Villegas-Gonzalez, M.A., Molina-Trinidad, E., Casas-Alancaster, N., & Revilla-Vazquez, A.L. (2011). Microneedles: a valuable physical enhancer to increase transdermal drug delivery. *Journal of Clinical Pharmacology*, 51, 7, (July 2011), pp. 964-977, ISSN 0091-2700.

Farokhzad, O.C., Dimitrakov, J.D., Karp, J.M., Khademhosseini, A., Freeman, M.R., & Langer, R. (2006). Drug delivery systems in urology - getting "smarter". *Urology*, 68, 3, (September 2006), pp. 463-469,ISSN 0090-4295.

Felten, M., Staroske, W., Jaeger, M.S., Schwille, P., & Duschl, C. (2008). Accumulation and filtering of nanoparticles in microchannels using electrohydrodynamically induced vertical flows. *Electrophoresis*, 29, 14, (July 2008), pp. 2987-2996, ISSN 0173-0835.

Ha, J.W., & Yang, S.M. (1999). Breakup of a multiple emulsion drop in a uniform electric field. *Journal of Colloid and Interface Science*, 213, (May 1999), pp. 92–100, ISSN: 0021-9797.

Hamouda, T., Hayes, N.M., Cao, Z.Y., Tonda, R., Johnson, K., Wright, D.C., Brisker, J., & Baker, J.R. (1999). A novel surfactant nanoemulsion with broad-spectrum sporicidal activity against bacillus species. *Journal of Infectious Diseases*, 180, 6, (December 1999), pp. 1939-1949, ISSN 0944-5013.

Hershock, D. (2006). Medical and Surgical Therapy for Gastric Cancer, In: Endoscopic Oncology, Faigel, D.O. and Kochman, M.L., pp. 173-181, Humana Press, Retrieved from <http://www.crcnetbase.com/search/advanced>.

Jensen, K. & Lee, A. (2004). The science & applications of droplets in microfluidic devices - Foreword. *Lab On A Chip*, 4, 4, (n.d.), pp. 31N-32N, ISSN 1473-0189.

Kang, Y. & Li, D. (2009). Electrokinetic motion of particles and cells in microchannels. *Microfluidics and Nanofluidics*, 6, 4, (April 2009), pp. 431-460, ISSN 1613-4982.

Kim, H., Luo, D.W., Link, D., Weitz, D.A., Marquez, M., & Cheng, Z.D. (2007). Controlled production of emulsion drops using an electric field in a flow-focusing microfluidic device. *Applied Physics Letters*, 91, 13, (September, 2007), pp.133106, ISSN 0003-6951.

Kim, S.H., Kim, J.W., Cho, J.C., & Weitz, D.A. (2011). Double-emulsion drops with ultra-thin shells for capsule templates. *Lab Chip*, 2011, 11, 18, (2011), pp. 3162-3166, ISSN 1473-0197.

Kiss, N., Brenn, G., Pucher, H., Wieser, J., Scheler, S., Jennewein, H., Suzzi, D., & Khinast, J. (2011). Formation of O/W emulsions by static mixers for pharmaceutical applications. *Chemical Engineering Science*, 66, 21, (November 2011), pp. 5084-5094, ISSN: 0009-2509.

Kirby, B.J., & Hasselbrink, E.F. (2004). Zeta potential of microfluidic substrates: 2. Data for polymers. *Electrophoresis*, 25, 2, (January 2004) pp. 203–213, ISSN: 0173-0835.

Lee, K., Lee, C.Y., & Jung, H. (2011). Dissolving microneedles for transdermal drug administration prepared by stepwise controlled drawing of maltose. *Biomaterials*, 32, 11, (April 2011), pp. 3134-3140, ISSN 0142-9612.

Lee, W., Walker, L.M., & Anna, S.L. (2009). Role of geometry and fluid properties in droplet and thread formation processes in planar flow focusing. Physics of Fluids, 21, 3, (March 2009), ISSN 1070-6631.

Levi, J.A., Dalley, D.N., & Aroney, R.S. (1979). Improved combination chemotherapy in advanced gastric-cancer. *British Medical Journal*, 2, 6203, (December 1979), pp. 1471-1473, ISSN 0959-8138.

Liao, C.Y., & Su, Y.C. (2010). Formation of biodegradable microcapsules utilizing 3D, selectively surface-modified PDMS microfluidic devices. *Biomedical Microdevices*, 12, 1, (February 2010), pp. 125-133, ISSN: 1387-2176.

Lima, J.L.F.C., Santos, J.L.M., Dias, A.C.B., Ribeiro, M.F.T, & Zagatto, E.A.G. (2004). Multi-pumping flow systems: an automation tool. *Talanta*, 64, 5, (December 2004), pp. 1091–1098, ISSN: 0039-9140.

Lin, J.T.Y., & Yeow, J.T.W. (2007). Enhancing dielectrophoresis effect through novel electrode geometry. *Biomedical Microdevices*, 9, 6, (December 2007), pp. 823-831, ISSN 1387-2176.

Lv, Y.G., Liu, J., Gao, Y.H., & Xu, B. (2006). Modeling of transdermal drug delivery with a microneedle array. Journal of Micromechanics and Microengineering, 16, 11, (November 2006), pp. 2492-2501, ISSN 0960-1317.

Martin-Banderas, L., Flores-Mosquera, M., Riesco-Chueca, P., Rodriguez-Gil, A., Cebolla, A., Chavez, S., & Ganan-Calvo, A.M. (2005). Flow Focusing: A Versatile Technology to Produce Size- Controlled and Specific-Morphology Microparticles. *Small*, 1, 7, (July 2005), pp. 688-692. ISSN: 1613-6810.

Mejia, A.F., He, P., Luo, D.W., Marquez, M., & Cheng, Z.D. (2009). Uniform discotic wax particles via electrospray emulsification, *Journal of Colloid and Interface Science*, 334, 1, (June 2009), pp. 22-28, ISSN: 0021-9797.

Mezzenga, R., Schurtenberger, P., Burbidge, A., & Michel, M. (2005), Understanding foods as soft materials. *Nature Materials*, 4, 10, (October 2005), pp. 729-740, ISSN: 1476-1122.

Migalska, K., Morrow, D.I.J., Garland, M.J., Thakur, R., Woolfson, A.D., &, Donnelly, R.F. (2011). Laser-engineered dissolving microneedle arrays for transdermal macromolecular drug delivery. *Pharmaceutical Research*, 28, 8, (August 2011), pp. 1919-1930, ISSN 0724-8741.

Minerick, A.R. (2008). The rapidly growing field of micro and nanotechnology to measure living cells. *AIChE Journal*, 54, 9, (September 2008), pp. 2230-2237, ISSN 0001-1541.

Moon, S., & Lee, S.S. (2003). Fabrication of microneedle array using inclined LIGA process. Transducer'03 12th International Conference on Solid-State Sensors, Actuators and Microsystems, 2, (n.d.), pp. 1546-1549.

National Cancer Institute. (n.d.). Stomach Cancer, In: *National Institute of Health*, September 2010, Available from: <http://www.cancer.gov/cancertopics/types/stomach>.

Nguyen, N.T., & Wereley, S.T. (2006). *Fundamentals and Applications of Microfluidics* (2nd Edition), Artech House, ISBN 1580539726, Boston.

Pacek, A.W., Chamsart, S., Nienow, A.W., & Bakker, A. (1999). The influence of impeller type on mean drop size and drop size distribution in an agitated vessel. *Chemical Engineering Science* 54, 19 (October 1999), pp. 4211-4222, ISSN: 0009-2509.

Pai, S.A., Rivankar, S.H., & Kocharekar, S.S. (2003). Parenteral cisplatin emulsion. United States Patent, US 6,572,884, date of filling (June 14, 2002), date of issue(June 3, 2003).

Pethig, R. (2010). Dielectrophoresis: status of the theory, technology, and applications. *Biomicrofluidics*, 4, 3, (September 2010), ISSN 1932-1058.

Power, D.G., Kelsen, D.P., & Shah, M.A. (2010). Advanced gastric cancer - slow but steady progress. Cancer Treatment Reviews, 36, 5, (August 2010), pp. 384-392, ISSN 0305-7372.

Rivera, F., Vega-Villegas, M.E., & Lopez-Brea, M.F. (2007). Chemotherapy of advanced gastric cancer. *Cancer Treatment Reviews*, 33, 4, (June 2007), pp. 315-324, ISSN 0305-7372.

Santini, J.T., Cima, M.J., & Langer, R.S. (1999). A controlled-release microchip. *Nature*, 397, 6717, (January 1999), pp. 335-338, ISSN0028-0836.

Santini, J.T., Richards, A.C., Scheidt, R.A., Cima, M.J., & Langer, R.S. (2000). Microchip technology in drug delivery. *Annals of Medicine*, 32, 6, (September 2000), pp. 377-379, ISSN 0785-3890.

Seo, M., Paquet, C., Nie, Z., Xu, S.Q., & Kumacheva, C. (2007). Microfluidic consecutive flow-focusing droplet generators. *Soft Matter*, 3, 8, (May 2007), pp. 986–992, ISSN: 1744-683X.

Soller, E., Murray, C., Maoddi, P., & Di Carlo, D. (2011). Rapid prototyping polymers for microfluidic devices and high pressure injections. *Lab on a chip, 22,* (November 2011), pp. *3752-65,* ISSN: 1473-0189.

Srivastava, S.K., Gencoglu, A., & Minerick, A.R. (2011). DC insulator dielectrophoretic applications in microdevice technology: a review. *Analytical and Bioanalytical Chemistry*, 399, 1, (January 2011), pp. 301-321, ISSN 1618-2642.

Thiele, J., Steinhauser, D., Pfohl, T., & Foster, S. (2010). Preparation of monodisperse block copolymer vesicles via flow focusing in microfluidics. *Langmuir*, 26, 9, (May 2010), pp. 6860-6863, ISSN 0743-7463.

Utada. A.S., Lorenceau, E., & Link, D.R. (2005), Monodisperse double emulsions generated from a microcapillary device. *Science*, 308, 5721, (April 2005), pp. 537-541, ISSN: 0036-8075

Vergauwe, N., Witters, D., Ceyssens, F., Vermeir, S., Verbruggen, B., Puerr, R., & Lammertyn, J. (2011). A versatile electrowetting-based digital microfluidic platform for quantitative homogeneous and heterogeneous bio-assays. *Journal of Micromechanics and Microengineering*, 21, 5, (May 2011), ISSN 0960-1317.

Wang, C., Wang, X., & Jiang, Z. (2011). Dielectrophoretic driving of blood cells in a microchannel. *Biotechnology and Biotechnological Equipment*, 25, 2, (May 2011), pp. 2405-2411, ISSN 1310-2818.

Wang, P.M., Cornwell, M., Hill, J., & Prausnitz, M.R. (2006). Precise microinjection into skin using hollow microneedles. *Journal of Investigative Dermatology*, 126, 5, (May 2006), pp. 1080-1087, ISSN 0022-202X.

Weng, C.H., Huang, T., Huang, C., Yeh, C., Lei, H., & Lee, G. (2011). A suction-type microfluidic immunosensing chip for rapid detection of the dengue virus. *Biomedical Microdevices*, 13, 3, (June 2011), pp. 585-595, ISSN 1387-2176.

Wibowo, C., & Ng, K.M. (2001). Product-Oriented Process Synthesis and Development: Creams and Pastes. *AIChE Journal*, 47, 12, (December 2001), ISSN: 0001-1541.

Xiao, Z.G., & Young, E.F.Y. (2011). Placement and routing for cross-referencing digital microfluidic biochips. *IEEE Transactions on Computer-Aided Design of Integrated Circuits and Systems*, 30, 7, (July 2011), pp. 1000-1010, ISSN 0278-0070.

Zagnoni, M., Baroud, C.N., & Cooper, J.M. (2009). Electrically initiated upstream coalescence cascade of droplets in a microfluidic flow. *Physical Review E*, 80, 4, (October 2009), pp. 046303, ISSN: 1539-3755.

Zagnoni, M., Lain, G.L., & Cooper, J.M. (2010). Electrocoalesence mechanisms of microdroplets using localized electric fields in microfluidic channels. *Langmuir*, 26, 18, (September 2010), pp. 14443-14449, ISSN 0743-7463.

Zahn, J.D., Deshmukh, A., Pisano, A.P., & Liepmann, D. (2004). Continuous on-chip micropumping for microneedle enhanced drug delivery. *Biomedical Microdevices*, 6, 3, (September 2004), pp. 183-190, ISSN 1387-2176.

Zhu, H., Mayandad, S., Coskun, A.F., Yagidere, O., & Ozcan, A. (2011). Optofluidic fluorescent imaging cytometry on a cell phone. *Analytical Chemistry*, 83, 17, (September 2011), pp. 6641-6647, ISSN 0003-2700.

10

Microfluidizer Technique for Improving Microfiber Properties Incorporated Into Edible and Biodegradable Films

Márcia Regina de Moura[1,2], Fauze Ahmad Aouada[2,3], Henriette Monteiro Cordeiro de Azeredo[4] and Luiz Henrique Capparelli Mattoso[2]
[1]*Nanomedicine and Nanotoxicology Laboratory, IFSC, University of São Paulo, São Carlos/SP,*
[2]*National Laboratory of Nanotechnology for Agribusiness (LNNA), EMBRAPA-CNPDIA, São Carlos/SP,*
[3]*IQ, UNESP – State University of São Paulo, Araraquara/ SP,*
[4]*Embrapa Tropical Agroindustry - CNPAT, Fortaleza/CE, Brazil*

1. Introduction

Microfluidics is a quite new technology platform that classically requires a multi-disciplinary approach for building effective systems through the interface of physics, chemistry, engineering and biochemistry. These devices usually consist of macro-structures that are fabricated in the micrometer-scale and can be designed to produce diagnostically useful systems for potential point-of-care measurements. Systems based upon microfluidics operation possess a unique set of potential advantages, such as reduction in reagent cost, enhancement of assay speed, potential for mass production of devices at low cost and the ability to integrate several processing steps into a single system (i.e., size control) [Situma et al., 2006; Wu & Nguyen, 2005].

Cellulose, the most abundant organic polymer in the biosphere, is a polydisperse linear homopolymer consisting of β-1-4 glucopyranose units. The properties of cellulose including good mechanical properties, low density, biodegradability, and availability from renewable resources have become increasingly important and have contributed to a rising interest in this material [S. Y. Lee et al., 2009]. Native cellulose is generally known to be fibrillar and crystalline and the cellulose fibrils play a significant role in contributing to the high strength of plant cell walls [Lu & Hsieh, 2010; Zuluaga et al., 2009].

Cellulose derivatives have been widely studied, and there are numerous industrial applications in fiber, film and gel based materials. Cellulose ethers are strongly studied in barrier packaging. As they are often water-soluble [Moura et al., 2008], they are mostly used as coating [Coma et al., 2001] or in the preparation of edible films with efficient oxygen barrier properties [Kamper & Fennema, 1984]. Hydroxypropyl methylcellulose (HPMC) is a most commonly used cellulose ether. It is approved for food applications by FDA (21 CFR 172.874) and the EU (EC, 1995). The films obtained from HPMC are resistant to oils and fats, flexible, transparent, odorless, and tasteless but tend to have moderate strength [Bilbao-Sainz et al., 2011]. Other cellulose ether most utilized for edible films is carboxymethylcellulose (CMC), a sodium salt of carboxymethyl ether cellulose. It is synthesized by swelling cellulose with sodium hydroxide and alkali-catalyzed reaction of cellulose with chloroacetic acid [Sudhakar et al., 2006]. CMC has a high water bonding capacity, good compatibility with skin and mucous membrane, is widely available and cheap [Ramli & Wong, 2011].

Natural cellulose fibers have been used as a reinforcement material. One disadvantage is the lack of strength of interfacial bonding. Natural fibers include those of vegetable origin constituted of cellulose, a polymer of glucose bound to lignin with varying amounts of other natural materials. The interest of using cellulose fibers for polymer matrix reinforcing has rapidly grown in the last decade because of the above mentioned advantages and because the natural fibers are obtained from annually renewable resources [Azeredo, 2009].

Edible coatings such as those formed by wax on various fruit surfaces have been used for centuries to prevent moisture loss and to create a shiny fruit surface. Several edible materials have had their film-forming properties studied to produce edible films and coatings to be used in food packaging, not completely replacing petroleum-derived plastics, but rather improving their efficiency, thus reducing the amount of synthetic polymers required for each application. Since edible biopolymers do not have satisfactory barrier and mechanical properties to packaging application, the addition of reinforcement material is helpful to make these materials more adequate to industrial application.

Polymer-based composite materials reinforced with organic-based fillers have received significant interest. Especially, many synthetic polymeric materials have been substituted by biopolymers combined with natural reinforcing fillers to improve their mechanical and barrier properties. In this chapter, the principal idea is to analyze the importance of microfluidics technic in produce cellulose microfibers for composite materials.

2. Microfluidic technology

2.1 Principles and advantages

Microfluidic technologies are indicated to manipulate small quantities of liquids or fluids usually through channels with at least one dimension smaller than 1 mm, for emulsion formation, mixing and dispersion [Skurtys & Aguilera, 2008]. This leads to greatly reduced reagent consumption and exhibits intrinsically efficient heat and mass transfer due to high surface area-to-volume ratios [Hung & Lee, 2007; Il Park et al., 2010].

Microfluidizer high shear fluid processors are unique in their ability to achieve uniform particle size reduction, bottom-up crystallization and efficient cell disruption - enabling innovative companies to develop nano-enabled medicines, chemicals and consumer products that change the world. Product enters the system via the inlet reservoir and is powered by a high-pressure pump into the interaction chamber at speeds up to 400 m/s. It is then effectively cooled, if required, and collected in the output reservoir. The exclusive fixed-geometry interaction chambers are the heart of our technology, and combines with a constant pressure pumping system to produce unparalleled results. By reducing particles to the nano-level more efficiently, customers use less energy to achieve particle size results that are, on average, half the size of even the most effective homogenizer outputs.

Lee et al., 2011 defined the aim of microfluidic mixing schemes as a device to enhance the mixing efficiency and to achieve a thorough and rapid mixing of multiple samples in microscale devices, being the sample mixing essentially achieved by enhancing the diffusion effect between the different species flows. Also, the diminutive scale of the flow channels in microfluidic systems increases the contact area between the species to be mixed, and this factor is one of the most efficient means of enhancing the diffusive mixing effect. Other important alternative to increase the contact area between the mixing species into the microfluidic devices is designing the microchannel configurations allowing that the species are folded multiple times as they flow along the mixing channel.

Microfluidic process has become an attractive technology for numerous applications [Atalay et al., 2011; Cho et al., 2011; Napoli et al., 2011; Schirhagl et al., 2010; Thompson et al., 2010; Zhang et al., 2011] due to unique characteristics, such as: i) efficient and rapid mix leading to rapid chemical reaction; ii) homogeneous reaction environments; iii) continuously varied reaction conditions and iv) precise time intervals to add reagents during reaction [deMello, 2006; DeMello & DeMello, 2004]. Therefore, the application of the microfluidic devices improves the control of the synthesis parameters and thus the nanoparticle sizes and properties [Il Park et al., 2010] and optimizes the miniaturization of the microstructures.

2.2 Utilization of microfluidizer systems for different applications

Microfluidizer system belongs to microfluidic process which has been extensive proposed as a green alternative technology in miniaturization from macro to micro/nano structure dimensions.

Some recent applications using microfluidizer technology are discussed as following. Nik et al., 2010 applied the microfluidics technology to obtain oil-in-water emulsions stabilized with whey protein isolate or with β-lactoglobulin or α-lactalbumin (individually or combined). The experimental set consisted of the mixture pre-homogenized with Ultra-Turrax mixer for 2 min followed by homogenization with four passes through a microfludizer (M-110EH Microfluidizer Processor, Microfluidics, Newton, MA, USA) at 350 kPa. In these experiments, bleached soybean oil was used as oil source.

In the work described by Cavender & Kerr, 2011, they investigated how microfluidization affects the physical and sensory properties of ice cream made from mixes which contain either xanthan gum or locust bean gum (LBG). The mixes were stirred for 1 min and processed at 220–250 MPa using a high pressure processing system (Model M140-K, Microfluidics) fitted with a diamond interaction chamber (Model G10Z, Microfluidics). They showed that by treating full-fat ice cream mix with microfluidization, one can affect both sensory and physical properties of the finished product, and further, those changes differ based on the gum used. The authors also concluded that consumers preferred the firmness and creaminess of ice cream made from microfluidized mixes with LBG.

Liu et al., 2011 investigated the aggregation changes of whey protein induced by high-pressure microfluidization (HPM) treatment by using M-7125 Microfluidizer with pressure range of 400,000 psi. The whey protein concentrate (WPC) at 1 mg/mL was treated under pressures of 40, 80, 120 and 160 MPa, being each solution repeated 3 times at each pressure. The average particle size of whey protein decreased as the pressure increased from 407 nm (untreated whey - 0 MPa) to 196 nm when the Microfluidizer pressure was increased to 160 MPa. Functional properties (solubility, foaming, and emulsifying properties) of WPC ultrafiltered from fluid whey were evaluated. As main results, they showed significant modifications in the solubility (30% to 59%) and foaming properties (20% to 65%) of WPC with increasing pressure, suggesting that HPM treatment of WPC is appropriate for applications in selected dairy products.

Chen et al., 2011 studied the effect of the sonolysis, microfluidization and shearing treatments on the degradation kinetics of chitosan solutions prepared by dissolving chitosan in an acetic acid buffer (0.2 mol L^{-1} acetic acid/0.1 mol L^{-1} sodium acetate, pH = 4.3). In the microfluidization treatment, 300 mL of chitosan solution was placed in a water bath at controlled temperature of 0 ± 1, 30 ± 1 and 50 ± 1°C and treated with a microfluidizer (M-100Y Cell Disruption, Microfluidics) at pressures of 82.7 and 117.2 MPa for 5, 10, 15, 20 and 25 passes. SEC-HPLC method was used to determine the molecular weight of chitosan and from this technique to accompany the chitosan degradation. The results showed that among three physical methods studied, the microfluidization treatment results in the highest efficiency when higher chitosan solution concentrations were utilized. Also, the degradation mechanism of the chitosan is different for each of the physical methods; in the microfluidization method, the mechanism is entanglement and stretch plus cavitation.

Rao and McClements, 2011 established experimental conditions to prepare food-grade stable microemulsions, nanoemulsions or emulsions using sucrose monopalmitate (SMP) as a surfactant and lemon oil as an oil phase. After a prehomogenization using a high-speed blender for 2 min at ambient temperature, the oil and aqueous phases were passed through a high pressure homogenizer for 3 passes at 9,000 psi (Model M-110L Microfludizer Processor, Microfluidics). According to the authors, emulsions ($r > 100$ nm) or nanoemulsions ($r < 100$ nm) were formed at low surfactant-to-oil ratios depending of homogenization conditions; and microemulsions were formed at higher ratios. In other work, Rao and McClements, 2012 prepared a nanoemulsion with mean droplet diameters of 105±15 nm (pH = 7) from stock emulsion formed by lemon oil (10 wt%), SMP (1wt%) and buffer solution (89 wt%) by using of the same microfluidization way.

Bonilla et al., 2012 reported the influence of homogenization treatment of the essential oil type (basil and thyme) on the physical properties of chitosan-based film-forming dispersions. Two different homogenization treatments were used. In the first one, the film-forming dispersions were prepared by using an Ultraturrax rotor-stator homogenizer at 21,500 rpm for 4 min. The second treatment consisted of the association of the first treatment and microfluidization at 165 MPa in a single pass by means of a Microfludizer® M110-P processor. The authors concluded that microfluidization at 165 MPa affected the properties of the emulsions and films based on chitosan and basil/thyme essential oils. Also, a reduction was observed in oil droplet size and viscosity, which promoted the adsorption of chitosan on the oil-water interface.

2.3 Our contribution to microfluidics technology

Our contribution to microfluidics technology focuses specifically on the miniaturization of the cellulose microfibers to obtain nanofibers to be incorporated into polymeric matrices for application as nano-reinforcement to edible films and polysaccharide hydrogels.

2.3.1 Miniaturization of cellulose microfibers for application in polymeric matrices

2.3.1.1 HPMC films

Recently, our research group [Moura et al., 2011] employed the microfluidics technology to decrease the size of cellulose microfibers and to study the effect of addition of such fibers on the properties of hydroxypropyl methylcellulose (HPMC) films with desired properties to be applied in food packaging. The particle size of fibers and mechanical properties, water vapor and oxygen permeabilities, total pore volume, and light and electron microscopy micrographs of films were analyzed. Data were analyzed by 2-sample t-Student tests, and one-way ANOVA with Tukey's multiple comparison tests at 95% confidence level using Minitab version 14.12.0 statistical software (Minitab Inc., State College, PA, USA).

To deagglomerate the commercial cellulose fibers (trade name CF1), and to reduce their median particle size, a Microfluidizer® processor model M-110EH-30 was used, with a pressure of 20,000 psi. The most important results of this research will be now discussed. Fig. 1 shows the influence of the amount of passes through the Microfluidizer® on the size values of cellulose fibers. A Partica LA-910 laser scattering particle size distribution analyzer was used to obtain the particle size distribution, assuming particles shaped as spheres. A gradual decrease in the fiber size was observed with an increase in the number of passes of the fiber suspension through the Microfluidizer®. This trend may be explained by the fact that the processor equipment maximizes the energy-per-unit fluid volume, resulting in uniform submicron particle and droplet sizes. The CF1 fibers originally had an average size of 37.5 ± 2.5 μm. When 7, 10, and 20 passes of the fiber solution were done in the Microfluidizer® equipment the size values decreased to 6.8 ± 0.6, 5.2 ± 0.5 and 1.6 ± 0.9 μm, respectively.

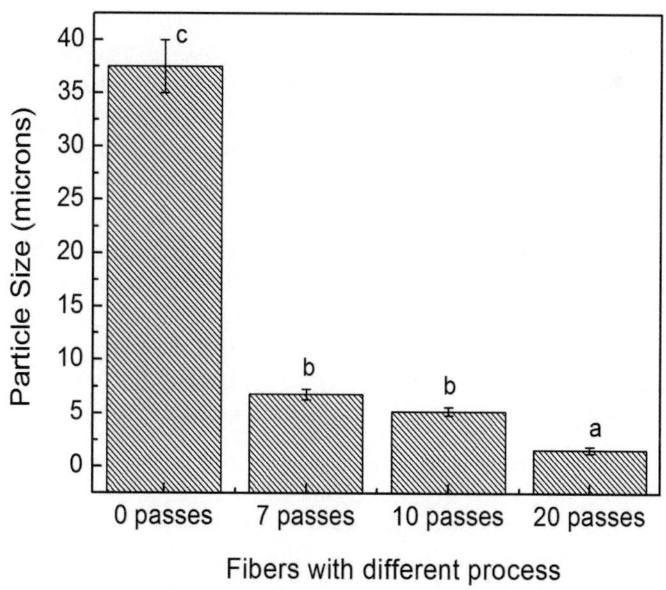

Fig. 1. Particle size of hydroxypropyl methylcellulose/cellulose fiber film-forming solutions affected by increasing number of passes in a Microfluidizer®. [a,b,c] Different letters indicated significant difference at $P < 0.05$ [Moura et al., 2011].

Fig. 2 shows that the size of the crystalline cellulose fibers, by birefringence under crossed polarizers, diminishes in size considerably following 7 passes in the Microfluidizer®. The fibers appear to continue to reduce in size following subsequent passes through the Microfluidizer®.

Fig. 3 shows the scanning electron microscopy micrographs of cross sections of HPMC films with crystalline cellulose fibers added. In Fig. 3a, the white spots throughout the HPMC film and on the surface are thought to be pores. In Fig. 3b HPMC film are viewed with the back-scattered electron detector to provide a slightly different view of the pores. The dark appearance of the pores indicates that they have less density than the surrounding material, thus they are likely to be voids or empty spaces. Fig. 3c and d shows HMPC with 1% cellulose fiber (CF1) with zero passes through the microfluidizer®. The CF1 are embedded in a smooth matrix of HPMC. Close examination of a fiber (Fig. 3d) shows that the fiber is made up of many smaller fibers. Fig. 3e and f shows the HMPC with CF1 films made with solutions after 7 passes through the microfluidizer®. The film has rough areas where small fibers have been incorporated as well as very smooth areas of HMPC. Some of the fibers are still in clumps. Fig. 3g and h shows the HMPC with CF1 films made with solutions after 10 passes while Fig. 3i and j corresponds to films made with solutions after 20 passes. The fibers might be smaller and better-dispersed with increased numbers of passes through the Microfluidizer®.

Transparent, flexible, homogeneous, surface smooth films without pores and cracks were obtained after drying the film-forming solutions containing fibers and HPMC. Incorporation of cellulose fibers in the films improved their mechanical and barrier properties significantly. The percentage of elongation of the HPMC films changed when particle size was decreased as shown in Table 1. The increase of the elongation improved the tenacity of the films. The elastic modulus of the HPMC films increases with addition of the fibers and did not present significant variation with different sizes of fibers. In addition, the elasticity of the films was preserved with addition of microfibers. The suitable use of packaging is also strongly dependent on its favorable mechanical and barrier properties. The addition of microfibers to HPMC films results in significant improvements in film mechanical properties. Table 1 also shows the effects of size on tensile strength of microfibers/HPMC films. When CF1 fibers with 37.5 μm were included in the HPMC films, the tensile strength of the film was 18.4 ± 1.0 MPa. For films containing fibers with 6.8 μm, TS increased from 18.4 to 42.7 MPa. When the fiber size was decreased from 6.8 to 1.6 μm, TS increased to 70.0 MPa. This enhancement is attributed to the increased strength and bonding of smaller fibers with the HPMC matrix compared to HPMC matrix with CF1 without microfluidization. This reinforcing effect is attributed to the better dispersion into the HPMC matrix of the fibers with small size because of increased contact surface area between HPMC-microfluidized cellulose promoted by increased hydrogen bonding of cellulose fibers with the HPMC matrix.

The knowledge of gas permeability is essential for the application of polymers as oxygen barrier food packaging materials. The important factors that affect the permeability are its dependence on relative orientation of the fibers in the matrix and the state of aggregation and dispersion of these fibers in the film matrix [Bharadwaj, 2001]. Table 2 shows the values of water vapor permeability of HPMC film with fibers at different sizes at 25 °C. In general, it was observed that the presence of microfluidized fibers in the HPMC films decreased the values of WVP. For instance, the incorporation of fibers with around 1.6 μm decreased the WVP from 0.894 to 0.455 ± 0.010 g mm kPa^{-1} h^{-1} m^{-2}. So, the water vapor permeability decreases with decreasing fiber sizes. The oxygen permeability of fibers/HPMC films (Table 2) were also analyzed, since they are important characteristics to be considered for a finished packaging material to increase the shelf life of foods [Miller & Krochta, 1997].

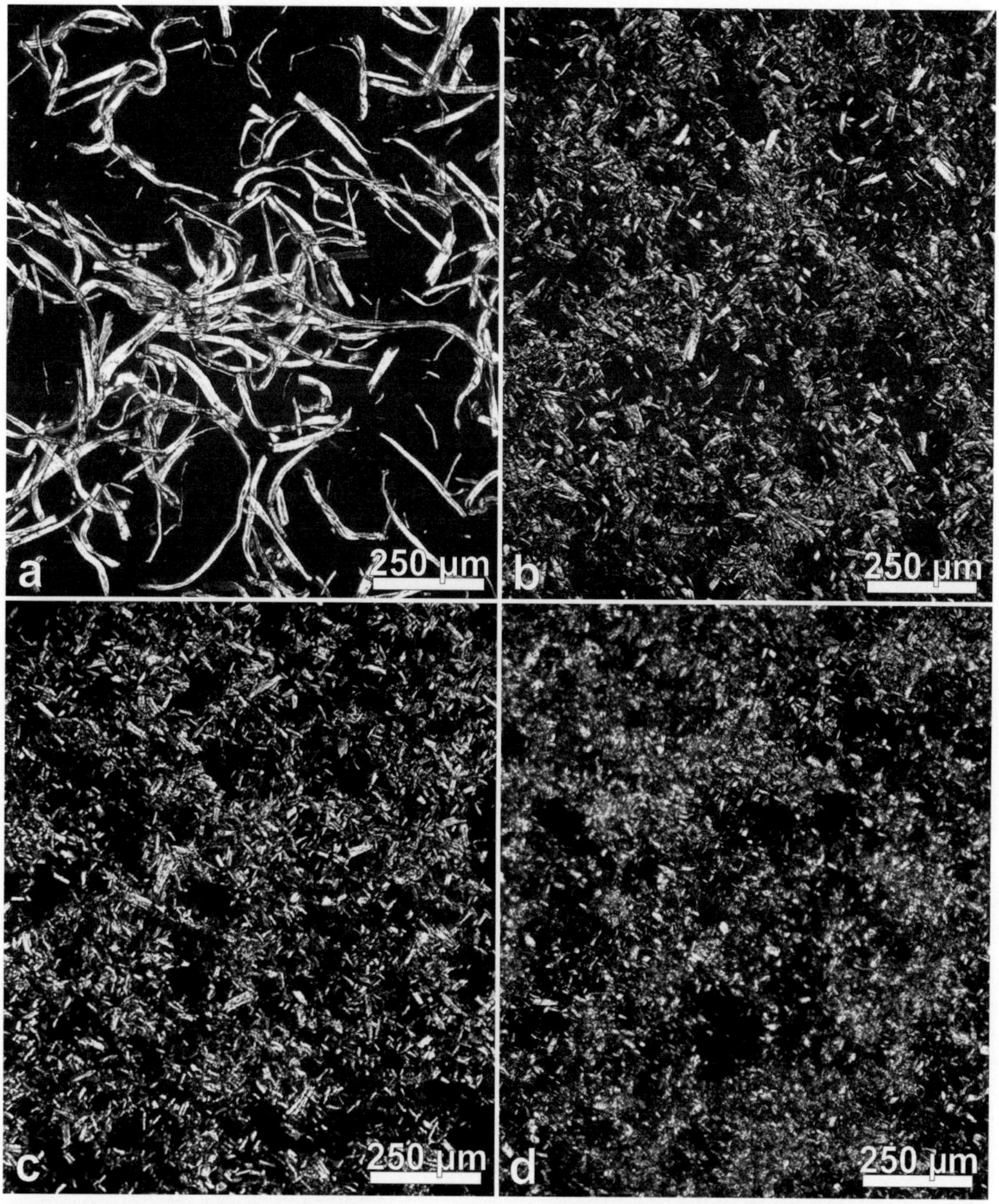

Fig. 2. Hydroxypropyl methylcellulose films incorporated with 1% cellulose fiber viewed through crossed polarizers to show the birefringence of the cellulose fibers in the matrix of the films. (a) zero passes; (b) after seven passes; (c) after ten passes and (d) after twenty passes through the Microfluidizer® [Moura et al., 2011].

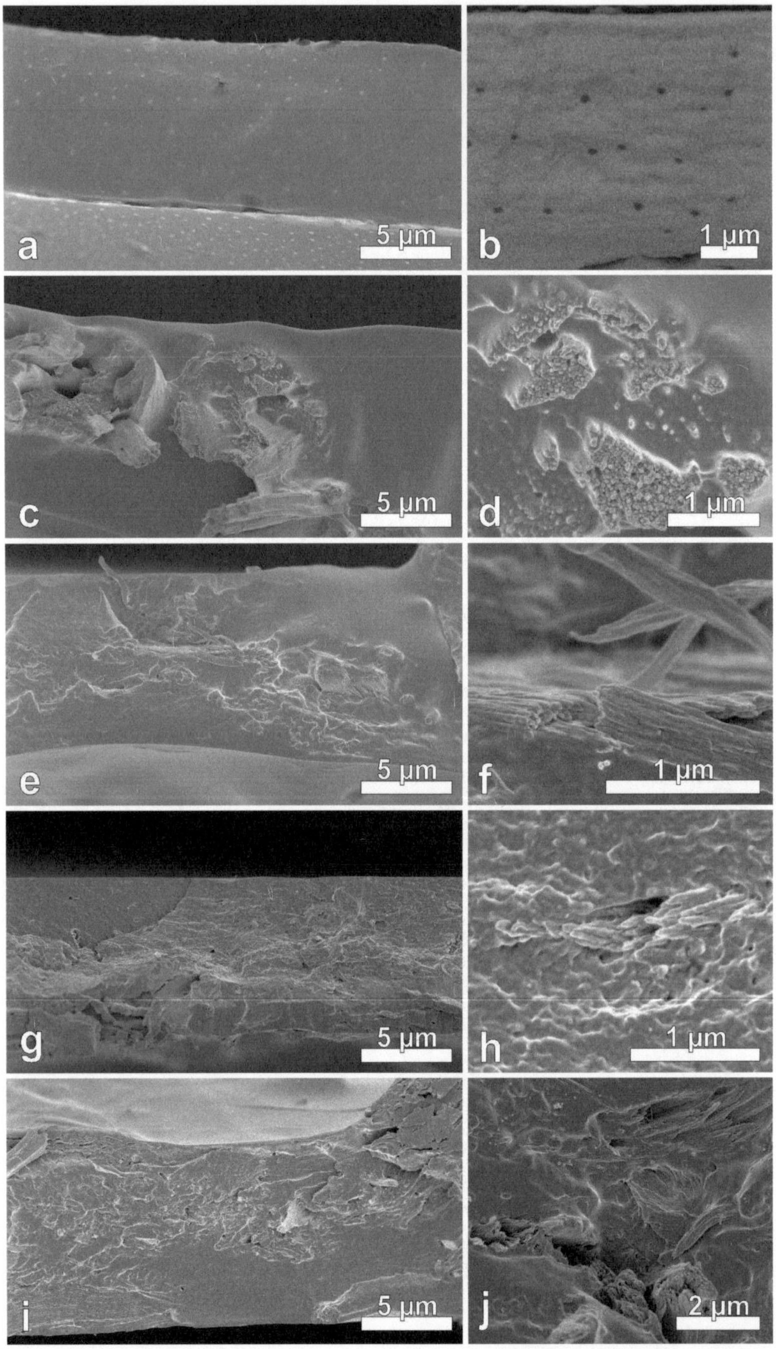

Fig. 3. Cross sections of hydroxypropyl methylcellulose (HPMC) films viewed using scanning electron microscopy (SEM). (a) HPMC film; (b) HPMC film viewed with a backscattered electron detector; (c and d) HMPC with 1% cellulose fiber (CF1) with zero passes; (e and f) HMPC with CF1 after seven passes; (g and h) HMPC with CF1 after ten passes, and; (i and j) after twenty passes through the Microfluidizer® [Moura et al., 2011].

Type of HPMC film	Thickness (μm)	Tensile strength (MPa)	Elastic modulus (MPa)	Elongation (%)
Zero passes (without CF1)	34 ± 2[a]	28.3 ± 1.0[b]	900 ± 34[b]	8.1 ± 0.7[b]
Zero passes (1% CF1)	57 ± 4[b]	18.4 ± 1.5[a]	783 ± 67[a]	5.7 ± 1.1[a]
After 7 passes (1% CF1)	36 ± 4[a]	42.7 ± 1.9[c]	1359 ± 110[c]	13.6 ± 3.6[c]
After 10 passes (1% CF1)	39 ± 2[a]	52.3 ± 1.3[d]	2273 ± 204[d]	7.5 ± 1.7[b]
After 20 passes (1% CF1)	39 ± 2[a]	70.0 ± 3.2[e]	1876 ± 162[d]	10.6 ± 4.6[c]

[a,b,c,d,e] Different letters within a column indicated significant difference at $P < 0.05$.

Table 1. Effect of cellulose fibers (CF1) reduced in sizes after successive passes in a Microfluidizer® on thickness, elastic modulus, and elongation of hydroxypropyl methylcellulose (HPMC) films [Moura et al., 2011].

Our findings showed that the oxygen permeability (O_2P) of films containing fibers was markedly low, this feature being probably associated with the more compact structure of these materials determined by the dispersion of fibers in the HPMC matrix. The important result is that after the Microfluidizer® treatment of fibers, without chemical treatment, the O_2P values to the fibers/HPMC matrix decreased to 101.2 ± 0.1; 92.0 ± 0.2 and 85.0 ± 0.2 cm^3 μm^{-2} d^{-1} kPa^{-1} for fibers processed with 7 passes; 10 passes and 20 passes, respectively.

Type of HPMC film	Water vapor permeability (g mm kPa^{-1} h^{-1} m^{-2})	Oxygen permeability (cm^3 μm^{-2} d^{-1} kPa^{-1})
Zero passes (without CF1)	0.89 ± 0.03[c]	182.4 ± 0.4[d]
Zero passes (1% CF1)	0.95 ± 0.03[c]	172.2 ± 0.3[d]
After 7 passes (1% CF1)	0.49 ± 0.01[b]	101.2 ± 0.1[c]
After 10 passes (1% CF1)	0.46 ± 0.02[a]	92.0 ± 0.2[b]
After 20 passes (1% CF1)	0.46 ± 0.02[a]	85.0 ± 0.2[a]

[a,b,c,d] Different letters within a column indicated significant difference at $P < 0.05$.

Table 2. Water vapor and oxygen permeability of hydroxypropyl methylcellulose (HPMC) film with cellulose fiber (CF1) reduced in sizes after successive passes in a Microfluidizer® [Moura et al., 2011].

In summary, our study demonstrated that the addition of fibers processed by microfluidics results in improvement of mechanical and barrier properties of the HPMC films. In special, the observed reductions in WVP and O_2P permeability are promising as a means to improve quality and shelf life of food or pharmaceutical products coated with these types of films.

2.3.1.2 Polysaccharide hydrogels

In other work developed by our group [Aouada et al., 2011], cellulose nanofibers (CNFibers) were extracted by acid hydrolysis from commercial fibrous cellulose powder (trade name CF11) and the effect of their incorporation on the mechanical, hydrophilic, thermal, morphological, microscopic and structural polyacrylamide-methyl cellulose (PAAm-MC) hydrogel properties was studied.

Firstly, CF11 was deagglomerated by using Microfluidizer processor (model M-110EH-30) at 20,000 psi, in order to facilitate the acid hydrolysis of the microfiber due to its better water dispersion. After sulfuric acid hydrolysis, a highly stable suspension of hydrolyzed cellulose nanofibers was obtained from CF11. The opalescent suspension was the first indication of cellulose nanofiber extraction. The efficiency of the acid hydrolysis treatment was confirmed by the transmission electronic microscopy (TEM) technique (Figure 4) in which nanostructures with nanometer dimensions can be visualized. So, the network-structured cellulose nanocrystals were observed to exhibit micrometer-scale dimensions along both their length and width.

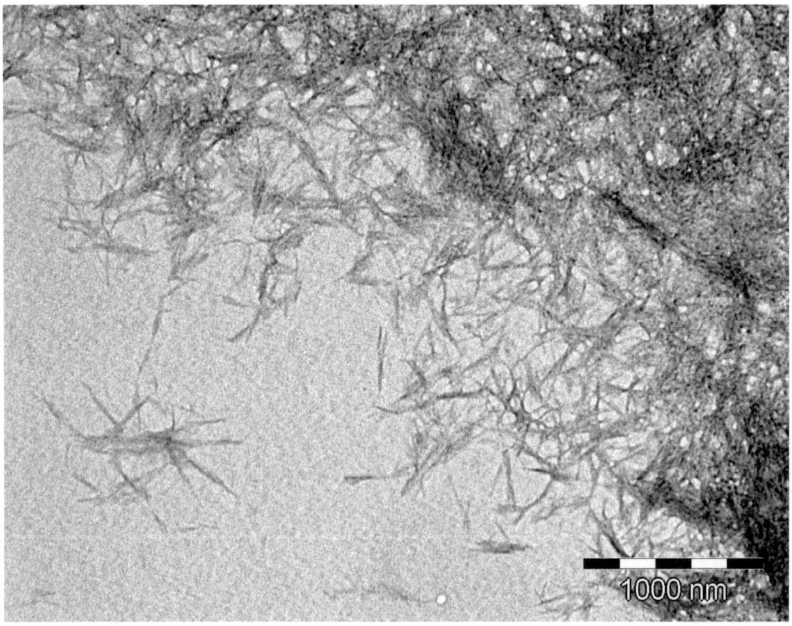

Fig. 4. TEM image of cellulose nanofiber after the acid hydrolysis process [Aouada et al., 2011].

In this research, we can highlight that the Microfluidizer process is very efficient artifice to help obtain of the nanofibers with narrow size distribution. In general, the presence of the nanofibers improved mechanical and structural network properties without negatively impacting their thermal properties. For instance, the value of maximum compressive stress increased from 2.1 to 4.4 kPa when the cellulose nanofiber was incorporated into PAAm-MC hydrogel.

Here, we only show the effect of the nanofibers on the morphologic properties of the hydrogels and more results of the improvement of the PAAm-MC hydrogels by incorporation of the cellulosic specimens are showed in Aouada et al., 2011. Fig. 5a depicts the SEM micrograph of a PAAm-MC hydrogel swollen in water, whereby a highly porous structure with well-defined shapes exhibiting some spread in pore size was observed. The presence of cellulose specimens causes a significant decrease in pore size, and the formation of three-dimensional well-oriented pore structure can be observed (Fig. 5b). This trend is possibly related to the fact that the cellulose nanofibers are distributed around and inside the three-dimensional porous material. This effect is better demonstrated by an analysis of further enlarged SEM micrographs (Fig. 6).

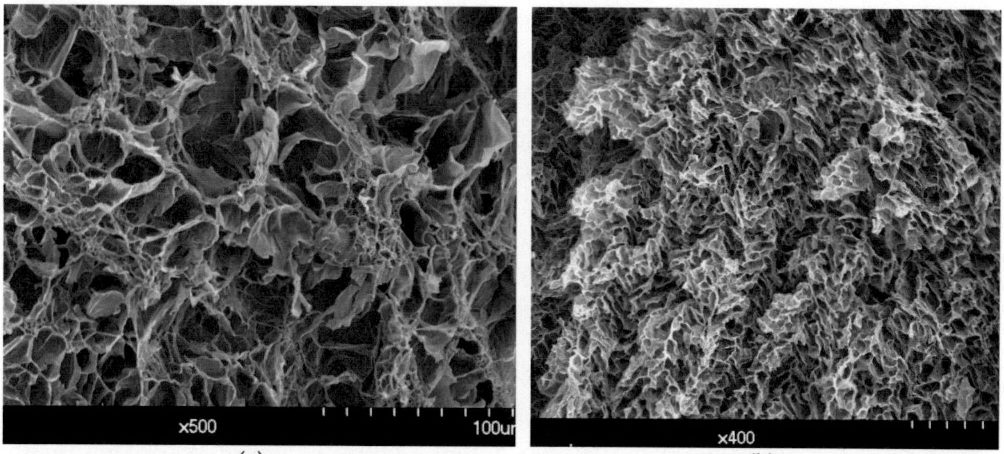

(a) (b)

Fig. 5. SEM micrographs of (a) PAAm-MC at 500 x magnification and (b) PAAm-NanFib hydrogels at 400 x magnification [Aouada et al., 2011].

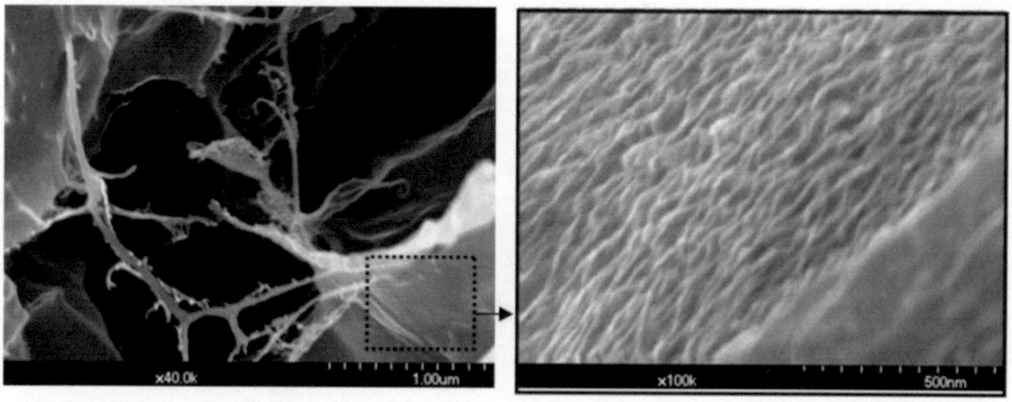

Fig. 6. Enlarged SEM micrographs of PAAm-NanFib hydrogels [Aouada et al., 2011].

3. Composite edible films

Nowadays, most materials used for food packaging are non-degradable, representing a serious environmental problem. Biopolymers materials have been increasingly exploited to develop edible and biodegradable films as an effort to extend shelf life and improve quality of food products while reducing packaging waste. However, the use of biopolymers has been limited because of problems related to performance, such as brittleness, poor gas and moisture vapor barrier, low degradation temperature, and cost.

Polymer composites are mixtures of polymers with inorganic or organic fillers with certain geometries (fibers, flakes, spheres, particulates). Several composites have been developed by adding reinforcing compounds to polymers to enhance their thermal, mechanical and barrier properties. Materials reinforced with macroscopic fillers usually contain defects and poor interactions at the filler-matrix interface, which become less important as the particles of the reinforcing component are smaller [Ludueña et al., 2007]. The addition of reinforcements to biopolymers opens new possibilities for improving the properties and applicability of biopolymers as food packaging materials.

Edible films and coatings may be heterogeneous in nature, consisting of a blend of polysaccharides, protein, and/or lipids. The main objective of producing composite films is to improve the permeability or mechanical properties as dictated by the need of a specific application. These heterogeneous films are applied either in the form of an emulsion, suspension, or dispersion of the non-miscible constituents, or in successive layers (multilayer coatings or films), or in the form of a solution in a common solvent [Bourtoom, 2008]. Recently, many researchers have extensively explored the development of composite films based on the diverse works.

Composite edible films and coatings can enhance food quality, safety and stability. They can control mass transfer between components within a product, as well as between product and environment. They can improve performance of the product through the addition of antioxidants, antimicrobial agents, and other food additives. Composite films can be formulated to combine the advantages of each component. Biopolymers, such as proteins and polysaccharides, provide the supporting matrix for most composite films, and generally offer good barrier properties to gases, with hydrocolloid components providing a selective barrier to oxygen and carbon dioxide [Baldwin et al., 1997; Drake et al., 1991; Guilbert, 1986; Kester & Fennema, 1986; Wong et al., 1992]. Lipids provide a good barrier to water vapor [Nisperos-Carriedo, 1994], while plasticizers are necessary to enhance flexibility and improve mechanical properties. Monitoring the film formulation allows adapting the mechanical and barrier properties of these materials to the desired application, improving the efficacy of preservation for packaged foods.

3.1 Preparation and characterization

Composite films are prepared by using 2 or more hydrocolloids. This requires dissolution of biopolymer molecules. Some polysaccharides require solubilization at a higher temperature, while others require dissolution in a pH-regulated medium (e.g., chitosan).

Composite suspensions can be used to obtain either coatings or films. In the case of coatings, surface tension and rheological behavior of the suspension are important factors that will affect suspension spread ability and coating adhesion. In formulation of composite biopolymer films, it is important to characterize the miscibility of biopolymers and interactions that may occur between them, since these attributes ultimately influence film microstructure.

Films can be prepared by casting, extrusion or lamination, which are similar processes used in the synthetic polymer industry [Stepto & Tomka, 1987].

Casting is a common, small-scale production method used to obtain biodegradable films. In this technique, a portion of the film suspension is poured onto a surface (e.g., acrylic plates), and then dried in a ventilated oven or at room conditions. This simple technique produces films that can be easily removed from plates, and tolerates films of variable thickness to be obtained by varying the weight of film suspension applied and the area of the plate onto which films are cast (Fig. 7).

Fig. 7. Casting of polysaccharide films at room temperature.

The casting technique is frequently used in systems containing micro- and nanomaterials. In the study performed by Moura et al., 2009, the hydroxypropyl methylcellulose (HPMC) films with CS-TPP nanoparticles were obtained by addition of 3.0 g of HPMC in 100 mL of nanoparticle solution (recently synthesized) under magnetic stirring for 12 h. After the solutions were prepared, the flasks were kept closed during 6 h to prevent microbubble formation in the films. The solutions were then poured in a glass plate (30 x 30 cm) covered with Mylar (Polyester film, DuPont, Hopewell, Va., USA) for film casting preparation. The solutions were cast at a wet thickness of 0.5 mm onto plates using casting bars and the plates were placed on a leveled surface at room temperature and let dry for 24 h. After drying, the films were removed and conditioned in sealed plastic bags stored at room temperature.

Fig. 8 shows the effect of CS/TPP nanoparticles on tensile strength of HPMC films. There were significant differences in tensile strength, demonstrating that the incorporation of nanoparticles improved the resistance of the film.

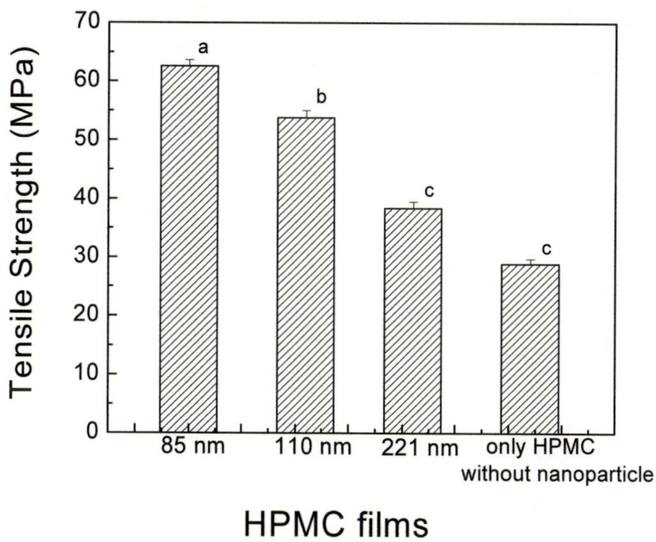

Fig. 8. Effect of nanoparticles on tensile strength of HPMC films. Columns show the means and error bars indicate the standard deviations. Different letters within a column indicated significant difference at $P < 0.05$ [Moura et al., 2009].

The morphology of the composite HPMC films containing nanoparticles (221 nm) prepared by casting was analyzed through scanning electron microscopy (SEM). Control film prepared from a solution containing only 3% w/v of HPMC in water (Fig. 9a), exhibited a high degree of porosity evenly distributed throughout the film. Compaction of HPMC film was observed when chitosan nanoparticles were added as shown in Fig. 9b. This compacting further decreased WVP and increased tensile properties of films. This result confirms the efficiency of the method to prepare cast films with properties suitable for various applications.

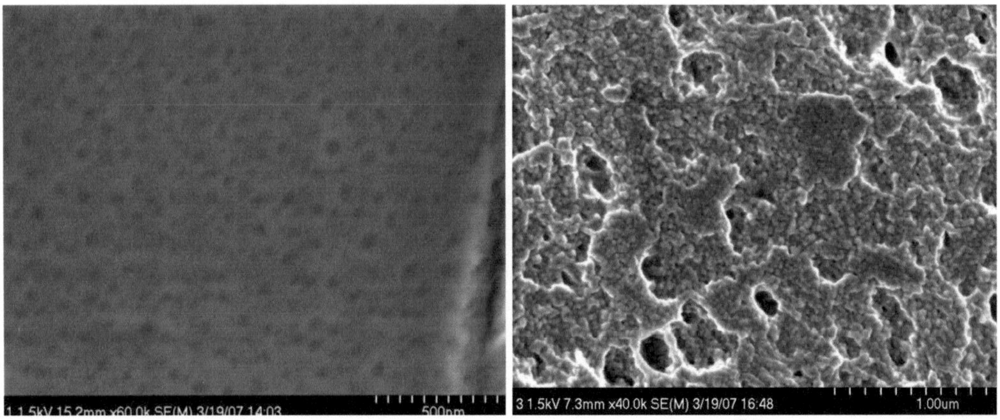

(a) (b)

Fig. 9. Scanning electron microphotographies of (a) HPMC films and (b) HPMC films with CS-TPP nanoparticles of 221 nm [Moura et al., 2009].

The main attributes involved in characterizing composite films are: optical properties, water-solubility, thickness, thermal behavior, barrier properties (water vapor and oxygen permeabilities) and mechanical properties.

All edible composite films should have the following properties as they are removed from the substrate drying: self-standing, peelable, colorless, flexible, easily handled, and with a homogenous, smooth surface. These characteristics should be the first analyzed after detached from casting substrate.

Solubility in water is an important property of edible films. Potential applications may require water insolubility to enhance product integrity and water resistance. However, in some cases, water solubility before consumption of the product might be beneficial [Pérez-Gago et al., 1999]. Film solubility in water was a very important characteristic of the polysaccharide films, because they must fully disintegrate upon immersion in aqueous solution. Turhan & Sahbaz, 2004 studied various aspects of methylcellulose (MC) films, including the solubility to evaluate the effects of film-forming solution composition and plasticizers on these properties.

The solubility of MC depends on its degree of substitution (DS), and MC used in this research (DS = 1.9) was reported as water-soluble. Films prepared only with water or water/ethanol dissolved slowly, whereas plasticized film pieces dissolved rapidly after coming into contact with water. The low dissolution rate indicated the high cohesion of MC matrix via numerous hydrogen bonds between MC chains.

It is important to know the behavior of any packaging material when exposed to changing temperatures during storage. The glass-rubber transition temperature (T_g) of a packaging material is an important parameter which controls mechanical properties, matrix chains dynamics, and swelling behavior. The glass transition is marked by a substantial change in molecular chain mobility. Below the T_g, the polymer chains show a minimum mobility and a low free volume resulting in great fragility and low diffusion properties; above T_g, the mobility sharply increases, and the mechanical and barrier properties of the polymer are impaired [S. L. Lee et al., 2008]. T_g increasing effects of cellulose micro- or nanofibers on polymer films have been reported [Alemdar & Sain, 2008; Anglès & Dufresne, 2000], probably because of the restricted mobility of polymer chains in the vicinity of the interfacial area.

The barrier properties of a food packaging material are related to the transport of gases or vapors through the material, influencing the preservation of the food quality. Oxygen permeation through the package may cause oxidation in lipid foods that result in off-flavors (the so-called oxidative rancidity), color changes and nutrient losses.

All the properties that describe the reactions of a material to application of forces are named mechanical properties. The kind of force involved, its time of application, magnitude, direction etc. are important to define different properties. Particularly important for packaging materials are tensile properties. Zimmermann et al., 2004 reported that the addition of cellulose fibrils to polymer matrices increased their tensile strength and elastic modulus.

4. Microfibers reinforced films and edible films

In an overview study, Sanchez-Garcia et al., 2010a described some important works related to addition of cellulose fibers and nanowhiskers for mechanical, thermal stability and barrier performance improvements of different matrices [Fendler et al., 2007; Kvien et al., 2005; Orts et al., 2005; Petersson et al., 2007; Tserki et al., 2006].

Other authors have proposed the addition of microfibers from different sources as reinforcement agent for improvement of the final film properties. The study realized by Dogan and McHugh, 2007 was the first to investigate the use of different size fillers for the purpose of preparing edible composite films. In this pioneering work, the inclusion of MCC (microcrystalline cellulose) fillers in different sizes on water vapor permeability and mechanical properties of HPMC based films was investigated. Measurements of the light scattering particle size analyzer indicated that the mean particle sizes of MCC were 0.5, 1.5, and 3.0 µm and the images from scanning electronic microscopy indicated that these MCC had rod-like shape. According to the authors, a decrease in diffusion coefficient is expected with the addition of cellulose fibrils because the diffusion of water in the films depends on the available pathways for water molecules. However, there was no observed noticeable difference in water vapor permeability of these composite films when compared to neat HPMC films. On the other hand, by analyzing the elongation and tensile strength, it was possible to see clearly the reinforcing effect of MCC fillers. For instance, tensile strength of films increased around 10% with the addition of 1.5 - and 3.0-µm MCC fibers to over 100% with the addition of 0.5 µm MCC to HPMC films. The authors also relates the importance of the edible films and coatings in foods as materials that increase shelf-life and improve sensory characteristics of foods by avoiding deterioration of food components and therefore promoting preservation of the final product. In addition, the composite materials with improved properties made from polymers reinforced with organic-based filler materials have received significant interest starting from the mid-1990s, when Favier et al., 1997 showed that the addition of 3% to 6% crystalline cellulose in a copolymer acrylate latex film increased dynamic modulus by more than 3-fold.

The addition of cellulose fibers to enhance the mechanical and moisture barrier properties of the starch films was reported by Muller et al., 2009. Composite films were prepared by casting technique from a suspension containing cassava starch, glycerol, water and cellulose fiber with 1.2 mm in length and 0.1 mm in diameter. The authors observed that the incorporation of cellulose fibers reduced the films moisture uptake due to the lower water affinity of cellulose fibers compared to starch. Also, films reinforced with fibers presented higher values of tensile strength and elasticity modulus, and lower values of elongation at break when compared to non-reinforced films. For example, the incorporation of 0.10 and 0.50 g fiber/g starch increased the tensile strength of reinforced films 6.7 and 18 times respectively, showing the reinforcement effect of the fibers addition. The addition of the fiber suspension also decreased the water vapor permeability of the starch films because of the low hygroscopicity of cellulose fibers. In fact, the water vapor permeability depends on both water diffusivity in the polymeric matrix and the solubility coefficient of water in the film [Larotonda et al., 2005; Muller et al., 2007].

Mango purees edible films reinforced by Azeredo et al., 2009 with cellulose nanofibers (CNF). CNF were effective in increasing tensile strength, and even more noticeable in enhancing elastic modulus, which increased more than 100% with a CNF loading of 10 g/100 g matrix. The remarkable effect of CNF on modulus was ascribed to the formation of a fibrillar network within the matrix, the fibers being probably linked through hydrogen bonds [Helbert et al., 1996]. The addition of CNF was also effective to improve water vapor barrier of the films.

Azeredo et al., 2010 reported effects of CNF in enhancing tensile strength, elastic modulus and water vapor barrier of chitosan films. CNF showed no specific orientation when observed by atomic force microscopy (AFM), which suggests an exfoliated nanocomposite, with good dispersion of CNF in the matrix, which explains the good performance of the nanocomposite films.

Sanchez-Garcia et al., 2010b reported enhanced water vapor barrier of carrageenan films due to incorporation of cellulose whiskers (CW), prepared by acid hydrolysis of highly purified cellulose microfibers. The nanobiocomposites containing 3 wt % of CW exhibited 71% reduction in water vapor permeability, which was mainly attributed to a filler-induced water solubility reduction. However, TEM and water vapor permeability data suggest that increasing the nanofiller loading in excess of 3 wt % leads to agglomeration of CW due to hydrogen-bonding-induced self-association.

By adding of the jute micro/nanofibrils (JNF) to starch/polyvinyl alcohol (PVA), Das et al., 2011 prepared and characterized different biocomposite films reinforced with 5, 10 and 15 wt.% JNF produced by solution casting method and using glycerol as plasticizer. Tensile properties results indicated that the tensile strength of the biocomposite films (5, 10 and 15 wt.% filler loaded) increased by 51%, 130% and 197% respectively in comparison to the unreinforced one. The authors attributed this increase to very fine nature of JNF and due to effective stress transfer at the interface between the matrix and JNF. The uniform dispersion of the JNF in the matrix and effective intercomponent bonding significantly increased the thermal stability of these biocomposites. Another important result proceeding of the fiber incorporation is that the moisture uptake decreased in all the biocomposite films compared to the unreinforced one. Possibly, this decreasing is related to lowering in the amount of free hydroxyl groups of the matrix which took part in hydrogen bonding with the JNF and the incorporation of JNF provided a stabilization effect to the matrix by forming a three dimensional cellulosic network which strongly restricted the dissolution of the matrix in water. Finally, according to the authors, the 10 wt.% JNF loaded films exhibited the best combination of moisture uptake behaviour, mechanical, thermal, and morphologic properties.

5. Final remarks

Researches on biopolymer based edible and biodegradable films have been a tendency for food packaging in times of great concern about the huge waste accumulation derived from the extensive use of non-biodegradable packaging materials. However, biopolymers have limited barrier and mechanical properties, requiring reinforcements to improve their performance. Cellulose micro- and nanofibrils have been extensively studied as reinforcement materials for biopolymers, since their good mechanical properties, biodegradability and compatibility with most biopolymers favor such applications. This kind of improvement in biopolymer based materials is of utmost importance when trying to replace (at least partially) the conventional petroleum-based polymers by biodegradable materials in packaging sector. Microfluidization has been shown as a useful technology to improve the applicability of cellulose microfibers as reinforcements to biobased packagings, enhancing cellulose interactions with the matrices and the resulting performance of the resulting composite materials.

6. Acknowledgment

MCT/FINEP, FAPESP, CNPq, Capes, Embrapa.

7. References

Alemdar, A. & Sain, M. (2008). Biocomposites from wheat straw nanofibers: Morphology, thermal and mechanical properties. *Composites Science and Technology*, Vol. 68, No.2, (February 2008), pp. 557-565, ISSN 0266-3538.

Anglès, M. N. & Dufresne, A. (2000). Plasticized starch/tunicin whiskers nanocomposites. 1. Structural analysis. *Macromolecules,* Vol.33, No.22, (October 2000), pp. 8344-8353, ISSN 0024-9297.

Aouada, F. A.; Moura, M. R. de; Orts, W. J. & Mattoso, L. H. C. (2011). Preparation and characterization of novel micro- and nanocomposite hydrogels containing cellulosic fibrils. *Journal of Agricultural and Food Chemistry,* Vol.59, No.17, (September 2011), pp. 9433-9442, ISSN 0021-8561.

Atalay, Y. T.; Vermeir, S.; Witters, D.; Vergauwe, N.; Verbruggen, B.; Verboven, P.; Nicolai, B. M. & Lammertyn, J. (2011). Microfluidic analytical systems for food analysis.

Trends in Food Science & Technology, Vol.22, No.7, (July 2011), pp. 386-404, ISSN 0924-2244.

Azeredo, H. C. M. (2009). Nanocomposites for food packaging applications. *Food Research International*, Vol.421, No.9, (November 2009), pp. 1240-1253, ISSN 0963-9969.

Azeredo, H. C. M.; Mattoso, L. H. C.; Avena-Bustillos, R. J.; Ceotto Filho, G.; Munford, M. L.; Wood, D. & McHugh, T. H. (2010). Nanocellulose reinforced chitosan composite films as affected by nanofiller loading and plasticizer content. *Journal of Food Science*, Vol.75, No.1, (January - February 2010), pp.N1-N7, ISSN 0022-1147.

Azeredo, H. C. M.; Mattoso, L. H. C.; Wood, D.; Williams, T. G.; Avena-Bustillos, R. J. & McHugh, T. H. (2009). Nanocomposite edible films from mango puree reinforced with cellulose nanofibers. *Journal of Food Science*, Vol.74, No.5, (September - October 2009), pp.N31-N35, ISSN 0022-1147.

Baldwin, E. A.; Nisperos-Carriedo, M. O.; Hagenmaier, R. D. & Baker, R. A. (1997). Use of lipids in coatings for food products. *Food Technology*, Vol.51, No.6, (June 1997), pp. 56-64, ISSN 0015-6639.

Bharadwaj, R. K. (2001). Modeling the barrier properties of polymer-layered silicate nanocomposites. *Macromolecules*, Vol.34, No.26, (December 2001), pp. 9189-9192, ISSN 0024-9297.

Bilbao-Sainz, C.; Bras, J.; Williams, T.; Sénechal, T. & Orts, W. (2011). HPMC reinforced with different cellulose nano-particles. *Carbohydrate Polymers*, Vol.86, No.4, (October 2011), pp. 1549-1557, ISSN 0144-8617.

Bonilla, J.; Atarés, L.; Vargas, M. & Chiralt, A. (2012). Effect of essential oils and homogenization conditions on properties of chitosan-based films. *Food Hydrocolloids*, Vol.26, No.1, (January 2012), pp. 9-16, ISSN 0268-005X.

Bourtoom, T. (2008). Edible films and coatings: characteristics and properties. *International Food Research Journal*, Vol.15, No.3, pp. 237-248, ISSN 22317546.

Cavender, G. A. & Kerr, W. L. (2011). Microfluidization of full-fat ice cream mixes: effects of gum stabilizer choice on physical and sensory changes. *Journal of Food Process Engineering*, In Press, doi:10.1111/j.1745-4530.2011.00650.x, ISSN 1745-4530.

Chen, R. H.; Huang, J. R.; Tsai, M. L.; Tseng, L. Z. & Hsu, C. H. (2011). Differences in degradation kinetics for sonolysis, microfluidization and shearing treatments of chitosan. *Polymer International*, Vol.60, No.6, (March 2011), pp. 897-902, ISSN 0959-8103.

Cho, D.; Matlock-Colangelo, L.; Xiang, C.; Asiello, P. J.; Baeumner, A. J. & Frey, M. W. (2011). Electrospun nanofibers for microfluidic analytical systems. *Polymer*, Vol.52, No.15, (July 2011), pp. 3413-3421, ISSN 0032-3861.

Coma, V.; Sebti, I.; Pardon, P.; Deschamps, A. & Pichavant, F. H. (2001). Antimicrobial edible packaging based on cellulosic ethers, fatty acids, and nisin incorporation to inhibit Listeria innocua and Staphylococcus aureus. *Journal of Food Protection*, Vol.64, No.4, (April 2001), pp. 470–475, ISSN 0362-028X.

Das, K.; Ray, D.; Bandyopadhyay, N. R.; Sahoo, S.; Mohanty, A. K. & Misrad, M. (2011). Physico-mechanical properties of the jute micro/nanofibril reinforced starch/polyvinyl alcohol biocomposite films. *Composites Part B: Engineering*, Vol.42, No.3, (April 2011), pp. 376-381, ISSN 1359-8368.

deMello, A. J. (2006). Control and detection of chemical reactions in microfluidic systems. *Nature*, Vol.442, No.7107, pp. 394-440, ISSN 0028-0836.

DeMello, J. & DeMello, A. (2004). Microscale reactors: nanoscale products. *Lab on a Chip*, Vol.4, No.2, pp. 11N-15N, ISSN 1473-0189.

Dogan, N. & McHugh, T. H. (2007). Effects of microcrystalline cellulose on functional properties of hydroxy propyl methyl cellulose microcomposite films. *Journal of Food Science*, Vol.72, No.1, (January - February 2007), pp. E16-E22, ISSN 0022-1147.

Drake S. R.; Cavalieri, R. & Kupferman, E. M. (1991) Quality attributes of Anjou pears after different wax drying and refrigerated storage. *Journal of Food Quality*, Vol.14, No.6, (December 1991), pp. 455-465, ISSN 0146-9428.

Favier, V.; Canova, G. R.; Shrivastava, S. C.; Cavaille, J. Y. (1997). Mechanical percolation in cellulose whisker nanocomposites. *Polymer Engineering and Science*, Vol.37, No.10, (October 1997), pp. 1732-1739, ISSN 0032-3888.

Fendler, A.; Villanueva, M. P.; Gimenez, E. & Lagaron, J. M. (2007). Characterization of the barrier properties of composites of HDPE and purified cellulose fibers. *Cellulose*, Vol.14, No.5, (October 2007), pp. 427-438, ISSN 0969-0239.

Guilbert, S. (1986). Technology and application of edible protective films. In: *Food Packaging and Preservation. Theory and Practice*, M. Mathlouthi (Ed.), 371-394, Elsevier Applied Science Publishing Co., ISBN 0853344132, London, England.

Helbert, W.; Cavaillé, J. Y. & Dufresne, A. (1996). Thermoplastic nanocomposites filled with wheat straw cellulose whiskers. Part I: processing and mechanical behavior. *Polymer Composites*, Vol. 17, No.4, (August 1996), pp. 604–611, ISSN 1548-0569.

Hung, L.-H. & Lee, A. P. (2007). Microfluidic devices for the synthesis of nanoparticles and biomaterials. *Journal of Medical and Biological Engineering*, Vol.27, No.1, pp. 1-6, ISSN 1609-0985.

Il Park, J.; Saffari, A.; Kumar, S.; Gunther, A.; Kumacheva, E.; Clarke, D. R.; Ruhle, M. & Zok, F. (2010). Microfluidic synthesis of polymer and inorganic particulate materials. *Annual Review of Materials Research*, Vol.40, pp. 415-443, ISSN 1531-7331.

Kamper, S. L. & Fennema, O. (1984). Water vapor permeability of an edible, fatty acid, bilayer film. *Journal of Food Science*, Vol.49, No.6, pp. 1482–1485, ISSN 0022-1147.

Kester, J. J. & Fennema, O. R. (1986). Edible films and coatings: a review. *Food Technology*, Vol.40, No.12, (December 1986), pp. 47-59, ISSN 0015-6639.

Kvien, I.; Bjørn, S. T. & Oksman, K. (2005). Characterization of cellulose whiskers and their nanocomposites by atomic force and electron microscopy. *Biomacromolecules*, Vol.6, No.6, (November -December 2005), pp. 3160-3165, ISSN 1525-7797.

Larotonda, F. D. S.; Matsui, K. N.; Sobral, P. J. A. & Laurindo, J. B. (2005). Hygroscopicity and water vapor permeability of Kraft paper impregnated with starch acetate. *Journal of Food Engineering*, Vol.71, No.4, pp. 394–402, (December 2005), ISSN 0260-8774.

Lee, C. Y.; Chang, C. L.; Wang, Y. N. & Fu, L. M. (2011). Microfluidic mixing: A Review. *International Journal of Molecular Sciences*, Vol.12, No.5, (May 2011), pp. 3263-3287, ISSN 1422-0067.

Lee, S. L.; Yam, K. L. & Piergiovanni, L. (2008). *Food Packaging Science and Technology*. Boca Raton: CRC Press, ISBN 9780824727796, Boca Raton, United States of America. pp. 43-77.

Lee, S. Y.; Mohan, D. J.; Kang, I. A.; Doh, G. H.; Lee, S. & Han, S. O. (2009). Nanocellulose reinforced PVA composite films: effects of acid treatment and filler loading. *Fibers and Polymers*, Vol.10, No.1, (February 2009), pp. 77-82, ISSN 1229-9197.

Liu, C. -H.; Zhong, J. -Z.; Liu, W.; Tu, Z. -C.; Wan, J.; Cai, X. -F. & Song, X. -Y. (2011). Relationship between functional properties and aggregation changes of whey protein induced by high pressure microfluidization. *Journal of Food Science*, Vol.76, No.4, (April 2011), pp. 341-347, ISSN 0022-1147.

Lu, P. & Hsieh, Y. L. (2010). Preparation and properties of cellulose nanocrystals: Rods, spheres, and network. *Carbohydrate Polymers*, Vol.82, No.2, (September 2010), pp. 329-336 ,0144-8617.

Ludueña, L. N.; Alvarez, V. A. & Vasquez, A. (2007). Processing and microstructure of PCL/clay nanocomposites. *Materials Science and Engineering: A - Structural Materials Properties Microstructure and Processing*, Vol.460-461, (July 2007), pp. 121–129, ISSN 0921-5093.

Miller, K. S. & Krochta, J. M. (1997). Oxygen and aroma barrier properties of edible films: A review. *Trends in Food Science & Technology*, Vol.8, No.7, (July 1997), pp. 228-237, ISSN 0924-2244.

Moura, M. R. de; Aouada, F. A.; Avena-Bustillos, R. J.; McHugh, T. H.; Krochta, J. M. & Mattoso, L. H. C. (2009). Improved barrier and mechanical properties of novel hydroxypropyl methylcellulose edible films with chitosan/tripolyphosphate nanoparticles. *Journal of Food Engineering*, Vol.92, No.4, (June 2009), pp. 448-453, ISSN 0260-8774.

Moura, M. R. de; Avena-Bustillos, R. J.; McHugh, T. H.; Krochta, J. M. & Mattoso, L. H. C. (2008). Properties of novel hydroxypropyl methylcellulose films containing chitosan nanoparticles. *Journal of Food Science*, Vol.73, No.7, (September 2088), pp. N31–N37, ISSN 0022-1147.

Moura, M. R. de; Avena-Bustillos, R. J.; McHugh, T. H.; Wood, D. F.; Otoni, C. G. & Mattoso, L. H. C. (2011). Miniaturization of cellulose fibers and effect of addition on the mechanical and barrier properties of hydroxypropyl methylcellulose films. *Journal of Food Engineering*, Vol. 104, No.1, (May 2011), pp. 154-160, ISSN 0260-8774.

Muller, C. M. O.; Laurindo, J. B. & Yamashita, F. (2009). Effect of cellulose fibers addition on the mechanical properties and water vapor barrier of starch-based films. *Food Hydrocolloids*, Vol.23, No.5, (July 2009), pp. 1328-1333, ISSN 0268-005X.

Muller, C. M. O.; Yamashita, F. & Laurindo, J. B. (2007). Evaluation of the effect of glycerol and sorbitol concentration and water activity on the water barrier properties of cassava starch films through a solubility approach. *Carbohydrate Polymers*, Vol.72, No.1, (April 2008), pp. 82-87, ISSN 0144-8617.

Napoli, M.; Atzberger, P. & Pennathur, S. (2011). Experimental study of the separation behavior of nanoparticles in micro- and nanochannels. *Microfluidics and Nanofluidics*, Vol.10, No.1, (January 2011), pp. 69-80, ISSN 1613-4982.

Nik, A. M.; Wright, A. J. & Corredig, M. (2010). Surface adsorption alters the susceptibility of whey proteins to pepsin-digestion. *Journal of Colloid and Interface Science*, Vol.344, No.2, (April 2010), pp. 372-381, ISSN 0021-9797.

Nisperos-Carriedo, M. O. (1994) Edible coatings and films based on polysaccharides. In: *Edible Coatings and Films to Improve Food Quality*, J. M. Krochta, E. A. Baldwin, M. O. Nisperos-Carriedo, (Eds.), 305-330, Technomic Publishing Co., ISBN 1566761131, Lancaster, United States of America.

Orts, W. J.; Shey, J.; Imam, S. H.; Glenn, G. M.; Guttman, M. E. & Revol, J. F. (2005). Application of cellulose microfibrils in polymer nanocomposites. *Journal of Polymers and the Environment*, Vol.13, No.4, (October 2005), pp. 301-306, ISSN 1566-2543.

Pérez-Gago, M. B.; Nadaud, P.; Krochta, J. M. (1999). Water vapor permeability, solubility, and tensile properties of heat-denatured versus native whey protein films. *Journal of Food Science*, Vol.64, No.6, (November-December 199), pp. 1034-1037, ISSN 0022-1147.

Petersson L.; Kvien I. Oksman K. (2007). Structure and thermal properties of poly(lactic acid)/cellulose whiskers nanocomposite materials. *Composites Science and Technology*, Vol.67, No.11-12, (September 2007), pp. 2535-2544, ISSN 0266-3538.

Ramli, N. A. & Wong, T. W. (2011). Sodium carboxymethylcellulose scaffolds and their physicochemical effects on partial thickness wound healing. *International Journal of Pharmaceutics*, Vol.403, No.1-2, (January 2011), pp. 73-82, ISSN 0378-5173.

Rao, J. & McClements, D. J. (2011). Food-grade microemulsions, nanoemulsions and emulsions: Fabrication from sucrose monopalmitate & lemon oil. *Food Hydrocolloids*, Vol.25, No.6, (August 2011), pp. 1413-1423, ISSN 0268-005X.

Rao, J. & McClements, D. J. (2012). Lemon oil solubilization in mixed surfactant solutions: Rationalizing microemulsion & nanoemulsion formation. *Food Hydrocolloids*, Vol.26, No.1, (January 2012), pp. 268-276, ISSN 0268-005X.

Sanchez-Garcia, M. D.; Hilliou, L. & Lagaron, J. M. (2010b). Morphology and water barrier properties of nanobiocomposites of κ/ι-hybrid carrageenan and cellulose nanowhiskers. *Journal of Agricultural and Food Chemistry*, Vol.58, No.24, (December 2010), pp. 12847–12857, ISSN 1520-5118.

Sanchez-Garcia, M. D.; Lopez-Rubio, A. & Lagaron, J. M. (2010a). Natural micro and nanobiocomposites with enhanced barrier properties and novel functionalities for food biopackaging applications: An overview. *Trends in Food Science & Technology*, Vol.21, No.11, (November 2010), pp. 528-536, ISSN 0924-2244.

Schirhagl, R.; Seifner, A.; Husain, F. T.; Cichna-Markl, M.; Lieberzeit, P. A. & Dickert, F. L. (2010). Antibodies and their replicae in microfluidic sensor systems-label free quality assessment in food chemistry and medicine. *Sensor Letters*, Vol.8, No.3, (June 2010), pp. 399-404, ISSN 1546-198X.

Situma, C.; Hashimoto, M. & Soper, S. A. (2006). Merging microfluidics with microarray based bioassays. *Biomolecular Engineering*, Vol.23, No.5, (October 2006), pp. 213–231, ISSN 1389-0344.

Skurtys, O. & Aguilera, J. M. (2008). Applications of microfluidic devices in food engineering. *Food Biophysics*, Vol.3, No.1, (March 2008), pp. 1-15, ISSN 1557-1858.

Stepto, R. F. T. & Tomka, I. (1987). Injection molding of natural hydrophilic polymers in the presence of water. *Chimia*, Vol.41, No.3, (March 1987), pp. 76-81, ISSN 0009-4293.

Sudhakar, Y.; Kuotsu, K. & Bandyopadhyay, A. K. (2006). Buccal bioadhesive drug delivery a promising option for orally less efficient drugs. *Journal of Controlled Release*, Vol.114, No.1, (August 2006), pp. 15-40, ISSN 0168-3659.

Thompson, J. A.; Du, X.; Grogan, J. M.; Schrlau, M. G. & Bau, H. H. (2010). Polymeric microbead arrays for microfluidic applications. *Journal of Micromechanics and Microengineering*, Vol.20, No.11, (November 2010), article number 115017, ISSN 0960-1317.

Tserki, V.; Matzinos, P.; Zafeiropoulos, N. E. & Panayiotou, C. (2006). Development of biodegradable composites with treated and compatibilized lignocellulosic fibers. *Journal of Applied Polymer Science*, Vol.100, No.6, (March 2006), pp. 4703-4710, ISSN 0021-8995.

Turhan, K. N.; Sahbaz, F. (2004). Water vapor permeability, tensile properties and solubility of methylcellulose-based edible films. *Journal of Food Engineering*, Vol.61, No.3, (February 2004), pp. 459-466, ISSN 0260-8774.

Wong, D. W. S.; Gastineau, F. A.; Gregorski, K. S.; Tillin, S. J. & Pavlath, A. E. (1992). Chitosan-lipids films: microstructure and surface energy. *Journal of Agricultural and Food Chemistry*, Vol.40, No.4, (April 1992), (April 1992), pp. 540-544, ISSN 0021-8561.

Wu, Z. & Nguyen, N. T. (2005). Hydrodynamic focusing in microchannels under consideration of diffusive dispersion: theories and experiments. *Sensors and Actuators B – Chemical*, Vol.107, No.2, (June 2005), pp. 965-974, ISSN 0925-4005.

Zhang, H. W.; Betz, A.; Qadeer, A.; Attinger, D. & Chen, W. (2011). Microfluidic formation of monodispersed spherical microgels composed of triple-network crosslinking. *Journal of Applied Polymer Science*, Vol.121, No.5, (September 2011), pp. 3093-3100, ISSN 0021-8995.

Zimmermann, T.; Pöhler, E. & Geiger, T. (2004). Cellulose fibrils for polymer reinforcement. *Advanced Engineering Materials*, Vol.6, No.9, (September 2004), pp.754-761, ISSN 1527-2648.

Zuluaga, R.; Putaux, J. L.; Cruz, J.; Velez, J.; Mondragon, I. & Ganan, P. (2009). Cellulose microfibrils from banana rachis: Effect of alkaline treatments on structural and morphological features. *Carbohydrate Polymers*, Vol.76, No.1, (March 2009), pp. 51-59, ISSN 0144-8617.